Elementary Algebra

by Toby Wagner

Elementary Algebra

ISBN: 978-1-943536-29-0

Edition 1.0 Fall 2017

© 2017, Chemeketa Community College. All rights reserved.

Chemeketa Press

Chemeketa Press is a nonprofit publishing endeavor at Chemeketa Community College. Working together with faculty, staff, and students, we develop and publish affordable and effective alternatives to commercial textbooks. All proceeds from the sale of this book will be used to develop new textbooks. For more information, please visit chemeketapress.org.

Publisher: Tim Rogers

Managing Editor: Steve Richardson

Production Editor: Brian Mosher

Manuscript Editors: Steve Richardson, Matt Schmidgall

Design Editors: Ronald Cox IV, Kristen MacDonald

Cover Design: Ronald Cox IV, Faith Martinmaas, Shaun Jaquez, Kristi Etzel

Interior Design: Ronald Cox IV, Kristi Etzel

Layout: Noah Barrera, Matthew Sanchez, Faith Martinmaas, Emily Evans, Steve Richardson, Kristi Etzel, Cierra Maher, Candace Johnson

Additional contributions to the design and publication of this textbook come from the students and faculty in the Visual Communications program at Chemeketa.

Chemeketa Math Faculty

The development of this text and its accompanying MyOpenMath classroom has benefited from the contributions of many Chemeketa math faculty in addition to the author, including:

Ken Anderson, Lisa Healey, Kelsey Heater, Kyle Katsinis, Tim Merzenich, Nolan Mitchell, Chris Nord, Martin Prather, and Rick Rieman

Text Acknowledgment

This book was originally developed using materials from *Elementary Algebra*, by Wade Ellis and Dennis Burzynski, which has been made available under a Creative Commons Attribution 2.0 license and may be downloaded for free from legacy.cnx.org/content/col10614/1.3/.

Printed in the United States of America.

Contents

CHAPTER 1

Solving Linear Equations and Inequalities

Solving equations is at the heart of algebra. Because of this, it seems fitting to begin our study of elementary algebra by learning how to solve equations algebraically. In Chapter 1, we will focus on solving linear equations, which are typically less complicated than non-linear equations. We will also learn how to solve linear inequalities. As we progress through the chapter, we will be using our skills to help us solve application problems, which are commonly known as "word problems" or "story problems."

In this chapter, you'll study the following topics:

1.1 Solving Linear Equations

Overview

Here's a problem:

> The area of a rectangle is the same as the value found by multiplying the length and the width. A rectangle that is 14 feet long has an area of 91 square feet. What is the width of the rectangle?

This problem states that two things are the same — the area of the rectangle and the value found by multiplying the length and width. Mathematicians write many sentences like this, though usually with mathematical notation instead of words. In math, *equations* are used to communicate sameness. Equations are the most common sentences in math.

When you are finished with this section, you will be able to:

- ◆ Identify various types of equations
- ◆ Understand the meaning of solutions and equivalent equations
- ◆ Use addition and subtraction to solve 1-step equations
- ◆ Use multiplication and division to solve 1-step equations
- ◆ Solve application problems involving 1-step equations

At the end of this section, we will write and solve an equation to find the width of the rectangle in the problem above. In the meantime, let's learn more about equations.

A. Types of Equations

An **equation** is a mathematical sentence that asserts that two things are the same or *equal*. An equals sign (=) means "is the same as." It's important, though, to understand that an equation only *asserts* that two things are the same. As you'll soon see, this doesn't guarantee that the statement is actually true.

Some equations are always true. These equations are called **identities**. Identities are equations that are true for all acceptable values of the variable. Here are some examples of identities:

$5x = 5x$ is true for all acceptable values of x because 5 times any number is always the same as 5 times that same number.

$y + 1 = y + 1$ is true for all acceptable values of y because any number plus 1 is always the same as that same number plus 1.

$2 + 5 = 7$ is true, and no substitutions are necessary because this equation doesn't use any variables.

About Acceptable Values

For the equations in the first few chapters of this book, the variables can be replaced with any real number. All real numbers are therefore *acceptable values* of the variable. Later, you will see equations with more complicated expressions in which some numbers will not be used to replace the variable.

Some equations are never true. These equations are called contradictions. Contradictions are equations in which the expression on the left side and the expression on the right side are never equal, no matter what value is substituted for the variable.

$x = x + 1$ is never true for any acceptable value of x because no real number is equal to itself plus 1.

$0 \cdot k = 14$ is never true for any acceptable value of k because the product of 0 and any real number is always 0. It can never be 14.

$2 = 1$ is never true because 2 can never equal 1.

The truth of some equations depends upon the number value chosen for the variable. These equations are called conditional equations. Conditional equations are true when using at least one replacement value for the variable and false when using at least one different replacement value for the variable.

$x + 6 = 11$ is true only if x is replaced with 5 ($x = 5$). It is false if x is replaced with any other number.

$y - 7 = -1$ is true only if $y = 6$, and it's false if y is replaced with any other number.

A conditional equation with one variable is a linear equation if the highest power of the variable is 1. For example, $t + 5 = 12$ is a linear equation because the only variable that appears is being raised to the power of 1 ($t = t^1$). However, the equation $m^2 + 6m = 16$ is *not* a linear equation because it contains a variable that is being raised to the power of 2. In Chapter 2, we will see that if an equation is linear, then the graph of that equation is a straight line.

The following examples will show you different types of equations and explain why each is an identity, a contradiction, or conditional equation.

▼ Example 1

Decide whether each equation is an *identity*, *contradiction*, or a *conditional equation*.

1. $x - 4 = 6$　　　　　　2. $n - 2 = n - 2$　　　　　　3. $a + 5 = a + 1$

Solutions

1. $x - 4 = 6$ is a *conditional equation* because it's true only on the condition that $x = 10$.

2. $n - 2 = n - 2$ is an *identity* because it is true for all values of n. For example, if $n = 5$, then $5 - 2 = 5 - 2$ is true. And if $n = -7$, then $-7 - 2 = -7 - 2$ is true. No matter what number we substitute for n, the equation will always be true.

3. $a + 5 = a + 1$ is a *contradiction* because every value of a produces a false statement. For example, if $a = 8$, then $8 + 5 = 8 + 1$ is false. And if $a = -2$, then $-2 + 5 = -2 + 1$ is false.

Practice A

Now it's your turn to classify a new set of equations. For each of the following equations, decide whether the equation is an *identity*, a *contradiction*, or a *conditional equation*. If you think that the equation is conditional, figure out the value of the variable that will make the equation true. When you are done, turn the page and check your solutions.

1. $x + 1 = 10$

2. $y - 4 = 7$

3. $5a = 25$

4. $\frac{x}{4} = 9$

5. $\frac{18}{b} = 6$

6. $y - 2 = y - 2$

7. $x + 4 = x - 3$

8. $x + x + x = 3x$

9. $8x = 0$

10. $m - 7 = -5$

B. Solutions

The equals sign of an equation indicates that the value represented by the expression to the *left* of the equals sign is the same as the value represented by the expression to the *right* of the equals sign.

This number	is the same as	this number.
x	=	6
$x + 2$	=	8
$x - 1$	=	5

Figure 1. Equivalent Equations

When we use variables like x or y, the values for those variables that will make an equation true are called solutions of the equation. An equation is solved when all of its solutions have been found.

Sometimes a set of equations will share the exact same solution. When that happens, the equations are called equivalent equations. For example, these equations are equivalent:

$$2x + 1 = 7 \qquad 2x = 6 \qquad x - 1 = 2$$

The only substituted value for x that makes each equation true is 3, so the solution for all three equations is the same, $x = 3$.

Take a look at the table in Figure 1. You'll notice that 6 is the solution for all three equations. That means that all three equations are equivalent equations. The first equation, $x = 6$, explicitly states the solution.

When solving equations algebraically, our goal is to end up with an equivalent equation in which the variable is isolated — that is, it stands by itself — on one side of the equation. It is customary to write the solution as an equation that has the variable by itself on one side of the equals sign and the value on the other side. For example, for the equation $x + 3 = 5$ we would write the solution as $x = 2$.

C. Using Addition and Subtraction to Solve 1-Step Equations

Look again at the table in Figure 1. Notice that the second equation, $x + 2 = 8$, can be obtained by adding 2 to both sides of the first equation, $x = 6$. Also notice that the third equation, $x - 1 = 5$, is generated by subtracting 1 from both sides of the first equation. This illustrates the following properties of equality:

1. We can create an equivalent equation by adding the same number to both sides of the equation. In other words, the equation $a = b$ is equivalent to $a + c = b + c$. This is called the addition property of equality.

2. We can also create an equivalent equation by subtracting the same number from both sides of the equation. This means that the equation $a = b$ is equivalent to $a - c = b - c$. This is called the subtraction property of equality.

We can use these properties to isolate a variable like x. Once x has been isolated, we say that the equation has been "solved for x." The following examples will show you how you can use the equality properties of addition and subtraction to solve an equation.

Example 2

1. Solve for x: $x + 7 = 10$

2. Solve for m: $m - 2 = -9$

Solutions

1. $x + 7 = 10$

$$x + 7 = 10 \quad \text{7 is being added to } x \text{ on the left side of the equation.}$$
$$\underline{-7 \qquad -7} \quad \text{Undo this operation by subtracting 7 from } both \text{ sides of the equation.}$$
$$x = 3 \quad \text{Now } x \text{ is isolated on the left side of the equation. The equation } x = 3$$

states the solution of the original equation.

Let's check the solution by substituting 3 for x in the original equation.

$$x + 7 = 10 \qquad \text{Start with the original equation and substitute 3 for } x.$$
$$3 + 7 = 10 \qquad \text{Simplify the left side.}$$
$$10 = 10 \checkmark \qquad \text{This is correct.}$$

2. $m - 2 = -9$

$$\begin{array}{rl} m - 2 = & -9 \\ +2 & +2 \\ \hline m = & -7 \end{array}$$

2 is being subtracted from m on the left side of the equation.

Undo the subtraction by adding 2 to *both* sides.

Let's check this solution by substituting -7 for m in the original equation.

$$\begin{array}{rl} m - 2 = -9 & \text{Take the original equation and replace } m \text{ with } -7. \\ -7 - 2 = -9 & \text{Finish by simplifying.} \\ -9 = -9 \ \checkmark \end{array}$$

In the previous example, the instructions indicated which variable to solve for. However, if an equation only contains one letter, we don't really need to be told which variable to solve for — there's only one option. So from now on, if you're instructed to solve an equation that only contains one letter, the instructions will simply tell you to "solve."

Example 3

1.   Solve: $y - 2.181 = -16.915$

2.   Solve: $p + \frac{3}{8} = \frac{7}{12}$

Solutions

1. $y - 2.181 = -16.915$. This time, see if you can figure out how this works on your own.

$$\begin{array}{rl} y - 2.181 = & -16.915 \\ +2.181 & +2.181 \\ \hline y = & -14.734 \end{array}$$

Now check the solution.

$$\begin{array}{rl} y - 2.181 = & -16.915 \\ -14.734 - 2.181 = & -16.915 \\ -16.915 = & -16.915 \ \checkmark \end{array}$$

2. $p + \frac{3}{8} = \frac{7}{12}$

$$\begin{array}{rl} p + \frac{3}{8} = & \frac{7}{12} \\ -\frac{3}{8} & -\frac{3}{8} \\ \hline p = & \frac{5}{24} \end{array}$$

Remember, when adding or subtracting fractions, the fractions must have the same number in the denominator: $\frac{7}{12} - \frac{3}{8} = \frac{14}{24} - \frac{9}{24} = \frac{5}{24}$

And now we check:

$$p + \frac{3}{8} = \frac{7}{12}$$

$$\frac{5}{24} + \frac{3}{8} = \frac{7}{12}$$

$$\frac{5}{24} + \frac{9}{24} = \frac{7}{12}$$

$$\frac{14}{24} = \frac{7}{12}$$

$$\frac{7}{12} = \frac{7}{12} \checkmark$$

Practice C

Solve these equations by finding equivalent equations in which the variable is isolated. You've seen how it's done in the examples above, and you've seen how to check your work, so you know what to do. When you are done, turn the page and check your solutions.

11. Solve: $y - 3 = 8$

12. Solve: $x + 9 = -4$

13. Solve: $g - 7.2 = 1.3$

14. Solve: $m - \frac{5}{4} = \frac{1}{6}$

D. Using Multiplication and Division to Solve 1-Step Equations

You remember that the equals sign indicates that the value represented by the expression on the *left* side of the equation is the same as the value represented by the expression on the *right* side. We solved equations by adding or subtracting the same number on *both* sides of the equation. Now it's time to look at two more properties of equality:

1. We can obtain an equivalent equation by dividing both sides of the equation by the same nonzero number. In other words, if $c \neq 0$, then $a = b$ is equivalent to $\frac{a}{c} = \frac{b}{c}$. This is called the **division property of equality**.

2. We can also obtain an equivalent equation by multiplying both sides of the equation by the same nonzero number. In this case, if $c \neq 0$, then $a = b$ is equivalent to $ac = bc$. This is called the **multiplication property of equality**.

Practice A — Answers

1. conditional, $x = 9$	**4.** conditional, $x = 36$	**7.** contradiction	**10.** conditional, $m = 2$
2. conditional, $y = 11$	**5.** conditional, $b = 3$	**8.** identity	
3. conditional, $a = 5$	**6.** identity	**9.** conditional, $x = 0$	

We can use these properties to solve an equation by undoing operations and isolating the variable on one side of the equation. To solve $ax = b$ for x, we divide both sides of the equation by a. To solve $\frac{x}{a} = b$ for x, we multiply both sides of the equation by a. Let's see how that works with some example equations.

> ### Example 4
>
> Solve: $5x = 35$
>
> $5x = 35$ x is being multiplied by 5 on the left side of the equation.
>
> $\frac{5x}{5} = \frac{35}{5}$ To undo that operation, divide both sides of the equation by 5. A fraction bar like this indicates division.
>
> $x = 7$ Finally, both sides of the equation are simplified: $5x$ divided by 5 is x and 35 divided by 5 is 7.
>
> Now we check our solution:
>
> $5x = 35$ Start with the original equation and substitute 7 for x.
>
> $5 \cdot 7 = 35$
>
> $35 = 35$ ✓

Some people use the divide sign ($\div$) to show division instead of a fraction bar:

$$\frac{5x}{\div 5} = \frac{35}{\div 5}$$
$$x = 7$$

While this isn't mathematically incorrect, the fraction bar is the preferred way to denote division.

> ### Example 5
>
> Solve: $\frac{x}{4} = 5$
>
> $\frac{x}{4} = 5$ On the left side of the equation, x is being divided by 4.
>
> $4 \cdot \frac{x}{4} = 4 \cdot 5$ Undo that operation by multiplying both sides by 4.
>
> $x = 20$ Finish by simplifying the equation: $4 \cdot \frac{x}{4} = \frac{4x}{4}$ and $\frac{4x}{4} = x$
>
> And, of course, we check the solution.
>
> $\frac{x}{4} = 5$ Take the original equation, and substitute 20 for x.
>
> $\frac{20}{4} = 5$ Simplify.
>
> $5 = 5$ ✓

Example 6

Solve: $\frac{2y}{9} = 3$

There are two ways to solve this equation. The first involves getting rid of the 9 and 2 separately.

$\frac{2y}{9} = 3$ On the left side of the equation, $2y$ is being divided by 9.

$9 \cdot \frac{2y}{9} = 9 \cdot 3$ Undo this operation by multiplying both sides of the equation by 9.

$2y = 27$ In the resulting equation, y is being multiplied by 2.

$\frac{2y}{2} = \frac{27}{2}$ To undo this operation, divide both sides of the equation by 2.

$y = \frac{27}{2}$ The solution is an improper fraction because the numerator is not less than the denominator. However, it is not reducible, because 27 and 2 don't have any common factors. Therefore, $\frac{27}{2}$ is a simplified solution.

Let's check this solution.

$\frac{2y}{9} = 3$ Start with the original equation and replace x with $\frac{27}{2}$.

$\frac{2\left(\frac{27}{2}\right)}{9} = 3$ Begin simplifying the left side of the equation by simplifying the numerator of the fraction: $2 \cdot \frac{27}{2} = \frac{54}{2} = 27$

$\frac{27}{9} = 3$ Now simplify.

$3 = 3 ✓$

The second way to solve this equation makes use of reciprocals.

$\frac{2y}{9} = 3$ To solve this equation we need to isolate y, so we want $1y$ to stand on its own on the left side of the equation. Multiplying a number by its reciprocal — in this case, multiplying $\frac{2}{9}$ by $\frac{9}{2}$ — gives a product of 1.

$\frac{9}{2} \cdot \frac{2y}{9} = \frac{9}{2} \cdot 3$ With that in mind, multiply both sides of the equation by $\frac{9}{2}$.

$1 \cdot y = \frac{27}{2}$ Finish by simplifying.

$y = \frac{27}{2}$

This approach also gives us the correct solution.

Practice C — Answers

11. $y = 11$ **12.** $x = -13$ **13.** $g = 8.5$ **14.** $m = \frac{17}{12}$

Notice that other forms of the solution, such as 13.5 and $13\frac{1}{2}$, are also mathematically correct. However, if the instructions do not call for the answer to be written in decimal or mixed number form, then $\frac{27}{2}$ is perfectly fine. In general, it's acceptable to give an improper fraction as a final answer as long as the fraction is not reducible. For example, $\frac{36}{32}$ would need to be reduced to $\frac{9}{8}$ in order to be an acceptable final answer.

Practice D

Use the equality properties of multiplication and division to undo associations and isolate the variables. When you are done, turn the page and check your solutions.

15. Solve $6a = 42$ for a.

16. Solve $-12m = 16$ for m.

17. Solve $\frac{y}{8} = -2$ for y.

18. Solve $6.42x = 1.09$ for x.

E. Application Problems

In mathematics, "word problems" or "story problems" are commonly called *application problems* because they require us to *apply* our mathematical knowledge in the process of solving a real-life problem.

To algebraically solve an application problem, we must first translate the problem into algebraic terms. This means looking closely at the wording of the problem to figure out the variables and constants and how they relate to each other. We then translate the verbal phrases and sentences into algebraic expressions and equations.

The following table is a mathematics dictionary that will help us decide which mathematical symbol to use to describe different written relationships.

Mathematics Dictionary

Symbol	Word or Phrases
+	Sum, sum of, added to, increased by, more than, plus, and
−	Difference, minus, subtracted from, decreased by, less, less than
·	Product, the product of, of, multiplied by, times
÷	Quotient, divided by, ratio of
=	Equals, is equal to, is, the result is, becomes
{variable}*	A number, an unknown quantity, an unknown, a quantity

*A variable is a letter or symbol that represents a number whose value is unknown or whose value can change.

The following examples illustrate how we can start with a written phrase or sentence and translate it into a mathematical expression or equation.

Example 7

Phrase or Sentence	Algebraic Expression or Equation
six plus a number	$6 + x$
a quantity minus eight	$n - 8$
Two times a number is ten.	$2 \cdot y = 10$
One half of a number is twenty.	$\frac{1}{2} \cdot z = 20$
If a number is divided by four, the result is twelve.	$\frac{c}{4} = 12$

In Example 7, the algebraic expressions and equations match the order of the phrases and sentences. The first item in the phrase or sentence matches the first item in the expression or equation. The second item in the phrase or sentence matches the second item in the expression or equation — and so on. However, there are times when the order of a phrase or sentence does not match the equivalent expression or equation.

Example 8

Phrase	Expression	Commentary
Fifteen is subtracted from a number	$x - 15$	The word "from" implies that we must start with something and then subtract 15.
Fifteen less than a number	$x - 15$	"less than" means the same thing as "is subtracted from"

Be careful with the phrase "less than." This shows up a lot in application problems, and students often translate it incorrectly. Consider this real-life scenario:

> Gerald has \$20, and Trina has three dollars less than him. To figure out how much money Trina has, we need to determine the value of "three less than twenty." 3 − 20 gives a value of −17, but that doesn't make sense. Trina doesn't have negative-seventeen dollars! So instead, we perform the operation 20 − 3, which gives the correct result. Trina has \$17.

With addition or multiplication, the order used in the algebraic expression doesn't matter because of the commutative properties of addition and multiplication. An operation is *commutative* if we can change the order of the numbers being operated on without changing the result:

$$a + b = b + a \qquad a \cdot b = b \cdot a$$

However, in situations involving subtraction and division, the order used definitely matters because these operations are not commutative. As we saw, 20 − 3 and 3 − 20 don't yield the same results.

Example 9

Operation	Phrase	Expression
addition	The sum of a number and eleven	$n + 11$ or $11 + n$
subtraction	The difference of a number and sixteen	$h - 16$
multiplication	The product of a number and three	$v \cdot 3$ or $3 \cdot v$
division	The quotient of a number and five	$\frac{x}{5}$

Practice E, Part 1

Translate the following sentences into algebraic equations. Then solve each equation. Use x for any unknown quantity. When you are done, turn the page and check your solutions.

19. Eleven plus a number results in thirty-eight.

20. Thirteen is nine less than a number.

21. The quotient of a number and six is twelve.

22. Four times a number is thirty two.

23. One third of a number is six.

24. The difference of a number and seven is negative-fifteen.

Five-Step Method for Solving Application Problems

Now that you've had some practice translating verbal phrases and sentences into algebraic expressions and equations, it's time to use your translation skills to tackle application problems. The following procedure is a useful approach to solving application problems:

1. Let x — or any letter — represent the unknown quantity.

2. Translate the words of the problem into numbers, variables, and mathematical symbols to form an equation.

3. Solve the equation.

4. Check to see if the solution makes sense in the original problem. Typically, we do this by replacing the variable in the equation from Step 2 and simplifying. It's also important to mentally check the answer within the context of the original application problem. If the answer doesn't make sense, we've either created an incorrect equation or solved the equation incorrectly. In this case, we have to reconsider our equation.

5. Translate the mathematical solution into a written statement for the word problem and include the correct units, when applicable.

In the past, you may have found application problems frustrating. Don't worry about that. Just try following the five-step method carefully. With practice, it gets easier.

Many people run into problems because of errors made in the first step. Always start by introducing a variable. Always. As you go through the other steps, remember what that variable represents.

Let's solve the application problem from the beginning of this section.

Example 10

A rectangle that is 14 feet long has an area of 91 square feet. What is the width of the rectangle?

Solution

Step 1:	Let x = the width of the rectangle.	We are given the length of the rectangle and the area. The width is unknown.
Step 2:	$14x = 91$	We know that for a rectangle, Length · Width = Area. In this problem, the length is 14, and the area is 91.
Step 3:	$14x = 91$	Now solve the equation using the properties in this chapter.

$$\frac{14x}{14} = \frac{27}{14}$$

$$x = 6.5$$

Step 4:	$14 \cdot 6.5 = 91 \checkmark$	
Step 5:	The width of the rectangle is 6.5 feet.	Then translate the solution into a written sentence.

Example 11

This year an item costs $44, which is $3 more than last year's price. What was last year's price?

Solution

Step 1:	Let x = last year's price.	That's what we don't know yet.
Step 2:	$x + 3 = 44$	$x + 3$ represents the $3 increase in price from last year.
Step 3:	$x + 3 = 44$	Now it's time to solve the equation.

$$\underline{-3 \qquad\quad -3}$$

$$x = 41$$

Step 4:	$41 + 3 = 44 \checkmark$	Finish by checking the solution, and translating it into a written statement."
Step 5:	Last year's price was $41.	

Practice D — Answers

15. $a = 7$　　　　16. $m = -\frac{4}{3}$　　　　17. $y = -16$　　　　18. $x = 0.17$ (rounded two decimal places)

Example 12

Six percent of a number is 54. What is the number?

Step 1:   Let x = the unknown number

Step 2: $.06x = 54$

To create this equation, we must use the numerical value of 6%, which is 0.06 or $\frac{6}{100}$. We'll use the decimal form and multiply x by .06.

Step 3: $.06x = 54$

$$\frac{.06x}{.06} = \frac{.06x}{.06}$$

$$x = 900$$

Now it's time to solve the equation.

Step 4: $.06(900) = 54 \checkmark$

Step 5:   The number is 900.

Practice E, Part 2

For the following practice problems, follow the same five-step process we used in the examples. When you are done, turn the page and check your solutions.

25. This year an item costs $79, a decrease of $5 from last year's price. What was last year's price?

26. Eight percent of a number is 36. What is the number?

27. Cassie's car gets 25 miles per gallon. On a recent trip, she used 4.8 gallons of gas. How many miles did she drive? (Note: Miles per gallon is found by dividing the number of miles traveled by the number of gallons of gas used.)

Practice E, Part 1 — Answers

19. Equation: $11 + x = 38$; Solution: $x = 27$

20. Equation: $13 = x - 9$; Solution: $x = 22$

21. Equation: $\frac{x}{6} = 12$; Solution: $x = 72$

22. Equation: $4x = 32$; Solution: $x = 8$

23. Equation: $\frac{1}{3}x = 6$; Solution: $x = 18$

24. Equation: $x - 7 = -15$; Solution: $x = -8$

Exercises 1.1

For the following exercises, classify each of the equations as an *identity*, *contradiction*, or *conditional equation*.

1. $m + 6 = 15$

2. $y - 8 = -12$

3. $x + 1 = x + 1$

4. $g + g + g + g = 4g$

5. $k - 2 = k - 3$

6. $x + 1 = 0$

For the following exercises, solve each of the conditional equations by isolating the variable. Use the properties of equality discussed above to undo operations.

7. $m - 2 = 5$

8. $k + 10 = 1$

9. $y + 6 = -11$

10. $y - 8 = -1$

11. $x + 14 = 0$

12. $m - 12 = 0$

13. $g + 164 = -123$

14. $h - 265 = -547$

15. $x + 17 = -426$

16. $h - 4.82 = -3.56$

17. $y + 17.003 = -1.056$

18. $k + 1.0135 = -6.0032$

19. $p + \frac{7}{20} = \frac{19}{20}$

20. $w - \frac{11}{48} = \frac{25}{48}$

21. $x - \frac{3}{4} = \frac{1}{6}$

22. $n + \frac{2}{9} = \frac{11}{12}$

23. $t + \frac{22}{15} = \frac{3}{10}$

24. $c - \frac{3}{2} = \frac{13}{21}$

25. $3x = 42$

26. $5y = 75$

27. $6x = 48$

28. $8x = 56$

29. $4x = 56$

30. $3x = 93$

31. $5a = -80$

32. $9m = -108$

33. $6p = -108$

34. $12q = -180$

35. $-4a = 16$

36. $-20x = 100$

37. $-6x = -42$

38. $-8m = -40$

39. $-3k = 126$

40. $-9y = 126$

41. $\frac{x}{6} = 1$

42. $\frac{a}{5} = 6$

43. $\frac{k}{7} = 6$

44. $\frac{x}{3} = 72$

45. $\frac{x}{8} = 96$

46. $\frac{y}{-3} = -4$

47. $\frac{m}{7} = -8$

48. $\frac{k}{18} = 47$

49. $\frac{f}{-62} = 103$

50. $3.06m = 12.546$

51. $5.012k = 0.30072$

52. $\frac{x}{2.19} = 5$

53. $\frac{y}{4.11} = 2.3$

54. $\frac{4y}{7} = 2$

55. $\frac{3m}{10} = -1$

56. $\frac{5k}{6} = 8$

57. $\frac{8h}{-7} = -3$

58. $\frac{-16z}{21} = -4$

For the following exercises, translate the sentence into an equation. Use x for any unknown quantity. Then solve the equation.

59. If thirteen is added to a number, the result is negative twenty.

60. The product of nine and a number is sixty-three.

61. Negative thirty-seven is eighteen less than a number.

62. Fifteen results from a number being divided by five.

63. The difference of a number and negative four is eleven.

64. If a number is multiplied by negative six, the result is negative twenty-eight.

65. Forty-five is twice a number.

66. The quotient of a number and sixteen is eight.

67. Eight times a number is equal to sixty-two.

68. The sum of a number and twenty-nine is negative twelve.

For the following exercises, use the five-step process to help you solve each problem.

69. This year, a teacher graded 73 fewer final exams than last year. If the teacher graded 483 final exams this year, how many final exams did the teacher grade last year?

70. Twenty percent of a number is 68. What is the number?

71. During the 2015-2016 NBA regular season, Steph Curry averaged 4.8 more points per game than LeBron James. If Steph Curry averaged 30.1 points per game, how many points per game did LeBron James average?

72. A proton is about 1837 times as heavy as an electron. If an electron weighs 2.68 units, how many units does a proton weigh?

73. Neptune is about 30 times as far from the sun as is the Earth. If it takes light 8 minutes to travel from the sun to the Earth, how many minutes does it take light to travel from the sun to Neptune?

Practice E, Part 2 — Answers

25.
Step 1 Let x = last year's price

Step 2 $x - 5 = 79$

Step 3
$$x - 5 = 79$$
$$\underline{+5 \qquad +5}$$
$$x = 84$$

Step 4 $84 - 5 = 79$ ✓

Step 5 Last year's price was $84.

26.
Step 1 Let x = the number

Step 2 $0.08x = 36$

Step 3
$$0.08x = 36$$
$$\frac{0.08x}{0.08} = \frac{36}{0.08}$$
$$x = 450$$

Step 4 $0.08(450) = 36$ ✓

Step 5 The number is 450.

27.
Step 1 Let x = the number of miles Cassie drove

Step 2 $25 = \frac{x}{4.8}$

Step 3
$$4.8 \cdot 25 = 4.8 \cdot \frac{x}{4.8}$$
$$120 = x$$

Step 4 $25 = \frac{120}{4.8}$ ✓

Step 5 She drove 120 miles.

74. A rectangle is 9.75 miles wide and has an area of 120.9 square miles. What is the length of the rectangle?

75. The radius of the sun is about 695,202 km (kilometers). That is about 109 times as long as the radius of the Earth. What is the radius of the Earth?

76. A statistician is collecting data to help him estimate the average income of accountants in California. He needs to collect 390 pieces of data and he is $\frac{2}{3}$ done. How many pieces of data has the statistician collected?

77. In the three football seasons spanning 2013-2015, the Washington Huskies won $\frac{8}{11}$ of the number of games that the Oregon Ducks won. If the Huskies won 24 games during this time period, how many games did the Ducks win?

1.2 Solving Multi-Step Equations

Overview

In the previous section, we used addition, subtraction, multiplication, and division to solve equations. In each case, we only used one step to solve an equation. For that reason, they were called *one-step equations*. However, we often need to solve more complicated equations that require two or more steps.

Here's a problem to consider:

> The length of a rectangular building is sixteen feet less than twice its width. The perimeter of the building is 220 feet. What are the dimensions of the building?

This problem is more complicated than the last one, and it requires multiple steps to solve. We'll revisit the problem after we become familiar with techniques used with solving multi-step equations.

As you study this section, you will learn how to:

- Use multiple steps in order to isolate a variable
- Simplify mathematical expressions before solving equations
- Solve equations with variable terms on both sides
- Recognize identities and contradictions
- Use the table feature of a graphing calculator to solve equations
- Solve application problems involving multi-step equations

A. Isolating the Variable

In solving equations, our goal is to isolate the variable by undoing any operations that are being performed on it. That's how we find the number value of the variable and solve the equation.

When we're presented with expressions that contain multiple operations, we must perform those operations in the correct order.

Order of Operations

Use the following order of operations to simplify mathematical expressions:

1. Simplify expressions inside of parentheses.
2. Evaluate exponents.
3. Multiply or divide, moving from left to right.
4. Add or subtract, moving from left to right.

If there are no parentheses or exponents, this is the correct order:

1. Multiplication and division come first.

2. Addition and subtraction then follow.

To *undo* operations between numbers and variables, we use the order of operations in reverse:

1. Undo addition and subtraction operations first.

2. Then undo multiplication and division operations.

This is similar to the processes of getting dressed and then undressing. If you put on socks and shoes, for example, you have to put the socks on first and then the shoes. Later, after a busy day, when it's time to take those off, the shoes have to come off first and then the socks.

Let's see how that works in the following examples.

Example 1

1. Solve: $4x - 7 = 9$

2. Solve: $\frac{3y}{4} - 5 = -11$

Solutions

1. $4x - 7 = 9$

$4x - 7 =$		9	7 is being subtracted from $4x$.
$+7$	$+7$		Undo the subtraction by adding 7 to both sides of the equation.
$4x =$		16	Now x is being multiplied by 4.
$\frac{4x}{4} =$		$\frac{16}{4}$	Undo this operation by dividing both sides of the equation by 4.
$x =$		4	Now x is isolated on the left side of the equation.

We can check our solution by replacing x with 4 in the original equation.

$4(4) - 7 = 9$ 4 times 4 is 16.

$16 - 7 = 9$ $16 - 7$ is 9.

$9 = 9 \checkmark$

2. $\frac{3y}{4} - 5 = -11$

Again, analyze the left side of the equation to determine what operations are at work:

♦ First, y is multiplied by 3, and the result is $3y$.

♦ Next, $3y$ is divided by 4, which makes $\frac{3y}{4}$.

♦ Finally, 5 is subtracted from $\frac{3y}{4}$.

Can you figure out which number we will get rid of first?

$\frac{3y}{4} - 5 = -11$ 5 is being subtracted from $\frac{3y}{4}$.

$\underline{\quad +5 \quad\quad\quad +5 \quad}$ Undo the subtraction by adding 5 to both sides of the equation.

$\frac{3y}{4} = -6$ 3y is being divided by 4.

$4 \cdot \frac{3y}{4} = 4 \cdot (-6)$ Undo the division by multiplying both sides by 4.

$3y = -24$ The variable y is being multiplied by 3.

$\frac{3y}{3} \quad \frac{-24}{3}$ Undo the multiplication by dividing both sides by 3.

$y = -8$ Now y is isolated on the left side of the equation.

Let's check our work by replacing y with -8 in the original equation.

$\frac{3y}{4} - 5 = -11$ First, substitute -8 for the variable.

$\frac{3(-8)}{4} - 5 = -11$ Next multiply.

$\frac{-24}{4} - 5 = -11$ Then divide.

$-6 - 5 = -11$ Then subtract.

$-11 = -11 \checkmark$

We used three steps to solve this equation. However, if we use the reciprocal approach discussed in the Section 1.1, we can solve the equation in just two steps.

$\frac{3y}{4} - 5 = -11$

$\underline{\quad +5 \quad\quad\quad +5 \quad}$

$\frac{3}{4}y = -6$

$\frac{4}{3} \cdot \frac{3}{4}y = \frac{4}{3} \cdot (-6)$

$y = -8$

Practice A

Now it's your turn to work with some multi-step equations. First undo addition and subtraction, and then undo multiplication and division. When you are done, turn the page and check your solutions.

1. Solve: $3y - 1 = 11$

2. Solve: $\frac{5m}{2} + 6 = 1$

B. Simplifying Expressions before Solving Equations

Sometimes when we solve an equation, we have to simplify the expressions on either side of the equals sign before we undo the operations and isolate the variable. One way to simplify an expression is to combine like terms to create a single term. For example, we can rewrite $3x + 2x$ as $5x$ and rewrite $-6 - 1$ as -7.

Another common skill is applying the distributive property to an expression in order to expand it. The distributive property states that

$$a(b + c) = ab + ac \qquad \text{and} \qquad a(b - c) = ab - ac.$$

For example:

$$3(5x + 9) = 15x + 27 \quad \text{and} \quad -2(7n - 12) = -14n + 24$$

Notice that a negative sign indicates that a quantity is being multiplied by -1:

$$-(3y - 2) \quad = \quad -1(3y - 2) \quad = \quad -3y + 2$$

Simplifying expressions helps us to minimize the number of steps we have to take when solving an equation. Let's take a look at how that works in some actual equations.

Example 2

Solve: $4x + 1 - 3x = (-2)(4)$

Solution

We begin by simplifying both sides of this equation.

$$
\begin{array}{rcl}
4x + 1 - 3x & = & (-2)(4) \\
\end{array}
$$
Combine like terms on the left side: $4x - 3x$ becomes x. Then simplify the numerical expression on the right side.

$$
\begin{array}{rcl}
x + 1 & = & -8 \\
\underline{-1} & & \underline{-1} \\
x & = & -9
\end{array}
$$
1 is being added to x.
Undo this operation by subtracting 1 from both sides.

We can check our work by replacing the variable x with -9 in the original equation and then simplifying. See if you can follow how this works without further explanation.

$$
\begin{array}{rcl}
4x + 1 - 3x & = & (-2)(4) \\
4(-9) + 1 - 3(-9) & = & -8 \\
-36 + 1 + 27 & = & -8 \\
-8 & = & -8 \ \checkmark
\end{array}
$$

Practice A — Answers

1. $y = 4$ 2. $m = -2$

☠ Warning! Incorrect Approach! ☠

When solving equations like this one, some people make the mistake of performing an operation twice on the *same* side of the equation and not doing anything on the other side. The incorrect approach below is an example of this error.

$$4x + 1 - 3x = -8$$
$$\underline{+\ 3x \qquad +\ 3x}$$

In this case, **$3x$ is added *twice* on the left side of the equation.**

$$7x + 1 = -8$$

Nothing has been done to the right side of the equation! If one side of an equation is modified mathematically, then the other side *must* be modified in exactly the same way.

Even if the remaining steps are done correctly, the final result will be incorrect.

By making sure both sides of an equation are simplified *before* beginning to solve the equation, we can guard against making mistakes like this.

Example 3

Solve: $7(m - 6) - 5m = -9 + 1$

Solution

$$7(m - 6) - 5m = -9 + 1$$

Begin by **applying the distributive property on the left side of the equation**. Rewrite $7(m - 6)$ as $7m - 42$. Then simplify the numerical expression on the right side.

$$7m - 42 - 5m = -8$$

Combine like terms on the left side.

$$2m - 42 = -8$$

Now we solve the equation. 42 is being subtracted from $2m$.

$$\underline{+\ 42 \qquad\quad +\ 42}$$

Undo this operation by **adding 42 to both sides**.

$$2m = 34$$

The variable m is being multiplied by 2.

$$\frac{2m}{2} = \frac{34}{2}$$

Undo that operation by **dividing both sides by 2**.

$$m = 17$$

We can now check the solution by replacing m with 17 in the original equation and simplifying.

$$7(17 - 6) - 5(17) = -9 + 1$$
$$7(11) - 85 = -8$$
$$77 - 85 = -8$$
$$-8 = -8 ✓$$

Notice that the variable m appeared twice in the original equation for Example 3. If it's possible to apply the distributive property when solving an equation in which the variable only appears *once*, then applying the distributive property is optional. Consider the next example.

Example 4

Solve: $3(c + 11) - 18 = -24$

Solution

The variable c appears only once in this equation. Therefore, we can solve the equation without applying the distributive property. Look at the operations on the left side of the equation:

- First, 11 is being added to c. This comes first because it's inside parentheses.

- Next, $(c + 11)$ is being multiplied by 3.

- Finally, 18 is being subtracted from $3(c + 11)$.

$$3(c + 11) - 18 = \quad -24 \qquad \text{The last operation on the left side is subtraction.}$$

$$\underline{ + 18 \quad + 18} \qquad \text{Undo the subtraction by adding 18 to both sides of the equation.}$$

$$3(c + 11) = \quad -6$$

$$\frac{3(c + 11)}{3} = \quad \frac{-6}{3} \qquad \text{Next, undo the multiplication by dividing both side by 3.}$$

$$c + 11 = \quad -2 \qquad \text{With the 3 gone, there's no need to enclose } c + 11 \text{ in parentheses.}$$

$$\underline{ - 11 \quad - 11} \qquad \text{Finally, undo the addition by subtracting 11 from both sides.}$$

$$c = \quad -13$$

Check to verify that the solution is correct.

$$3(-13 + 11) - 18 = -24$$

$$3(-2) - 18 = -24$$

$$-6 - 18 = -24$$

$$-24 = -24 \checkmark$$

Some people get confused with the approach we just used — particularly in step two where we divide both sides of the equation by 3 and remove the parentheses. If you prefer to begin by applying the distributive property and simplifying *before* you begin solving, that approach will also work well.

$$3(c + 11) - 18 = \quad -24 \qquad \text{Start by applying the distributive property on the left side.}$$

$$3c + 33 - 18 = \quad -24 \qquad \text{Simplify the left side.}$$

$$3c + 15 = \quad -24 \qquad \text{Now } c \text{ is being multiplied by 3, and 15 is added to the result.}$$

$$\underline{ -15 \quad -15} \qquad \text{To begin solving, undo the addition by subtracting.}$$

$$3c = \quad -39$$

$$\frac{3c}{3} = \quad \frac{-39}{3} \qquad \text{Finish by undoing the multiplication with division.}$$

$$c = -13$$

The first approach allows us to start solving the equation immediately, and we have to use three steps to solve it. The second approach requires a total of four steps — applying the distributive property, simplifying, and then two steps of solving. Both approaches give us the correct solution, so it's up to you to choose the approach you like best. However, if you like the first approach, remember that it only works if the variable appears just *once* in the equation.

There are some situations in which the distributive property is not required, but using it strategically can make the equation easier to solve. In the next example, we will encounter multiple fractions. When that happens, we can multiply both sides of the equation by the least common denominator (LCD) of the fractions. The LCD is the least common multiple of the denominators. When we do this, we create an equivalent equation that involves integers instead of fractions. Integers are positive or negative whole numbers, so this makes the rest of the problem go more smoothly.

Example 5

Solve: $\frac{3}{5}x - \frac{11}{10} = \frac{5}{2}$

Solution

Because the denominators are 5, 10 and 2, the LCD is 10. We begin by multiplying both sides of this equation by 10 and simplifying the results.

$$10 \cdot \left(\frac{3}{5}x - \frac{11}{10}\right) = 10 \cdot \frac{5}{2}$$ Multiply both sides of the equation by 10. Then apply the distributive property on the left side.

$$\frac{30}{5}x - \frac{110}{10} = \frac{50}{2}$$ Simplify both sides of the equation. The simplified expressions shouldn't have any fractions.

$$6x - 11 = 25$$

$$\underline{+11 \qquad +11}$$ Use addition to undo the subtraction.

$$6x = 36$$

$$\frac{6x}{6} = \frac{36}{6}$$ Use division to undo the multiplication.

$$x = 6$$

We'll check our work by replacing x with 6 in the original equation.

$$\frac{3}{5}(6) - \frac{11}{10} = \frac{5}{2}$$ The equation is written with 6 in place of x.

$$\frac{18}{5} - \frac{11}{10} = \frac{5}{2}$$ Rewrite $\frac{3}{5}(6)$ as $\frac{18}{5}$.

$$\frac{36}{10} - \frac{11}{10} = \frac{5}{2}$$ Rewrite $\frac{18}{5}$ as $\frac{36}{10}$ so that you can perform the subtraction on the left side of the equation.

$$\frac{25}{10} = \frac{5}{2}$$ Now simplify.

$$\frac{5}{2} = \frac{5}{2} \checkmark$$

Practice B

Now it's time for you to solve some equations. In some cases, you will need to begin by simplifying the expressions on either side of the equals sign — as we did in the examples — and then solve. When you are done, turn the page and check your solutions.

3. Solve: $16x - 3 - 15x = 8$

4. Solve: $5(n - 3) + 12 = 57$

5. Solve: $4(y - 5) - 3y = -1$

6. Solve: $\frac{3}{4}a + \frac{5}{6} = \frac{2}{3}$

C. Equations with Variable Terms on Both Sides

In many cases, the variable we are solving for appears on both sides of the equation. To solve an equation like this, we have to write an equivalent equation in which the term containing the variable only appears on one side. We can do that with the same techniques that we've used so far in this chapter.

Example 6

Solve: $12x - 4 = 10x + 8$

Solution

$$12x - 4 = 10x + 8$$
$$\underline{-10x \qquad -10x}$$ To remove the variable term on the right, subtract 10x from both sides.
$$2x - 4 = \qquad 8$$ Now we undo the operations and isolate x. We'll finish solving without further explanation. See if you can follow along.

$$2x - 4 = \qquad 8$$
$$\underline{+4 \qquad\quad +4}$$
$$2x = \qquad 12$$

$$\frac{2x}{2} = \frac{12}{2}$$

$$x = 6$$

Let's check our work by replacing x with 6 in the original equation and simplifying both sides.

$$12(6) - 4 = 10(6) + 8$$
$$72 - 4 = 60 + 8$$
$$68 = 68 \checkmark$$

When variable terms appear on both sides of the equation, we get to decide which variable term to get rid of. In the previous example, we got rid of the variable term on the right side of the equation. In the next example, after simplifying expressions, we will get rid of the variable term on the left side of the equation.

Example 7

Solve: $6(1 - 3x) + 1 = 2x - [3(x - 7) - 20]$

Solution

$6(1 - 3x) + 1 = 2x - [3(x - 7) - 20]$	Apply the distributive property to the parentheses on both sides of the equation.
$6 - 18x + 1 = 2x - [3x - 21 - 20]$	Simplify the expression on the left side and the expression inside the brackets on the right side.
$-18x + 7 = 2x - [3x - 41]$	Apply the distributive property to the brackets.
$-18x + 7 = 2x - 3x + 41$	Simplify the expression on the right side.
$-18x + 7 = -x + 41$	This time we'll get rid of the variable term on the left.
$\underline{+18x \qquad\qquad +18x}$	Add $18x$ to both sides.
$7 = 17x + 41$	We finish by undoing operations to isolate x.
$\underline{-41 \qquad\qquad -41}$	
$-34 = 17x$	
$\dfrac{-34}{17} = \dfrac{17x}{17}$	
$-2 = x$	

We won't work through the process of checking our solution here. However, if you replace x with -2 in the original equation, and use the correct order of operations to simplify both sides, you can verify that our answer is correct.

Some people prefer to write the final answer with the variable appearing on the left side of the equation, so that $x = -2$. This is fine. Equality is a symmetric relationship, which means that if $a = b$ then $b = a$.

We can summarize all that we've learned about solving multi-step equations in the box that follows.

Practice B — Answers

3. $x = 11$ 4. $n = 12$ 5. $y = 19$ 6. $a = -\dfrac{2}{9}$

Procedure for Solving Multi-Step Equations

If the variable only appears once in the equation, then we need to:

1. Simplify numerical expressions, if needed.

2. Undo operations to isolate the variable.

This is done in reverse order: The last operation that takes place on the side of the equation with the variable is the first operation to undo. Perform operations on *both* sides of the equation. If the variable appears more than once in the equation, then we need to:

1. Simplify the expressions on either side of the equation. This can include doing things like using the distributive property to get rid of parentheses or brackets, simplifying numerical expressions, and combining like terms.

2. If the variable still appears on both sides of the equation, use addition or subtraction to get rid of the variable term on one side of the equation.

3. Undo operations to isolate the variable.

Practice C

Now it's your turn to solve a few slightly more complicated equations. Isolate the variable on one side of the equation and find its number value. When you are done, turn the page and check your solutions.

7. Solve: $8a + 5 = 3a - 5$

8. Solve: $9y + 3 (y + 6) = 15y + 21$

9. Solve: $3k + 2[4(k - 1) + 3] = 63 - 2k$

D. Recognizing Identities and Contradictions

As you learned in the previous section, an identity is an equation that is always true, and a contradiction is an equation that is never true. If we attempt to solve an identity, all of the variables are eliminated and the resulting equation is a true statement. If we attempt to solve a contradiction, all of the variables are eliminated, and the resulting equation is a false statement.

You'll see how that works in this example.

Example 8

Solve the following equations.

1. Solve: $9x + 3(4 - 3x) = 12$

2. Solve: $-2(10 - 2y) - 4y + 1 = -18$

Solutions

1. $9x + 3(4 - 3x) = 12$

$$9x + 3(4 - 3x) = 12 \qquad \text{Apply the distributive property on the left side.}$$
$$9x + 12 - 9x = 12 \qquad \text{Now simplify the left side.}$$
$$12 = 12$$

The variable x has been eliminated, and the result is a true statement. This means the original equation is an identity.

2. $-2(10 - 2y) - 4y + 1 = -18$

$$-2(10 - 2y) - 4y + 1 = -18 \qquad \text{Apply the distributive property on the left side.}$$
$$-20 + 4y - 4y + 1 = -18 \qquad \text{Next combine like terms to simplify the left side.}$$
$$-19 = -18$$

The variable has been eliminated, but this time the result is a false statement. The original equation is a contradiction.

Be careful when combining like terms! When attempting to solve the first equation in Example 8, some people make the mistake of rewriting $9x - 9x$ as x. That's not correct because x is the same as $1x$. In fact, $9x - 9x = 0x$, and $0x$ is the same as 0.

Practice D

Now it's your turn. Classify each equation as an identity or a contradiction. When you are done, turn the page and check your solutions.

10. $6x + 3(1 - 2x) = 3$

11. $-8m + 4(2m - 7) = 28$

12. $3(2x - 4) - 2(3x + 1) + 14 = 0$

13. $-5(x + 6) + 8 = 3[4 - (x + 2)] - 2x$

E. Graphing Calculator: Using the Table Feature

Sometimes we can use tables to solve equations. In those cases, graphing calculators can quickly generate a table of values. When doing this, our goal is to create a table that gives us values for each side of the equation when the variable is replaced by various numbers. If we discover a value of the variable that causes both sides of the equation to take on the same value, we have the solution for the equation.

In essence, we are using trial-and-error to solve the equation. When working by hand, trial-and-error tends to be a *very* inefficient method. However, when we can make use of technology to do the math for us, this method works well.

The following example will provide instructions for how to work through this process with a TI-83 or TI-84 calculator.

Example 9

Use the table feature of a graphing calculator to solve: $3x + 52 = 7x - 12$

Solution

Step 1: First, we need our calculators to store both sides of this equation separately. With a TI-83 or TI-84, we can do this by accessing the [Y=] screen.

The [⌐] button is located on the left side of the top row of buttons. Once there, we enter 3x+52 for Y_1 and 7x−12 for Y_2. Enter the variable *x* by pressing the [X,T,Θ,*n*] button, which is located in the third row of buttons from the top.

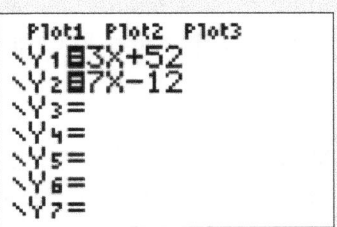

Step 2: We need to view a table that will give us values of Y_1 for the left side of equation and Y_2 for the right side.

With a TI-83 or TI-84, access the table screen by pressing [2nd] and then the [GRAPH] button, which is located on the right side on the top row of buttons. If your table is set up to begin with an *x*-value of 0, and if the value of *x* increases incrementally by 1s, the screen will look like this second one. If you need to adjust your table settings, press [2nd] and then press the [WINDOW] button located in the top row of buttons.

The second screen shows that if *x* is replaced by 0, the left side of the equation takes on a value of 52 and the right side of the equation takes on a value of −12. If *x* is replaced by 1, then the left side takes on a value of 55 while the right side takes on a value of −5. The replacement values continue with *x* = 3, *x* = 4, and so on.

Step 3: Using the arrow keys, we can scroll down the table until we view the screen show here.

This screen shows us that if *x* is replaced by 16, the left side of the equation takes on a value of 100 and the right side of the equation also takes on a value of 100. The *x*-value of 16 causes both sides of the equation to take on the same value. In other words, the solution for the equation is *x* = 16.

Practice C — Answers

7. $a = -2$ 8. $y = -1$ 9. $k = 5$

Using tables may seem like a convenient way to solve equations, but there are limitations. If the solution of an equation is an integer, we should be able to find it on a table. However, if the solution is a very large or a very small integer, like 586 or −752, then we will have to spend a lot of time scrolling through the values before we locate the solution. If the solution of an equation is *not* an integer, then finding it on a table becomes quite difficult — or in many cases, impossible. Despite limitations, however, tables can help us quickly check solutions we've found by hand.

Example 10

When solving the equation $-3[2x - (x + 5)] = x + 3(x + 4) - 6$ by hand, a student got an answer of $\frac{9}{7}$. Use the table feature of a graphing calculator to check this answer.

Begin by storing the left side of the equation as Y_1 and the right side of the equation as Y_2.

```
Plot1 Plot2 Plot3
\Y1■-3(2X-(X+5))
\Y2■X+3(X+4)-6
\Y3=
\Y4=
\Y5=
\Y6=
\Y7=
```

Because the solution we're checking is $\frac{9}{7}$, the table settings must be adjusted so that the x-values will be shown in increments of $\frac{1}{7}$. Remember, to access the table settings, press 2nd and then press the WINDOW button.

```
TABLE SETUP
 TblStart=0
 ⌂Tbl=1/7■
Indpnt: Auto Ask
Depend: Auto Ask
```

With newer calculators like the TI-84 Plus, the calculator mode can be set to display answers as fractions. In that case, the table will look like this when you check the solution.

X	Y₁	Y₂
6/7	87/7	66/7
1	12	10
8/7	81/7	74/7
9/7	78/7	78/7
10/7	75/7	82/7
11/7	72/7	86/7
12/7	69/7	90/7

X=9/7

With older calculators — or newer calculators that are not set to display answers as fractions — the table will look like this when you check the solution. Notice that finding the solution of $\frac{9}{7}$ would have been virtually impossible using the table feature. Before solving the problem, how could we possibly know to set our table up to go by increments of $\frac{1}{7}$? However, once we find the solution, checking it is easy.

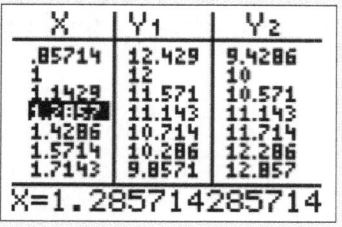

X	Y₁	Y₂
.85714	12.429	9.4286
1	12	10
1.1429	11.571	10.571
1.2857	11.143	11.143
1.4286	10.714	11.714
1.5714	10.286	12.286
1.7143	9.8571	12.857

X=1.285714285714

Practice E

To solve, use the table feature of a graphing calculator. When you are done, turn the page and check your solutions.

14. $-2x + 20 = -7x + 5$ **15.** $4x + 5(x - 4) = 3(x + 5) + 7$ **16.** $5.8x + 13.9 = 3.1x - 15.8$

17. $0.8(0.2x - 7.1) + 0.03x - 1.07 = -0.9(0.4x + 10.5) + 0.08x + 3.64$

F. Application Problems

Now that we've developed techniques for solving multi-step equations, it's time to take on the challenge of more complicated application problems. Before we do that, let's practice translating verbal sentences into algebraic equations.

Example 11

Translate each sentence into an algebraic equation. Then solve the equation.

1. Seven times the difference of a number and two is one hundred nineteen.

2. Twice the sum of the opposite of a number and nine is fifteen less than four times the number.

Solutions

1. Seven times the difference of a number and two is one hundred nineteen.

Based on what we saw in the previous section, we know that "the difference of a number and two" can be written as "$x - 2$." When multiplying this quantity by 7, we must use parentheses, $7 \cdot (x - 2)$ or $7(x - 2)$, and we can then write the equation like this: $7(x - 2) = 119$.

Now it's time to solve this equation. Because the variable only appears once, we'll begin solving immediately.

$$\frac{7(x-2)}{7} = \frac{119}{7}$$ Can you remember why the 7 comes off first?

$$x - 2 = 17$$
$$\underline{+ 2 \qquad + 2}$$
$$x = 19$$

In the expression on the left side of the original equation, the parentheses dictate that 2 is subtracted from x first. After that, 7 is multiplied. The 7 has to come off first because, as we discussed earlier, we undo operations in reverse order when solving equations.

Practice D — Answers

10. identity	**11.** contradiction	**12.** identity	**13.** contradiction

2. Twice the sum of the opposite of a number and nine is fifteen less than four times the number.

"Twice" means that we're multiplying by 2. "The sum of the opposite of a number and nine" is written as $-x + 9$. Because this entire expression is being multiplied by 2, we enclose it in parentheses: $2(-x + 9)$. "Fifteen less than four times the number" is written as $4x - 15$. With all of this in mind, we can write the equation $2(-x + 9) = 4x - 15$.

Because the variable appears more than once, we cannot begin solving immediately. We must begin by applying the distributive property in order to write the left side without parentheses.

$2(-x + 9)$ = $4x - 15$ Apply the distributive property on the left side.

$-2x + 18$ = $4x - 15$ We can eliminate the variable term on the left side by adding $2x$ to
$\underline{\quad +\, 2x}$ $\underline{\quad +\, 2x}$ *both* sides of the equation.

18 = $6x - 15$ It takes two more steps to isolate the variable.
$\underline{\quad +\, 15}$ $\underline{\quad +\, 15}$

33 = $6x$

$\dfrac{33}{6}$ = $\dfrac{6x}{6}$

$\dfrac{11}{2}$ = x

Practice E — Answers

The tables shown here have been generated by storing the left side of the original equation as Y_1 and the right side of the original equation as Y_2.

14.

X	Y₁	Y₂
-6	32	47
-5	30	40
-4	28	33
-3	26	26
-2	24	19
-1	22	12
0	20	5

X= -3

16.

X	Y₁	Y₂
-14	-67.3	-59.2
-13	-61.5	-56.1
-12	-55.7	-53
-11	-49.9	-49.9
-10	-44.1	-46.8
-9	-38.3	-43.7
-8	-32.5	-40.6

X= -11

15.

X	Y₁	Y₂
4	16	34
5	25	37
6	34	40
7	43	43
8	52	46
9	61	49
10	70	52

X=7

17.

X	Y₁	Y₂
-1	-6.94	-5.53
0	-6.75	-5.81
1	-6.56	-6.09
2	-6.37	-6.37
3	-6.18	-6.65
4	-5.99	-6.93
5	-5.8	-7.21

X=2

☠ Warning! Incorrect Approach! ☠

Some people attempt to get rid of the 2 as the first step in solving this equation. In general, you should not attempt to get rid of a number that is associated with a variable inside of parentheses.

$$7(x - 2) = 119$$
$$\underline{+2 \qquad \quad +2}$$
$$7x = 121$$

If we finish solving, we will get $x = \frac{121}{7}$. This is not the correct solution.

Now, let's solve some application problems. We'll use the five-step approach from Section 1.1.

Example 12

An astronomer notices that the star Sirius gives off 3.6 times as much energy as the star Arcturus. Together, they give off 55.844 units of energy. How many units of energy does each star give off?

Solution

Step 1: Let x = number of units of energy given off by Arcturus.	We have two unknowns, so it looks like we need two variables. However, the energy given off by Sirius is defined in terms of the energy given off by Arcturus. That means we only need one variable for Arcturus.
Let $3.6x$ = number of units of energy given off by Sirius.	The energy given off by Sirius is written as an expression with the same variable.
Step 2: $\quad x + 3.6x = 55.844$	The energy of Arcturus, x, plus the energy of Sirius, $3.6x$, equals 55.844
Step 3: $\quad x + 3.6x = 55.844$	Now simplify the left side of the equation and solve the equation for x, the energy given off by Arcturus.

$$4.6x = 55.844$$
$$\frac{4.6x}{4.6} = \frac{55.844}{4.6}$$
$$x = 12.14$$

Now that we know the value of x, the energy given off by Arcturus, we can calculate $3.6x$, the energy given off by Sirius.

$$3.6x = 3.6(12.14)$$
$$= 43.704$$

Step 4: 12.14 + 43.704 = 55.844 ✓   When we add the amount of energy from both stars, we see that our equation is correct.

Step 5:   The star Arcturus gives off 12.14 units of energy. The star Sirius gives off 43.704 units of energy.   We can then translate our mathematical solution into two complete sentences.

Here's another problem that requires us to find two different unknowns.

Example 13

Two consecutive even integers add up to 432. What are the two numbers?

Solution

In order to solve this problem, we must know that consecutive even integers differ by 2.

Step 1:   Let x = the smaller even integer. Let $x + 2$ = the larger even integer.   The larger even integer is 2 more than the smaller even integer.

Step 2: $x + x + 2$ = 432   The smaller integer, x, plus the larger integer, $x + 2$, equals 432.

Step 3: $x + x + 2$ = 432   Now solve to find the value of the smaller integer.

$$2x + 2 = 432$$
$$\underline{-2 \qquad -2}$$
$$2x = 430$$
$$\frac{2x}{2} = \frac{430}{2}$$
$$x = 215$$

Now find the value of the larger integer.

$$x + 2 = 215 + 2$$
$$= 217$$

Step 4: 215 + 217 = 432 ✓   The solutions work. However, 215 and 217 are *odd* integers, not even integers. So the mathematical solution is not an acceptable answer to the original question!

Step 5:   This problem has no solution. There are no two consecutive *even* integers that add up to 432.

It's tempting to think that working through a problem and finding no valid solution is an exercise in futility. However, proving that a problem has no solution can be just as important as solving problems for which valid solutions exist.

We'll finish by revisiting the question posed at the beginning of this section. This problem also requires us to write and solve an equation in order to find one missing value and then use the result to find another missing value.

Example 14

The length of a rectangular building is sixteen feet less than twice the width. The perimeter of the building is 220 feet. What are the dimensions of the building?

Solution

For problems like this one, an optional — but often helpful — step is to make use of a small diagram within our problem-solving process.

Step 1:   Let w = the width of the rectangle.

The first sentence in the problem describes the length of the rectangle in terms of the width. However, because it does not tell us the width, we let w represent the width.

Let $2w - 16$ = the length of the rectangle.

Using the first sentence in the problem, we can write an expression representing the length of the rectangle.

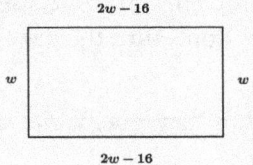

Optionally, we can draw a small diagram of a rectangle and label the sides with the algebraic expressions we came up with in Step 1.

Step 2: $w + 2w - 16 + w + 2w - 16 = 220$

We know that the perimeter of a rectangle is found by adding up the lengths of all four sides. We also know, from the problem, that the perimeter is 220.

$$6w - 32 = 220$$

Before solving our equation, we simplify the left side by combining like terms. Remember, w is the same as $1w$.

Step 3:

$$6w - 32 = 220$$

Now it's time to solve the equation.

$$\underline{+32 \qquad +32}$$

$$6w = 252$$

$$\frac{6w}{6} = \frac{252}{6}$$

$$w = 42$$

Now that we know w, the width, is 42, we can find the value of the length, $2w - 16$

$$2w - 16 = 2(42) - 16$$

$$= 68$$

Step 4: $42 + 68 + 42 + 68 = 220$ ✓   Adding up all four sides gives us the correct perimeter

Step 5: The building has a width of 42 feet and a length of 68 feet.

Practice F

Try your hand at the following application problems. For problems 20 to 22, use the same five-step approach that we've used in the examples. When you are done, turn the page and check your solutions.

18. If five is subtracted from eight times a number, the result is fifty-one. Write and solve an equation to help you find the number.

19. Three times the sum of a number and eight is equal to six times the difference of twice the number and fourteen. Write and solve an equation to help you find the number.

20. Doug's garden produces 5.8 times as many vegetables as Kelsey's garden. Together, the gardens produce 102 pounds of vegetables. How many pounds of vegetables does each garden produce? Answer in a complete sentence.

21. The sum of two consecutive odd integers is 772. What are the two integers? Hint: remember that consecutive odd integers differ by 2.

22. An isosceles triangle has two equal sides called legs, and a third side called the base. If the length of the base is 4 inches less than the twice the length of a leg, and the perimeter of the triangle is 28 inches, find the length of the base and legs.

Exercises 1.2

For the following problems, solve each conditional equation. If the equation is not conditional, then classify the equation as an *identity* or a *contradiction*.

1. $3x + 1 = 16$

2. $6y - 4 = 20$

3. $4a - 1 = 27$

4. $3x + 4 = 40$

5. $2y + 7 = -3$

6. $8k - 7 = -23$

7. $5x + 6 = -9$

8. $7a + 2 = -26$

9. $10y - 3 = -23$

10. $14x + 1 = -55$

11. $\frac{x}{9} + 2 = 6$

12. $\frac{m}{7} - 8 = -11$

13. $\frac{y}{4} + 6 = 12$

14. $\frac{x}{8} - 2 = 5$

15. $\frac{m}{11} - 15 = -19$

16. $\frac{k}{15} + 20 = 10$

17. $6 + \frac{k}{5} = 5$

18. $1 - \frac{n}{2} = 6$

19. $\frac{7x}{4} + 6 = -8$

20. $\frac{-6m}{5} + 11 = -13$

21. $\frac{3k}{14} + 25 = 22$

22. $3(x - 6) + 5 = -25$

23. $16(y - 1) + 11 = -85$

24. $6x + 14 = 5x - 12$

25. $23y - 19 = 22y + 1$

26. $-3m + 1 = 3m - 5$

27. $8k + 7 = 2k + 1$

28. $12n + 5 = 5n - 16$

29. $2(x - 7) = 2x + 5$

30. $-4(5y + 3) + 5(1 + 4y) = 0$

31. $3x + 7 = -3 - (x + 2)$

32. $4(4y + 2) = 3y + 2[1 - 3(1 - 2y)]$

33. $5(3x - 8) + 11 = 2 - 2x + 3(x - 4)$

34. $12 - (m - 2) = 2m + 3m - 2m + 3(5 - 3m)$

35. $-4 \cdot k - (-4 - 3k) = -3k - 2k - (3 - 6k) + 1$

36. $3[4 - 2(y + 2)] = 2y - 4[1 + 2(1 + y)]$

37. $-5[2m - (3m - 1)] = 4m - 3m + 2(5 - 2m) + 1$

38. $\frac{2}{5}x + \frac{7}{10} = -\frac{1}{2}$

39. $-\frac{3}{8}n + \frac{7}{12} = \frac{1}{6}$

40. $\frac{3}{2}k - 4 = \frac{5}{6}k - \frac{7}{9}$

41. $\frac{11}{20}w + \frac{7}{8} = \frac{3}{4}w - \frac{1}{10}$

For the following exercises, use the table feature on a graphing calculator to solve each equation. Each solution will be an integer.

42. $7.6x - 8.3 = 2.8x + 15.7$

43. $9.1x - 2.3 = 5.9x - 5.5$

44. $1.39 + 0.41x = 3.51 - 0.12x$

45. $-2.91 - 1.2x = 4.02 - 1.97x$

46. $13(2x + 6) - 8x + 17 = -7(x - 5) - 15$

47. $5x + 4(6x - 9) = 10 - 2x - 7(-8 - 2x)$

48. $-0.9[5(0.4x + 0.3) - 0.1] = -0.3(2.1x - 3.6)$

49. $2.4[0.75(0.3x - 1.6)] = 0.15[2(0.4x - 0.4) + 1.2]$

Practice F — Answers

18. Equation: $8x - 5 = 51$; Solution: $x = 7$

19. Equation: $3(x + 8) = 6(2x - 14)$; Solution: $x = 12$

20. *Step 1:* Let x = pounds of vegetables from Kelsey's garden, and let $5.8x$ = pounds of vegetables from Doug's garden.

Step 2: $5.8x + x = 102$

Step 3:
$$5.8x + x = 102$$
$$6.8x = 102$$
$$\frac{6.8x}{6.8} = \frac{102}{6.8}$$
$$x = 15$$
$$5.8x = 5.8(15)$$
$$= 87$$

Step 4: $5.8(15) + 15 = 102$ and $87 + 15 = 102$ ✓

Step 5: Doug's garden produces 87 pounds of vegetables. Kelsey's garden produces 15 pounds.

21. *Step 1:* Let x = the smaller integer. Let $x + 2$ = the larger integer.

Step 2: $x + x + 2 = 772$

Step 3:
$$x + x + 2 = 772$$
$$2x + 2 = 772$$
$$\underline{-2 \qquad\qquad -2}$$
$$2x = 770$$
$$\frac{2x}{2} = \frac{770}{2}$$
$$x = 385$$
$$x + 2 = 385 + 2$$
$$= 387$$

Step 4: $385 + 387 = 772$ ✓

Step 5: The two odd integers are 385 and 387.

22. *Step 1:* Let x = the length of a leg. Let $2x - 4$ = length of the base.

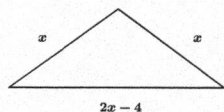

Step 2: $x + x + 2x - 4 = 28$

Step 3:
$$x + x + 2x - 4 = 28$$
$$4x - 4 = 28$$
$$\underline{+4 \qquad\qquad +4}$$
$$4x = 32$$
$$\frac{4x}{4} = \frac{32}{4}$$
$$x = 8$$
$$2x - 4 = 2(8) - 4$$
$$= 12$$

Step 4: $8 + 8 + 2(8) - 4 = 28$ and $8 + 8 + 12 = 28$ ✓

Step 5: Each leg is 8 inches long, and the base is 12 inches long.

For the following exercises, translate the sentence into an equation. Use x for any unknown quantity. Then solve the equation.

50. Three times a number is nine more than twice the number.

51. Five times a number is that number minus two.

52. Twice a number added to six results in thirty.

53. Four less than ten times a number results in sixty-six.

54. Seven more than some number is five more than twice the number.

55. Eleven fifteenths of the sum of a number and two is eight.

56. One tenth of a number is one less than that number.

57. Two more than twice a number is three less than half the number.

58. A number is equal to itself plus four times itself.

59. The sum of a number and seven is divided by two and the result is twenty-two.

60. The difference of ten times a number and one is divided by fourteen, and the result is two.

61. An unknown quantity is decreased by eleven. This result is then divided by fifteen. Now, one is subtracted from this result and five is obtained.

For the following exercises, use the five-step process to help you solve each problem.

62. If a quantity plus 85% more of the quantity is 62.9, what is the original quantity? This might help: Step 1: Let x = original quantity. Step 2: x (original quantity) + $0.85x$ (85% more) = 62.9

63. A company must increase production by 12% over last year's production. The new output will be 56 items. What was last year's output?

64. A company decides to increase production by $1\frac{1}{2}$ times last year's production. The new output will be 2885 items. What was last year's output?

65. A chemistry student has a beaker that contains 84 ml of an alcohol and water solution. Her lab directions tell her there is 4.6 times as much water as alcohol in the solution. How many milliliters of alcohol are in the solution? How many milliliters of water are in the solution?

66. A television commercial advertises that Battery A will last 20 hours longer than twice the life of Battery B. If consumer tests show that Battery A lasts for 725 hours, how many hours does Battery B last?

67. A 1000-ml flask containing a chloride solution will fill 3 beakers of the same size, and 210 ml of the solution will be left over. How many milliliters of the chloride solution will each beaker hold?

68. The sum of two consecutive even integers is 394. What are the even integers?

69. The sum of two consecutive odd integers is −28. What are the odd integers?

70. The sum of three consecutive odd integers is 495. What are the odd integers?

71. The sum of four consecutive odd integers is 260. What are the odd integers?

72. The sum of four consecutive even integers is 564. What are the even integers?

73. A rectangular building is 16 feet longer than it is wide. The perimeter of the building is 260 feet. Find the dimensions of the building.

74. The length of a rectangular picture frame is 15 cm less than twice the width. The perimeter of the frame is 108 cm. Find the dimensions of the frame.

75. Each leg of an isosceles triangle is twice the length of the base. The perimeter of the triangle is 105 inches. Find the length of each side of the triangle.

76. The length of the base of an isosceles triangle is 29 meters less than three times the length of each leg. The perimeter of the triangle is 26 meters. Find the length of each side of the triangle.

1.3 Solving Literal Equations

Overview

In the previous sections, we learned how to use algebra to solve equations containing just one variable. As you have seen, there are many applications of solving one-variable equations. However, there are also many situations that require us to work with multivariable equations.

Here's a multivariable problem:

Solve for b: $-2(10a - 3b) = 11b + 25$

We'll work through this problem later on. For now, though, notice that this equation has two variables, a and b. Because it contains more than one variable, we call it a *literal* equation. "Literal" means that something is made with letters — just like "literature." In this case, the letters are the variables.

In this section, you will learn how to:

- ◆ Recognize and solve literal equations
- ◆ Write linear equations in slope-intercept form
- ◆ Solve more complicated literal equations

A. Literal Equations

A literal equation is an equation involving more than one variable quantity. In other words, a literal equation contains more than one letter. A literal equation is solved for a particular variable when that variable appears by itself on one side of the equation, and the other side of the equation is an expression that does not contain that variable. Here are some examples:

$y = 2x + 7$	In this equation, y is isolated on the left side and doesn't appear on the right side. Therefore, this equation is solved for y.
$d = rt$	This equation is solved for d.
$m = 4m - 7n$	The variable m is by itself on the left side. However, it also appears in the expression on the right side. This equation is *not* solved for m.
$y + 1 = x + 4$	This equation is not solved for any particular variable yet because no variable has been isolated.

In many situations, literal equations relate two or more real-life quantities. Real-life equations like this are often called formulas. For example, the second equation, $d = rt$, is a formula that shows the relationship between d, the distance traveled, r, the average rate of travel, and t, the time spent traveling.

It's often necessary to solve a literal equation for a particular variable. To do that, we have to isolate the variable. We'll use the same techniques we've used earlier in this chapter for undoing operations.

Example 1

1. Solve for y: $y + m = s$

2. Solve for k: $k - 3q = -8q + 5$

3. Solve for r: $9t - 2r = 7$

Solutions

1. Solve for y: $y + m = s$

$$\begin{array}{rl} y + m = & s \\ \underline{-m \qquad -m} & \\ y = & s - m \end{array}$$

On the left side of the equation, m is being added to y.

We undo that operation by subtracting m from both sides.

Now y is isolated on the left side , so the equation is solved for y.

We can check our work by replacing y with the expression $s - m$ in the original equation:

$$\begin{array}{rl} y + m = s \\ s - m + m = s \\ s = s \end{array}$$

First we substitute $s - m$ for y.

We simplify the left side of the equation by combining like terms.

Even without knowing any number values for these variables, we see that our solution is correct.

2. Solve for k: $k - 3q = -8q + 5$

$$\begin{array}{rl} k - 3q = & -8q + 5 \\ \underline{+3q \qquad +3q} & \\ k = & -5q + 5 \end{array}$$

To isolate k, we undo the subtraction of $3q$ from k.

At the beginning of the problem, we did not think of 3 and q separately. Our goal was to isolate k, not q, so we didn't have to undo the operation between 3 and q. Because of that, we treated the expression $3q$ as a single entity.

☠ Warning! Incorrect Approach ! ☠

Here is one incorrect approach:

$$\begin{array}{rl} k - 3q = & -8q + 5 \\ \underline{+3 \qquad +3} & \\ k - q = & -8q + 5 \end{array}$$

This is incorrect because adding 3 to both sides does not undo the operation between 3 and q. 3 and q are being multiplied, and addition does not undo multiplication.

This is also an incorrect approach:

$$k - 3q = -8q + 5$$
$$k - \frac{3q}{3} \quad \frac{-8q}{3} + 5$$
$$k - q = -\frac{8}{3}q + 5$$

In this case, division is being used to undo multiplication. That is appropriate. However, only the terms containing q are being divided by 3. If both sides, in their entirety, had been correctly divided by 3, the equivalent equation would be $\frac{k}{3} - q = -\frac{8}{3}q + \frac{5}{3}$.

3. Solve for r: $9t - 2r = 7$

In this equation, it might be difficult to fully understand what's happening to the variable r, but we can rewrite the expression $9t - 2r$ as $9t + (-2r)$. Now it's easier to see what's happening:

- First, r is being multiplied by -2.

- Second, $9t$ is being added to the result.

$$9t - 2r = 7$$
$$\underline{\quad -9t \qquad -9t \quad}$$ Begin by subtracting $9t$ from both sides.
$$-2r = 7 - 9t$$
$$\frac{-2r}{-2} = \frac{7 - 9t}{-2}$$ Finish by dividing both sides by -2.
$$r = \frac{7 - 9t}{-2}$$

When solving equation 3 in Example 1, we got a negative number in the denominator. In general, mathematicians try to avoid having negative denominators in a final answer.

One way to avoid this is to write an equivalent fraction by moving the negative sign from the dominator out to the front of the full fraction:

$$\frac{7 - 9t}{-2} = -\frac{7 - 9t}{2}$$

We can also split the fraction into two smaller fractions and then simplify the smaller fractions:

$$\frac{7 - 9t}{-2} = \frac{7}{-2} - \frac{9t}{-2}$$
$$= -\frac{7}{2} + \frac{9}{2}t$$

When a fraction is split, each smaller fractions has the same denominator as the original fraction.

In some cases, it's desirable to have two separate terms instead of a single, more complicated fraction. Later on, we will be writing equations in slope-intercept form. When doing this, we want our solutions to appear as two separate terms.

Practice A

Now it's your turn to solve some literal equations. Pay close attention to which variable you are being asked to solve for. When you are done, turn the page and check your solutions.

1. Solve for f: $f + 2d = 5d$

2. Solve for k: $k + 3t - 4 = 8t + 13$

3. Solve for h: $3h - 9p = 11$

4. Solve for y: $7x - 4y = 15$

B. Writing Linear Equations in Slope-Intercept Form

When graphing a linear equation, it helps to write the equation in what is called **slope-intercept form**, which looks like this:

$$y = mx + b$$

An equation in slope-intercept form contains numbers representing slope, m, and y-intercept, b. For example, $y = 3x + 7$ is written in slope-intercept form. We will take a much closer look at slope and y-intercept in Chapter 2. For now, we will focus on solving for y and writing our solution with two separate terms on the right side of the equation. Keep in mind that the letter x should appear in the first term of the solution.

Example 2

Write the following equations in slope-intercept form:

1. $8x + 2y = 22$

2. $0.81x = 0.3y - 2.88$

3. $5x - \frac{3}{2}y = 12$

Solutions

1. $8x + 2y = 22$

$8x + 2y = \quad\quad 22$	$8x$ is being added to $2y$.
$\underline{-8x \quad\quad\quad\quad -8x}$	Undo that operation by subtracting $8x$ from both sides.
$2y = \quad 22 - 8x$	We want the term containing x to come first, so we switch the terms on the right side.
$\frac{2y}{2} = \frac{-8x + 22}{2}$	Now undo the multiplication of 2 and y by dividing.
$y = \frac{-8x + 22}{2}$	The big fraction needs to be written as two separate terms.
$y = -\frac{8}{2}x + \frac{22}{2}$	Finish by simplifying the solution.
$y = \quad -4x + 11$	

We have solved for y, and our solution consists of two terms. Furthermore, the x variable appears in the first of those two terms. The first equation is now in slope-intercept form.

2. $0.81x = 0.3y - 2.88$

$$
\begin{array}{rl}
0.81x = & 0.3y - 2.88 \\
\underline{+\,2.88} & \underline{\qquad +\,2.88} \\
0.81x + 2.88 = & 0.3y
\end{array}
$$
Add 2.88 to both sides.

$\dfrac{0.81x + 2.88}{0.3} = \dfrac{0.3y}{0.3}$ Divide both sides by 0.3.

$\dfrac{0.81x + 2.88}{0.3} = y$ Next, rewrite the big fraction as two separate terms.

$\dfrac{0.81x}{0.3} + \dfrac{2.88}{0.3} = y$ Simplify the solution.

$2.7x + 9.6 = y$ Finally, rewrite the equation so that y is on the left side.

$y = 2.7x + 9.6$

3. $5x - \dfrac{3}{2}y = 12$

$$
\begin{array}{rl}
5x - \dfrac{3}{2}y = & 12 \\
\underline{-\,5x} & \underline{\qquad -\,5x} \\
-\dfrac{3}{2}y = & 12 - 5x
\end{array}
$$
$5x$ is being added to $-\dfrac{3}{2}y$.
Undo that operation by subtracting $5x$ from each side.
Rewrite the expression on the right side so that the term containing x comes first.

$-\dfrac{3}{2}y = -5x + 12$ Now y is being multiplied by $-\dfrac{3}{2}$. The fastest way to undo this operation is by using the reciprocal method.

$-\dfrac{2}{3} \cdot \left(-\dfrac{3}{2}y\right) = -\dfrac{2}{3} \cdot (-5x + 12)$ Simplify the left side and apply the distributive property on the right side.

$y = \dfrac{10}{3}x - \dfrac{24}{3}$ Finish by simplifying the solution.

$y = \dfrac{10}{3}x - 8$

Practice B

Now that you've seen how this works, it's time to try it for yourself. Write each of these equations in slope-intercept form. When you are done, turn the page and check your solutions.

5. $6x + 3y = -12$

6. $12x - 16y = -48$

7. $-0.08x - 0.2y = -1.8$

8. $\dfrac{3}{4}x = -\dfrac{7}{8}y + 14$

C. Solving More Complicated Literal Equations

Let's take a look at some more complicated literal equations. As we do, remember that if the variable we are solving for appears on both sides of the equation, then we have to isolate the term containing the variable on just one side of the equation.

Example 3

Solve for n: $13n - 14 = 21m + 6n - 49$

Solution

$$13n - 14 = 21m + 6n - 49$$

| $-6n$ | $-6n$ | To isolate the term containing n on the left side of the equation, |
| $7n - 14 =$ | $21m - 49$ | subtract $6n$ from both sides. |

| $+14$ | $+14$ | Now that there's a single term containing n on the left side, |
| $7n =$ | $21m - 35$ | undo operations to isolate n. |

$$\frac{7n}{7} = \frac{21m - 35}{7}$$

$$n = \frac{21m}{7} - \frac{35}{7} \qquad \text{Finish by rewriting the right side of the equation as two terms}$$

$$n = 3m - 5 \qquad \text{and then simplify both terms.}$$

☠ Warning! Incorrect Approach! ☠

When simplifying an expression like the one below, some people make the sad mistake of only reducing the 21 and the 7:

$$\frac{21m - 35}{7} = \frac{3 \,\cancel{21}m - 35}{1 \,\cancel{7}}$$

$$= 3m - 35$$

This in not correct! Every term in the expression needs to be divided by 7, not just the 21.

When the variable we are solving for appears inside a set of parentheses, we have to decide whether the distributive property needs to be applied as part of the process of isolating the variable. Remember what we learned in the last section: when solving for a particular variable, we do *not* need to apply the distributive property if variable only appears *once* in the equation.

Practice A — Answers

1. $f = 3d$

2. $k = 5t + 17$

3. $h = \dfrac{11 + 9p}{3}$ Equivalently: $h = \dfrac{11}{3} + 3p$.

4. $y = -\dfrac{15 - 7x}{4}$ Equivalently: $y = -\dfrac{15}{4} + \dfrac{7}{4}x$.

Example 4

Solve for t: $4p(7t - 5p) = 32p$

Solution

The variable t only appears once in this equation, so we won't have to apply the distributive property in this situation.

$$\frac{4p(7t - 5p)}{4p} = \frac{32p}{4p}$$ Divide both sides by $4p$.

$$7t - 5p = 8$$ It takes two more steps to get t by itself.

$$\underline{+5p \qquad\quad +5p}$$

$$7t = 8 + 5p$$

$$\frac{7t}{7} \qquad \frac{8 + 5p}{7}$$

$$t = \frac{8 + 5p}{7}$$ Equivalently, we could write $t = \frac{8}{7} + \frac{5}{7}p$.

The numbers in the numerator of the big fraction, 8 and 5, do not have any common factors with the denominator, 7. Therefore, if we decide to write an equivalent equation with two separate terms instead of one big fraction, neither of the fractions in the equivalent equation can be reduced.

We'll finish by solving the problem presented at the beginning of this section. Remember, if a term containing the variable we are solving for appears inside a set of parentheses and another term containing the same variable appears elsewhere in the equation, we have to apply the distributive property to remove the parentheses. After that, we can combine like terms in order to consolidate the term that contains the variable we are solving for.

Example 5

Solve for b: $-2(10a - 3b) = 11b + 25$

Solution

Because the term $-3b$ appears inside of the parentheses and the term $11b$ appears elsewhere in the equation, we have to apply the distributive property in this example.

$$-2(10a - 3b) = 11b + 25$$ Start by applying the distributive property on the left side of the equation.

Practice B — Answers

5. $y = -2x - 4$ 7. $y = -0.4x + 9$

6. $y = \frac{3}{4}x + 3$ 8. $y = -\frac{6}{7}x + 16$

$$-20a + 6b = 11b + 25$$

$$\underline{-6b \qquad -6b}$$

$$-20a = 5b + 25$$

To consolidate the term containing b on the right side, subtract $6b$ from both sides of the equation.

$$\underline{-25 \qquad -25}$$

$$-20a - 25 = 5b$$

Now take two more steps to isolate b.

$$\frac{-20a - 25}{5} = \frac{5b}{5}$$

$$\frac{-20a - 25}{5} = b$$

Rewrite the left side as two separate terms.

$$\frac{-20}{5}a - \frac{25}{5} = b$$

Simplify the terms on the left side.

$$-4a - 5 = b$$

$$b = -4a - 5$$

Rewrite the equation with b on the left side.

In Example 5, both of the numbers in the numerator of the big fraction, -20 and -25, have a nontrivial factor in common with the number in the denominator, 5. Because of this, it is a good idea to write an equivalent equation with two separate terms instead of one big fraction so that the terms can be simplified completely.

Practice C

Now it's time for you to solve more complicated literal equations. When you see parentheses, review examples 4 and 5 to help you decide whether or not you should apply the distributive property. When you are done, turn the page and check your solutions.

9. Solve for g: $7g - 4v = 12g + 8$

10. Solve for w: $18k - 4w + 5 = 12k + 6w$

11. Solve for n: $8x(13n + 9x) = 24x$

12. Solve for h: $-5(8h - 3y) = -45$

13. Solve for d: $4(2c - d) = 3(5d + 3c)$

14. Solve for z: $4z - 2(7z + 9a) = -5a + 30$

Exercises 1.3

Decide which of the following equations have been solved for a variable and which equations have not yet been solved.

1. $y = 3x + 7$

2. $m = 2k + n - 1$

3. $4a = y - 6$

4. $hk = 2k + h$

5. $2a = a - b + 1$

6. $7m - 3p + 13 = n$

7. $\frac{13q - 5}{9} = d$

8. $1 - 5f - 2w = -z$

Solve each equation for the indicated variable. When you are asked to write an equation in slope-intercept form, you should solve for y and the variable x should appear in the first term of the solution.

9. Solve for n: $n + m = 4$

10. Solve for P: $P + 3Q - 8 = 0$

11. Solve for b: $a + b - 3c = d - 2f$

12. Solve for x: $x - 3y + 5z + 1 = 2y - 7z + 8$

13. Solve for c: $4a - 2b + c + 11 = 6a - 5b$

14. Solve for v: $\frac{v}{12} = g$

15. Solve for p: $pq = 7r$

16. Solve for n: $m2n = 2s$

17. Solve for b: $2.8ab = 5.6d$

18. Solve for p: $\frac{mnp}{2k} = 4$

19. Solve for b: $\frac{-8b}{3c} = -5\,a^2$

20. Solve for c: $\frac{3pc}{2m} = 2b$

21. Solve for Δ: $\frac{\square \cdot \Delta}{\Phi} = \nabla$

22. Solve for x: $5wx - 7 = 4h$

23. Solve for m: $8t - 12mk = 21$

24. Solve for z: $-9rz - 7ny = 40$

25. Solve for Φ: $3\square + 5\Phi\nabla = 2\Delta$

26. Write $8x + y = 11$ in slope-intercept form.

27. Write $3x - y = -7$ in slope-intercept form.

28. Write $15x + 3y = 17$ in slope-intercept form.

29. Write $-12x + 4y = 32$ in slope-intercept form.

30. Write $x + 2y + 10 = 0$ in slope-intercept form.

31. Write $-18x - 8y - 56 = 0$ in slope-intercept form.

32. Write $11x = 55y + 132$ in slope-intercept form.

33. Write $17.76x - 4.8y = 60$ in slope-intercept form.

34. Write $4x + \frac{8}{5}y = 24$ in slope-intercept form.

35. Write $-\frac{2}{3}x - \frac{4}{5}y = \frac{12}{35}$ in slope-intercept form.

36. Solve for h: $7h - 9w + 3 = 12h - 2$

37. Solve for a: $-21a + 15b = -18a + 3b + 27$

38. Solve for n: $4n + 7p - 8 = 2n - 11p + 3$

39. Solve for k: $8(3k + 4c) = -56c$

40. Solve for m: $3x(5m - 7x) = 45x$

41. Solve for q: $-2(q - 3r) = 5(2q - 6r)$

42. Solve for v: $17v = 4(3s - v) + 42$

43. Solve for g: $9f - 11g + 12 = 5g - 2(3f - 4g)$

44. Solve for n: $m = \frac{2n - h}{5}$

45. Solve for P: $t = \frac{Q + 6P}{8}$

46. Solve for c: $-13 = \frac{k - dc}{3} - 4$

47. Solve for j: $\Phi = \frac{\square + 9j}{\Delta}$

Practice C — Answers

9. $g = \frac{-4v - 8}{5}$ Equivalently: $g = -\frac{4}{5}v - \frac{8}{5}$

10. $w = \frac{6k + 5}{10}$ Equivalently: $w = \frac{3}{5}k + \frac{1}{2}$

11. $n = \frac{3 - 9x}{13}$ Equivalently: $n = \frac{3}{13} - \frac{9}{13}x$

12. $h = \frac{9 + 3y}{8}$ Equivalently: $h = \frac{9}{8} + \frac{3}{8}y$

13. $d = -\frac{c}{19}$

14. $z = \frac{-13a + 30}{10}$ Equivalently: $z = -\frac{13}{10}a - 3$

1.4 Solving Linear Inequalities

Overview

We have learned that an equation is a mathematical sentence stating that two things are the same. However, not all mathematical relationships are relationships of equality. We know, for example, that the number of human beings in Oregon is greater than 20. We also know that the average Oregonian consumes less than 10 grams of vitamin C every day.

Here's a problem for you:

> An oak tree that is 105 inches tall is growing next to a pine tree that is 69 inches tall. If the pine tree grows 22 inches per year and the oak tree grows 14 inches per year, in how many years will the pine tree be taller than the oak tree?

In this problem, we're asked to determine when one quantity, the height of the pine tree, will be greater than another quantity, the height of the oak tree. Because this is not a relationship of equality, we will be using an inequality to help us solve the problem. An inequality is a mathematical sentence that asserts that two things are *not* the same. (In some cases, an inequality may assert that two things are not necessarily the same.) In addition to telling us that two expressions are not equal, an inequality typically indicates which of the expressions is greater, and which is less.

When you have worked through this section, you will be able to:

- Know what inequalities are, and understand inequality notation
- Graph inequalities on a number line
- Use algebra to solve linear inequalities and compound inequalities
- Solve application problems involving inequalities

We'll solve the tree problem, and other application problems, at the end of this section

A. Introduction to Inequalities

There are four symbols commonly used to express relationships of inequality. When read from left to right, the symbols have the following meanings:

>	means "is strictly greater than"
<	means "is strictly less than"
≥	means "is greater than or equal to"
≤	means "is less than or equal to"

Operations vs. Inequalities

The words "more" and "less" indicate the operations of addition and subtraction:

Phrase	Expression
"3 more than a number"	$x + 3$
"11 less than a number"	$x - 11$

The words "is more" and "is less" are used in sentences to indicate relationships of inequality:

Sentence	Inequality
"3 is more than a number."	$3 > x$
"11 is less than a number."	$11 < x$

The expression $x > 12$, for example, means that x is strictly greater than 12. There are an infinite number of solutions for this inequality. Any number greater than 12 will make the inequality true when substituted for x. Solutions include such numbers as 15, 103, and 12.01.

Consider the inequality $7 \geq x$. This means that 7 is greater than or equal to x. Some find it easier to identify solutions for inequalities when the variable appears on the left side, so you can write an equivalent inequality by switching sides and reversing the inequality symbol to $x \leq 7$, which means that x is less than or equal to 7.

Rule for Writing Equivalent Inequalities

If the sides of an inequality are switched *and* the inequality symbol is reversed, then the resulting inequality is equivalent to the original. $A > B$ is equivalent to $B < A$.

A statement that uses the not-equal-to sign ($\neq$) also expresses inequality, but it doesn't indicate which expression is greater or less. For example, the statement $y + 1 \neq 5$ only tells us that the expression $y + 1$ is not equal to 5. It would be more useful to know which expression is greater, and which is less. For this reason, it's customary to only use the symbols $<$, $>$, $\geq$, and $\leq$ when writing inequalities.

Linear Inequalities in One Variable

The table below provides examples of linear inequalities that only involve one variable.

Notation	Sentence
$x \leq 12$	x is less than or equal to 12.
$x + 7 > 4$	The sum of x and 7 is greater than 4.
$y + 3 \geq 2y - 7$	The sum of y and 3 is greater than or equal to the difference of $2y$ and 7.
$P + 26 < 10(4P - 6)$	The sum of P and 26 is less than the product of 10 and the difference of $4P$ and 6.
$2r - \frac{9}{5} > 15$	The difference of $2r$ and $\frac{9}{5}$ is greater than 15.

These expressions look like the ones in the table, but they're not linear inequalities in one variable.

$x^2 < 4$ This expression has one variable, but the term x^2 is not linear because the variable is to the second power.

$x \le 5y + 3$ This expression is a linear inequality, but there are two variables. We call something like this a a linear inequality in two variables.

You've learned that a linear equation in one variable may have exactly one solution (conditional equation), an infinite number of solutions (identity), or no solution at all (contradiction). With a linear inequality in one variable, the possible number of solutions will either be an infinite number of solutions or no solution at all.

For example, consider the inequality $x \ge 9$. There are infinitely many solutions because x is equal to or greater than 9. That means that 9 and every number greater than 9 is a solution. On the other hand, consider the inequality $x + 6 < x$. In this case, there is no solution because if 6 is added to a number, the result cannot be less than the original number.

Number Line Graphs

The solutions of an inequality in one variable, when they exist, can be graphed on a number line. Remember, if solutions exist, there are infinitely many of them. We would never try to plot each solution as an individual point on the number line because it would take the rest of our lives, and we would still not be finished. Instead, we draw a bold line that represents all of the possible solutions of the inequality.

The boundary number, the starting point for the solutions, is marked by a dot. If the boundary number is included among the solutions, which happens when we use the symbols $\le$ and $\ge$, we use a solid dot. If the boundary number is not included among the solutions, which happens when we use the symbols $<$ and $>$, we use a hollow dot — also known as an open circle.

Consider this statement: $n \ge 3$. Figure 1 shows how we would graph this on a number line:

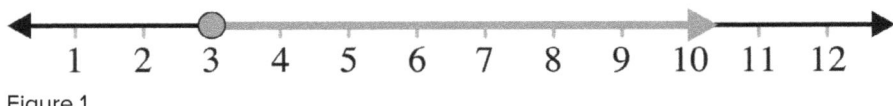

Figure 1.

In this case, every number greater than or equal to 3 is a solution. We draw a bold line that covers every number to the right of 3 on the number line. The arrow at the end of this line indicates that the solutions continue forever in that direction. In other words, the solution set doesn't just consist of numbers that we see on our paper. *Every* number that is greater than 3 is included among the solutions. Because the inequality is "greater than or equal to", the number 3 is included among the solutions. We show that with a solid dot.

Let's look at another example: $k < 7$. This inequality is graphed in Figure 2.

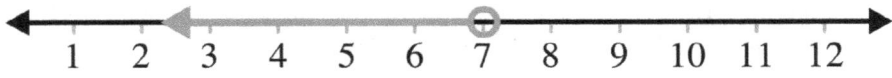

Figure 2.

In this case, every number less than 7 is a solution, so we indicate that with a bold line and an arrow that points to an infinite number of solutions in that direction. Because *k* is strictly less than 7, the number 7 is not included among the solutions. We indicate that exclusion with a hollow dot.

B. The Algebra of Linear Inequalities

We can solve linear inequalities using the same methods that we used to solve linear equations. However, there is an important rule pertaining to inequalities that we have to keep in mind when we do this:

> If both sides of an inequality are multiplied or divided by the same negative number, the inequality symbol must be reversed in order for the resulting inequality to be equivalent to the original inequality.

That may sound a little confusing at first, but it's fairly straightforward. Let's consider the numbers 3 and 7. We know that 3 < 7. Now consider the opposites of these numbers, −3 and −7. With these numbers, the correct inequality statement is −3 > −7. The inequality changes from "is less than" to "is greater than" when we consider opposites.

On a number line, the *opposite* of a number is found by locating the number that is the same distance away from zero on the opposite side of zero. The number 3, for example, is located three spaces to the *right* of zero, while the number −3 is located three spaces to the *left* of zero. Informally, we can say that in order to locate the opposite of 3 on a number line, we take the point representing 3 and "flip" it over to the other side of 0, making sure that the resulting point is still three spaces away from zero. Figure 3 shows what happens when the numbers 3 and 7 are "flipped" over zero in order to locate their opposites.

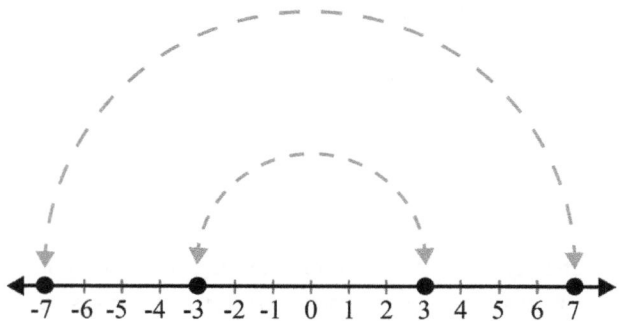

Figure 3. Flipping 3 and 7 to find their opposites on a number line.

Note how the process of "flipping" changes the orientation of the points. The point representing 3 is located to the *left* of the point representing the number 7. The point representing the number −3 is located to the *right* of the point representing the number −7. This explains why the symbol changed from < to > when we re-wrote the inequality statement using opposites.

Now, consider the fact that *every* time we multiply or divide a value by a negative number, the operation involves taking opposites. For example, $-2 \cdot 3$ can be thought of as the *opposite* of $2 \cdot 3$. Informally, we say that this process involves flipping to the other side of zero on the number line. When two distinct points on a number line are "flipped" to the other side of zero, their orientation changes.

The rules below summarize what we can do to an inequality in our quest to isolate the variable.

Rules for Forming Equivalent Inequalities

1. We can add or subtract the same number on both sides of the inequality.

2. We can multiply or divide both sides of the inequality by the same positive number.

3. We can multiply or divide both sides of the inequality by the same negative number *as long as we also reverse the inequality symbol.*

Let's take a look at how this works with some examples.

Example 1

Solve each inequality. Then graph the solution on a number line.

1. $3x > 15$ 2. $2y - 1 \leq 16$ 3. $-8x + 5 < 14$

Solutions

1. $3x > 15$

$\dfrac{3x}{3} > \dfrac{15}{3}$ We isolate x by dividing both sides by 3.

$x > 5$ Because 3 is a positive number, we don't reverse the inequality symbol.

Here is the graph for this solution:

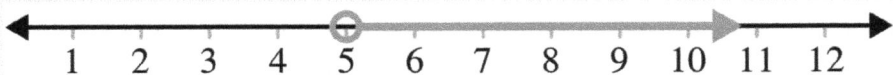

The hollow dot shows us that 5 is not included as a solution to the inequality.

2. $2y - 1 \leq 16$

$$2y - 1 \leq 16$$

$$\underline{+1 \qquad +1}$$

$2y \leq 17$ Add 1 to both sides. Addition and subtraction don't affect the inequality symbol.

$\dfrac{2y}{2} \leq \dfrac{17}{2}$ Divide both sides by 2. Because 2 is positive, the inequality symbol isn't affected.

$y \leq \dfrac{17}{2}$

The solution to problem 2 is written with an improper fraction, which is fine. However, because we're graphing the solution, it may be helpful to write the solution as a mixed number, $y \leq 8\frac{1}{2}$, or as a decimal, $y \leq 8.5$. Here is the graph for the solution:

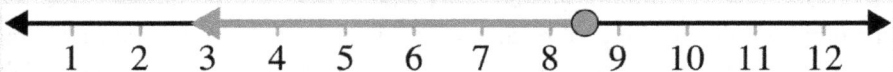

The solid dot shows us that $\frac{17}{2}$ *is* included as a solution to the inequality.

3. $-8x + 5 < 14$

$$
\begin{array}{rcl}
-8x + 5 & < & 14 \\
\underline{-5} & & \underline{-5} \quad \text{Subtract 5 from both sides.} \\
-8x & < & 9 \\
\frac{-8x}{-8} & > & \frac{9}{-8} \quad \text{Divide both sides by } -8. \text{ Because the final step involves division by a} \\
& & \text{negative number, we reverse the inequality symbol.} \\
x & > & -\frac{9}{8}
\end{array}
$$

If it's helpful, you can rewrite this solution as a mixed number or decimal.

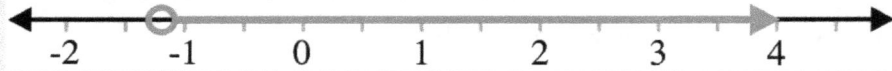

Example 2

Solve: $5 - 3(y + 2) < 6y - 10$. Then graph the solution on a number line.

Solution

This example includes a variable that appears on *both* sides of the inequality symbol.

$$
\begin{array}{rcl}
5 - 3(y + 2) & < & 6y - 10 \quad \text{On the left side, apply the distributive property.} \\
5 - 3y - 6 & < & 6y - 10 \quad \text{Then simplify.} \\
-3y - 1 & < & 6y - 10
\end{array}
$$

At this point, we can choose which side we want the variable term to be eliminated from. Let's look at both cases:

Case 1: Remove variable term from right side	*Case 2: Remove variable term from left side*
$-3y - 1 < 6y - 10$	$-3y - 1 < 6y - 10$
$\underline{-6y} \qquad \underline{-6y}$	$\underline{+3y} \qquad \underline{+3y}$
$-9y - 1 < -10$	$-1 < 9y - 10$
$\underline{+1} \qquad \underline{+1}$	$\underline{+10} \qquad \underline{+10}$
$-9y < -9$	$9 < 9y$
$\frac{-9y}{-9} > \frac{-9}{-9}$	$\frac{9}{9} < \frac{9y}{9}$
$y > 1$	$1 < y$

In Case 1, we are faced with dividing both sides by a negative number in the final solving step. This means that we have to reverse the inequality symbol.

In Case 2, the final step involves dividing both sides by a positive number, so we don't need to worry about reversing the inequality symbol. However, the variable ends up on the right side. This can make things confusing when graphing the solution.

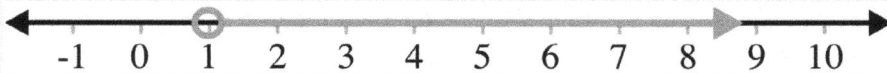

Example 3

Solve $\frac{2z + 7}{-4} \geq -6$. Then graph the solution on a number line.

Solution

$$\frac{2z + 7}{-4} \geq -6 \qquad 2z + 7\text{ is being divided by }-6.$$

$$-4 \cdot \frac{2z + 7}{-4} \leq -4 \cdot (-6) \qquad \text{Undo the division by } \textcolor{gray}{\text{multiplying both sides by }-4}. \text{ Don't forget to} \\ \textcolor{gray}{\text{reverse the inequality symbol.}}$$

$$2z + 7 \leq 24 \qquad \text{Then finish isolating the variable.}$$

$$\underline{-7 \qquad -7}$$

$$2z \leq 17$$

$$\frac{2z}{2} \leq \frac{17}{2}$$

$$z \leq \frac{17}{2}$$

Here is the graph for this solution:

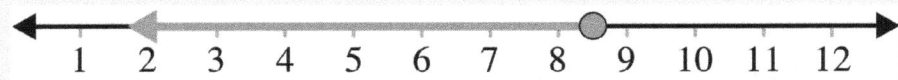

Practice B

Now it's time for you to try your hand at some practice problems. Solve the inequalities and then graph the solution on a number line. When you are done, turn the page and check your solutions.

1. $y - 6 \leq 5$

2. $x + 4 > 9$

3. $4x - 1 \geq 15$

4. $-5y + 16 \leq 7$

5. $7(4s - 3) < 26s + 8$

6. $5(1 - 4h) + 4 < (1 - h)2 + 6$

7. $18 \geq 4(2x - 3) - 9x$

8. $-\frac{3b}{16} \leq 4$

9. $\frac{-7z + 10}{-12} < -1$

10. $-x - \frac{2}{3} \leq \frac{5}{6}$

C. Compound Inequalities

Another type of inequality is the **compound inequality**. A compound inequality combines two statements of inequality with the word "and" or the word "or."

"And" Compound Inequalities

Typically, the solutions for an "and" compound inequality are all the numbers found between two boundary numbers. For example, we can say that x is greater than a and that x is less than b. The common way to write this statement is like this:

$$a < x < b$$

The statement $a < x$ can be restated as $x > a$ without changing the meaning. In the same way, we can read $a < x < b$ as "x is greater that a and less than b." It's also acceptable to say that "a is less than x and x is less than b."

Example 4

Draw graphs for the following "and" inequalities.

 1. $4 < x < 9$ **2.** $-2 < z \leq 0$ **3.** $5 < x + 6 < 8$

Solutions

1. $4 < x < 9$

The letter x is strictly greater than 4 and, at the same time, strictly less than 9. Therefore, x is some number strictly between 4 and 9. The numbers 4 and 9 are not included, so in the number line below, we use open circles at these points when graphing the inequality.

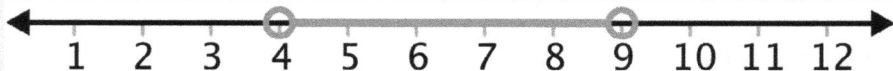

2. $-2 < z \leq 0$

The inequality states that z is strictly greater than -2 and less than or equal to 0. This means that z represents a number between -2 and 0. In this number line, 0 is included in the solutions to the inequality, while -2 is not. We use a solid dot for 0 and a hollow dot for -2 in the graph.

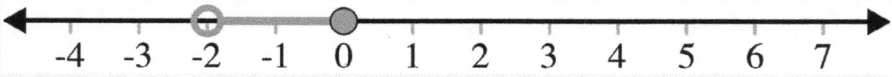

3. $5 < x + 6 < 8$

The expression $x + 6$ represents some number that is strictly greater than 5 and strictly less than 8. Therefore, $x + 6$ represents a number that is strictly between 5 and 8.

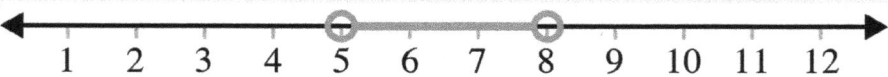

This is a number line graph of all possible values of $x + 6$. It's *not* a graph of all possible values of x.

"Or" Compound Inequalities

Typically, the solutions of an "or" compound inequality are found outside of two boundary numbers, not between them. For example, we can say that x is less than a or that x is greater than b:

$$x < a \quad \text{or} \quad x > b$$

Notice that with an "or" compound inequality, we write two simple inequalities and actually include the word "or." We'll see how this works in the next example.

Example 5

Draw graphs for the following "or" inequalities.

  1. $x < -7$ or $x > -2$ **2.** $x \le 14$ or $x > 16$

Solutions

1. $x < -7$ or $x > -2$

The letter x represents any number that is less than -7 or greater than -2. In our graph of these solutions, the numbers -7 and -2 are not included, so we use open circles at these points.

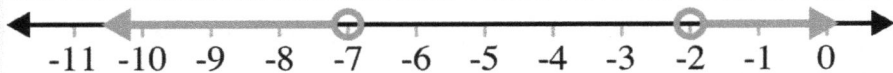

2. $x \le 14$ or $x > 16$

The letter x represents any number that is less than or equal to 14 or any number that is strictly greater than 16. Here4 is how we graph the solution:

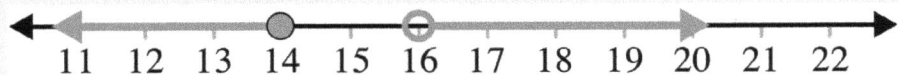

☠ Warning! Incorrect Approach! ☠

It may be tempting to write an "or" compound inequality in combined form. Many people try to write the inequality from Example 5 like this:

$$-7 > x > -2$$

This notation, however, is not mathematically correct. By writing the variable x only once and putting an inequality symbol on either side of it, we imply that both relationships of inequality are happening simultaneously. In this case, we are trying to say that x represents any value that is simultaneously less than -7 and greater than -2. But that's impossible. A number cannot be less than -7 and, at the same time, greater than -2.

When writing an "or" compound inequality, always write two simple inequalities and then put the word "or" between them.

Solving Compound Inequalities

Let's revisit the third inequality from Example 4, $5 < x + 6 < 8$. The statement tells us that the quantity $x + 6$ is strictly between 5 and 8. We would like to know for exactly which values of x the statement $5 < x + 6 < 8$ is true. We will solve this compound inequality by using the properties discussed throughout this chapter. However, because there are *three* parts to this compound inequality instead of two, we apply the rules to all three parts.

Example 6

Solve each compound inequalities. Then graph your solutions.

1. $5 < x + 6 < 8$.

2. $-3 < \dfrac{-2x - 7}{5} < 8$

Solutions

1. $5 < x + 6 < 8$

$\quad 5 < \quad x + 6 < \quad\; 8 \quad$ In the middle part, 6 is being added to x.

$\quad \underline{-6 \qquad -6 \qquad -6} \quad$ Undo this operation by subtracting 6 from all three parts..

$\quad -1 < \quad x \quad < \quad 2 \quad$ Now we know that x must be any number strictly between -1 and 2.

This is how to graph the solutions:

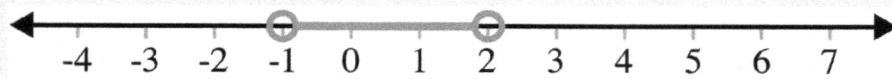

Practice B — Answers

1. $y \le 11$

2. $x > 5$

3. $x \ge 4$

4. $y \ge \dfrac{9}{5}$

5. $s < \dfrac{29}{2}$

6. $h > \dfrac{1}{18}$

7. $x \ge -30$

8. $b \ge -\dfrac{64}{3}$

9. $z < -\dfrac{2}{7}$

10. $x \ge -\dfrac{3}{2}$

2. $-3 < \dfrac{-2x-7}{5} < 8$

$-3 < \dfrac{-2x-7}{5} < 8$ Remember, do all of the solving steps to all three parts of the inequality.

$5 \cdot (-3) < 5 \cdot \dfrac{-2x-7}{5} < 5 \cdot 8$ The first step is to **multiply all three parts by 5**.

$-15 < -2x-7 < 40$ It takes two more steps to isolate the variable.

$\underline{\quad +7 \qquad\qquad +7 \qquad\qquad +7\quad}$

$-8 < -2x < 47$

$\dfrac{-8}{-2} > \dfrac{-2x}{-2} > \dfrac{47}{-2}$ Careful — we're **dividing by a negative number**.

$4 > x > -\dfrac{47}{2}$

It's customary to write "and" compound inequalities so that the lower boundary, in this case $-\frac{47}{2}$, comes first. We do this by switching sides and reversing the inequality symbols:

$4 > x > -\dfrac{47}{2}$ is rewritten as $-\dfrac{47}{2} < x < 4$

Writing the inequality like this also makes it a bit easier to graph:

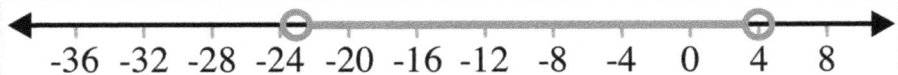

With an "or" compound inequality, solve each simple inequality separately.

Example 7

Solve $x + 5 < 8$ or $2x - 1 \geq 11$. Then graph the solution on a number line.

Solution

$x + 5 < 8$ We will begin by solving the first simple inequality.

$\underline{\quad -5 \qquad -5\quad}$

$x < 3$

$2x - 1 \geq 11$ Next we solve the second simple inequality.

$\underline{\quad +1 \qquad +1\quad}$

$2x \geq 12$

$\dfrac{2x}{2} \geq \dfrac{12}{2}$

$x \geq 6$

$x < 3$ or $x \geq 6$ Now we have the final compound inequality.

Here's how we graph this solution:

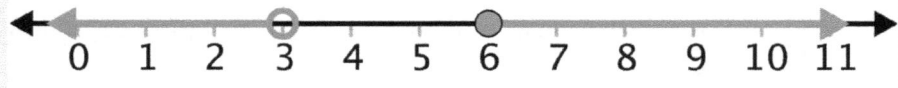

Practice C

For practice problems 11 through 18, your job is to solve each given compound inequality. After solving each inequality, graph the solution on a number line. For practice problems 19 and 20, use what you now understand about compound inequalities to answer the questions. When you are done, turn the page and check your solutions.

11. $4 < x - 5 < 12$

12. $-3 < 7y + 1 < 18$

13. $0 \leq 1 - 6x \leq 7$

14. $-5 \leq \frac{2x + 1}{3} \leq 10$

15. $9 < \frac{-4x + 5}{-2} < 14$

16. $n - 12 \leq -3$ or $2n > 26$

17. $0.7t - 1.3 < 0.8$ or $1.6t + 1.5 > 7.9$

18. $\frac{2}{3}x + \frac{1}{2} \leq \frac{13}{6}$ or $-\frac{7}{8}x \leq -\frac{14}{3}$

19. Explain why $4 < x < -1$ does not have a solution.

20. Explain why all real numbers are a solution of $m < 5$ or $m > 3$.

D. Application Problems

Some application problems require inequalities instead of equations. We can recognize these situations by paying careful attention to the words in those sentences. As we've seen, one important distinction to consider is the difference between phrases like "more than" and "is more than." It's amazing how inserting the word "is" can change the entire meaning of an expression.

The phrases "at least" and "at most" are also common in application problems. The words "at least" indicate that a *lower* boundary is being set. In other words, the acceptable values are greater than or equal to the given number. Similarly, the words "at most" indicate that an *upper* boundary is being set. In other words, the acceptable values are less than or equal to the given number.

Sentence	Inequality
A number is at least twelve.	$d \geq 12$
A number is at most sixty-one.	$v \leq 61$

In the next two examples, we will translate sentences into inequalities and then solve them.

Example 8

Find the acceptable values of the following numbers.

1. The sum of a number and thirteen is less than twenty-one.

2. Twice the difference of a number and ten is at least three times the number.

Solutions

1. The sum of a number and thirteen is less than twenty-one.

$$x + 13 < 21$$ The words "is less than" indicate inequality, so we use the $<$ symbol.

$$\underline{ - 13 \quad\quad - 13}$$

$$x < 8$$

Any number that is less than 8 is an acceptable value.

2. Twice the difference of a number and ten is at least three times the number.

$$2(x - 10) \geq 3x$$ The words "at least" indicate inequality, so we use the $\geq$ symbol.

$$2x - 20 \geq 3x$$

$$\underline{ - 2x \quad\quad - 2x}$$

$$-20 \geq x$$

It may be helpful to rewrite the solution as $x \leq -20$. Any number that is less than or equal to -20 is an acceptable value.

We'll finish off this section by looking at a couple of application problems. We will use the same five-step approach that we used in earlier sections.

Example 9

In the 2012 NFL season, Adrian Peterson was hoping to rush for at least 2,106 yards in order to set the single season rushing record. Going into the last game of the season, he had amassed 1,898 rushing yards. How many rushing yards did he need in the last game in order to set the record?

Solution

Step 1: Let x = the number of rushing yards in the last game.

Step 2: $x + 1,898 \geq 2,106$ The number of rushing yards in the last game, plus the number of rushing yards amassed during the rest of the season need to be at least 2,106. The words "at least" translate to $\geq$.

Step 3: $x + 1{,}898 \geq 2{,}106$

$\underline{\quad -1{,}898 \qquad -1{,}898\quad}$

$x \geq 208$

Step 4: $208 + 1{,}898 = 2{,}106$

$208.1 + 1{,}898 > 2{,}106$

$2{,}106.1 > 2{,}106 \checkmark$

It's difficult to check the solutions of an inequality because there are infinitely many of them. One strategy is to check 208, to verify equality, and then check a value slightly greater than 208, such as 208.1, to verify that the inequality holds true.

Step 5:   He needed to rush for at least 208 yards.

We'll finish by looking at the tree problem from the beginning of this section.

Example 10

An oak tree that is 105 inches tall is growing next to a pine tree that is 69 inches tall. If the pine tree grows 22 inches per year and the oak tree grows 14 inches per year, in how many years will the pine tree be taller than the oak tree?

Solution

The phrase "is taller than" means "is more than" when comparing heights. Therefore, we need to set up an inequality like this: {height of pine tree} > {height of oak tree}

Step 1:   Let x = the number of years that pass.

Step 2: $69 + 22x > 105 + 14x$ The pine tree starts out at a height of 69 inches and adds 22 inches of height per year: $69 + 22x$. The oak tree starts out at a height of 105 inches and adds 14 inches of height per year: $105 + 14x$.

Step 3: $69 + 22x > 105 + 14x$

$\underline{\quad -14x \qquad\qquad -14x\quad}$

$69 + 8x > 105$

$\underline{\quad -69 \qquad\qquad -69\quad}$

$8x > 36$

$\dfrac{8x}{8} > \dfrac{36}{8}$

$x > 4.5$

Step 4: $69 + 22(4.5) = 105 + 14(4.5)$ The solution indicates that any value greater than 4.5 is acceptable. Even though 4.5 is not an acceptable solution, check to see if replacing x with 4.5 causes both sides to be equal. Then check a value slightly greater than 4.5. We'll use 4.6.

$$168 = 168 \checkmark$$

$$69 + 22(4.6) > 105 + 14(4.6)$$

$$170.2 > 169.4 \checkmark$$

Step 5: After more than 4.5 years pass, the pine tree will be taller than the oak tree.

We can also check our solution by using the table feature of a graphing calculator — as you saw in Section 1.2. The left side of the inequality is stored as Y_1, and the right side is stored as Y_2. If the table settings are adjusted to make x increase by increments of 0.1, then the table will look like this.

X	Y1	Y2
4.2	161.4	163.8
4.3	163.6	165.2
4.4	165.8	166.6
4.5	168	168
4.6	170.2	169.4
4.7	172.4	170.8
4.8	174.6	172.2

X=4.5

Practice C — Answers

11. $9 < x < 17$

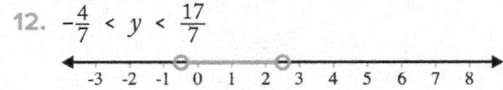

12. $-\frac{4}{7} < y < \frac{17}{7}$

13. $-1 \le x \le \frac{1}{6}$

14. $-8 \le x \le \frac{29}{2}$

15. $\frac{23}{4} < x < \frac{33}{4}$

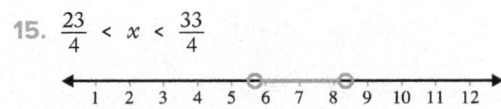

16. $n \le 9$ or $n > 13$

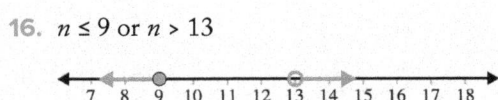

17. $t < 3$ or $t > 4$

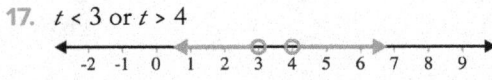

18. $x \le \frac{5}{2}$ or $x \ge \frac{16}{3}$

19. A number cannot be greater than 4 and, at the same time, less than −1.

20. Every number that is not a solution of $m < 5$ is a solution of $m > 3$. In other words, if a number is not less than 5, then it is automatically greater than 3.

Practice D — Answers

Practice D

Now it's time for you to practice solving word problems involving inequalities. For problems 23 and 24, use the five-step approach. When you are done, turn the page and check your solutions.

21. The difference of twice a number and eighteen is greater than seventy-two. Write and solve an inequality to help you find acceptable values of the number.

22. A number is at most four times the sum of the number and thirty-nine. Write and solve an inequality to help you find the acceptable values of the number.

23. Janice has determined that, in order for her business — Janice's Java Hut — to not lose money, the product of the number of employees and the hourly wage must be, at most, $204. If Janice's Java Hut has 12 employees, what is the hourly wage that Janice can pay her employees in order to not lose money?

24. Bovinia has a national debt of $2 trillion and Porcini has a national debt of $5.5 trillion. If Bovinia is adding $650 billion ($0.65 trillion) to its debt each year and Porcini is adding $250 billion ($0.25 trillion), in how many years will the national debt of Bovinia exceed — that is, be greater than — the national debt of Porcini?

Exercises 1.4

For the each of the following exercises, solve the inequality. Then graph the solution on a number line.

1. $x + 7 < 12$

2. $y - 5 \le 8$

3. $y + 19 \ge 2$

4. $x - 5 > 16$

5. $5x - 7 \le 8$

6. $9y - 12 \le 6$

7. $2z + 8 < 7$

8. $4x - 14 > 21$

9. $-5x \le 20$

10. $-8x < 40$

11. $-7z < 77$

12. $-3y > 39$

13. $\frac{x}{4} \ge 12$

14. $\frac{y}{7} > 3$

15. $\frac{2x}{9} \ge 4$

16. $\frac{5y}{2} \ge 15$

17. $\frac{10x}{3} \le 4$

18. $-\frac{5y}{4} < 8$

19. $-\frac{12b}{5} < 24$

20. $\frac{-6a}{7} \le -24$

21. $\frac{8x}{-5} > 6$

22. $\frac{14y}{-3} \ge -18$

23. $-\frac{21y}{8} < -2$

24. $-3x + 7 \le -5$

25. $-7y + 10 \le -4$

26. $6x - 11 < 31$

27. $3x - 15 \le 30$

28. $-2y + \frac{4}{3} \le -\frac{2}{3}$

29. $5(2x - 5) \ge 15$

30. $4(x + 1) > -12$

31. $6(3x - 7) \ge 48$

32. $3(-x + 3) > -27$

33. $-4(y + 3) > 0$

34. $-7(x - 77) \le 0$

35. $\frac{2n - 4}{3} < 6$

36. $\frac{2x + 14}{-5} < -8$

37. $\frac{-4p + 18}{6} \ge 5$

38. $\frac{6w - 18}{9} \ge 3$

39. $2x - 1 < x + 5$

40. $6y + 12 \leq 5y - 1$

41. $3x + 2 \leq 2x - 5$

42. $4x + 5 > 5x - 11$

43. $3x - 12 \geq 7x + 4$

44. $-2x - 7 > 5x$

45. $-x - 4 > -3x + 12$

46. $3 - x \geq 4$

47. $5 - y \leq 14$

48. $2 - 4x \leq -3 + x$

49. $3[4 + 5(x + 1)] < -3$

50. $2[6 + 2(3x - 7)] \geq 4$

51. $7[-3 - 4(x - 1)] < 91$

52. $-2(4x - 1) < 3(5x + 8)$

53. $-5(3x + 2) > -3(-x - 15) - 1$

54. $-.0091x \geq 2.885x - 12.014$

For each of the following exercises, solve the compound inequality. Then graph the solution on a number line.

55. $-6 < 2n - 14 < 8$

56. $17 \leq 3y + 2 \leq 32$

57. $15 < -2x < 22$

58. $-19 \leq 6 - 3n < 3$

59. $-\frac{13}{8} \leq 5x - \frac{3}{4} < \frac{17}{6}$

60. $0.48 < 0.32x + 0.16 < 1.68$

61. $4x + 7 < 17$ or $9x - 15 \geq 66$

62. $-19 + 2m < -47$ or $6m - 23 > -65$

63. $t - 31.7 \leq 14.8$ or $26.9 - t \leq -23.4$

64. $7x + 49 < 5x - 13$ or $-9x + 1 < -6x + 79$

For each of the following exercises, translate the sentence into an inequality. Then solve the inequality to find acceptable values of the number.

65. Twice a number plus one is greater than negative three.

66. The difference of thrice a number and eleven is less than the sum of the number and sixty-seven. Hint: "thrice" means "three times"

67. The sum of a number and twenty-five is at most the difference of fifty-seven and the number.

68. A number increased by ten is at least negative thirty-four.

69. Four times a number is subtracted from the sum of the number and fifty-one, and the result is greater than eighteen.

70. Negative three times a number is decreased by fifteen, and the result is greater than the sum of twice the number and forty.

71. The difference of a number and eight is doubled, and the result is at least as much as the result obtained by quadrupling the number and then adding seven. (Hint: "double" means "times two", and "quadruple" means "times four")

For each of the following exercises, use the five-step process to help you solve the problem.

72. In order to qualify for a bowling tournament, Joey must score at least 705 points over the course of three games. If Joey scored 256 points in the first game and 201 points in the second game, how many points must he score in the third game in order to qualify for the tournament?

73. Maggie is trying to make the cut in the Cherry City Golf Tournament. In order to make the cut, she must score, at most, 151 over the course of two days. If Maggie scored 73 on the first day, what score must she earn on the second day in order to make the cut?

74. Derek, the owner of Tattoo Inferno, has figured out that in order to not lose money, the product of the number of his employees and the hourly wage must be, at most, $312. Tattoo Inferno has sixteen employees. What is the hourly wage that Derek can pay them in order to not lose money?

75. Tommy has \$6,500 in debt and Juan has \$4,600 in debt. If Tommy pays off \$316 in debt each month and Juan pays off \$116 in debt each month, in how many months will the amount of Tommy's debt be less than the amount of Juan's debt?

76. A maple tree that is 206 inches tall is growing next to a cedar tree that is 90 inches tall. If the cedar tree grows 27 inches per year and the maple tree grows 11 inches per year, in how many years will the cedar tree be taller than the maple tree?

77. The area of a rectangular closet is found by multiplying the length by the width ($A = L \cdot W$). The length of the closet is 8 feet. What are the possible measures for the width of the closet if the area must be, at most, 48 square feet? (Hint: The area of any rectangle — or closet — must be more than 0 square feet, so set up an "and" compound inequality with a lower boundary of 0.)

21. $2x - 18 > 72$;   Solution: $x > 45$

22. $x \leq 4(x + 39)$;   Solution: $x \geq -52$

23. *Step 1:*   Let x = hourly wage.

   Step 2: $12x \leq 204$

   Step 3: $\dfrac{12x}{12} \leq \dfrac{204}{12}$

   $x \leq 17$

   Step 4: $12(17) = 204$ ✓

   $12(16.9) < 204$

   $202.8 < 204$ ✓

   Step 5:   Janice can pay her employees — at most — \$17 per hour.

24. *Step 1:*   Let x = the number of years that pass.

   Step 2: $2 + 0.65x > 5.5 + 0.25x$

   Step 3: $\underline{-0.25x} > \underline{-0.25x}$

   $2 + 0.4x > 5.5$

   $\underline{-2} \qquad \underline{-2}$

   $0.4x > 3.5$

   $\dfrac{0.4x}{0.4} > \dfrac{3.5}{0.4}$

   $x > 8.75$

   Step 4: $2 + 0.65(8.75) = 5.5 + 0.25(8.75)$

   $7.6875 = 7.6875$ ✓

   $2 + 0.65(9) > 5.5 + 0.25(9)$

   $7.85 > 7.75$ ✓

   Step 5:   If more than 8.75 years pass, the national debt of Bovinia will exceed the national debt of Porcini.

CHAPTER 2

Linear Models

In chapter 1, we learned the skills necessary to solve linear equations and inequalities. Most of our work involved equations with a single variable, although in section 1.3 we worked with literal equations — equations containing more than one variable.

It is common for two things to exhibit a mathematical relationship. For example, the amount of tuition paid by a college student is related to the number of credits for which he or she is enrolled. We can write models for these types of relationships in a number of ways, one of which is an equation containing two variables. It is important to understand all of the ways that mathematical models can be written, and to be able to convert models from one form to another. This is what we will be focusing on in Chapter 2.

In this chapter, you will study the following topics:

2.1 Introduction to Linear Models

Overview

The ability to model real-life situations mathematically is one of the most important things you will learn in this, and other mathematics courses.

Here's a problem for you:

> A person decides to save money in a jar at home. The person puts $200 in the jar, and every month adds $20 more. Identify an appropriate mathematical model for this situation.

As it turns out, this situation matches up with a linear model. Let's take a look at the different forms a linear model can take, and practice interpreting these models. After that, we will take another look at this problem. As you work through this section, you will learn to:

- Interpret a linear model given as a table, graph, equation, or in words
- Match a given situation with the correct table or graph

A. Interpreting Linear Models

You remember that a linear equation is an equation in which the highest power of the variable(s) is 1. In this chapter, we will explore relationships between two quantities, and our focus will be on relationships that can be illustrated with linear models.

When a linear relationship between two quantities is written as an equation, the equation contains two variables, and each variable has an exponent of 1. However, an equation is not the only way we can write a linear model. In this section, we will construct linear models in four different ways — using words, creating tables, writing equations, and drawing graphs. Before we begin constructing linear models on our own, we'll analyze and interpret linear models that have been created for us.

Example 1

Let's consider a store that sells jugs of milk that each contain one gallon. The store sells each jug for $4. Below are the different ways we can create a model that illustrates the relationship between gallons of milk and cost.

Words
The store sells gallons of milk for $4 each.

Table
If x represents the number of gallons of milk purchased and y represents the total cost, in dollars, then Figure 1 is the table.

x	y
1	4
2	8
3	12
4	16
5	20

Figure 1.

Equation

If x represents the number of gallons of milk purchased and y represents the total cost in dollars, then the equation is

$$y = 4x$$

Graph

Figure 2 is the graph of the equation.

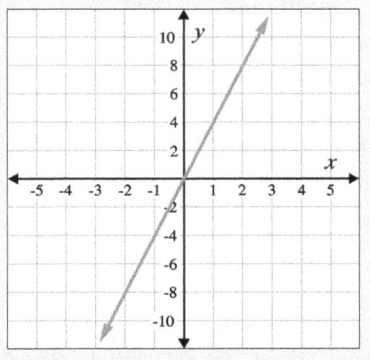

Figure 2. A graph of $y = 4x$.

There are advantages and drawbacks to each form of the model.

Words communicate relationships in everyday life, but if calculations are needed, we must understand the underlying mathematics. In order to determine the cost of 3 gallons of milk, for example, a person needs to understand that *multiplication* must be used: $4(3) = \$12$. The underlying math is easy to interpret in simple relationships like this, but it becomes difficult with complicated relationships.

Tables essentially eliminate the need to understand the underlying math. If we know that the numbers in the x-column correspond to gallons of milk and the numbers in the y-column correspond to cost in dollars, we can use the table in Example 1 to find the cost of 3 gallons of milk. We simply find the number 3 in the x-column and note the corresponding y-value, 12. Unfortunately, tables are limited in their scope. We can't use the table given in Example 1 to determine the cost of 587 gallons of milk.

Equations don't have the same limitations as tables, and unlike words, an equation explicitly shows people the underlying mathematics of a relationship. The equation $y = 4x$ shows people that 4 is *multiplied* by the number of gallons purchased in order to compute the cost. In order to compute the cost of 3 gallons of milk, we replace x with 3. This gives $y = 4(3)$, which can be simplified to read $y = 12$, so the cost is \$12. Unfortunately, equations often generate results that are not practical in real life.

For example, we could use this equation to compute that the cost of 2.7 gallons of milk is \$10.80 or that the cost of −2 gallons of milk is −\$8. Neither of these results can occur when purchasing milk that is sold in 1-gallon jugs. A person can either buy 2 or 3 gallons of milk — not 2.7 gallons — and purchasing negative-2 gallons of milk doesn't make sense in real life. When a model gives results that aren't possible in real life, we call it **model breakdown**.

Graphs give a visual representation of the values computed by an equation. To use the graph in Example 1 to calculate the cost of 3 gallons of milk, we find the number 3 on the x-axis, locate the point on the graph directly above this, and then determine which number on the y-axis lines up with this point, as you see in Figure 3. Based on the graph, when the number of gallons purchased (x) is 3, the cost in dollars (y) is 12.

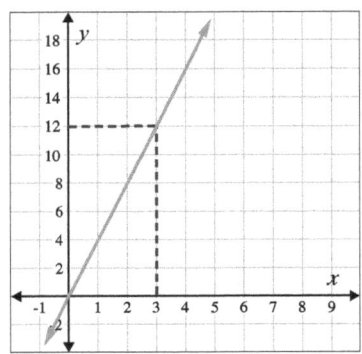

Figure 3.

In many cases, it's faster to use a graph to find a particular result than to use an equation. However, graphs have the same drawbacks as tables and equations. We can't use the graph in Example 1 to determine the cost of 587 gallons of milk, and most points on the graph correspond to values that are decimals or negative numbers. Despite these drawbacks, we often use graphs to present information in real life because they allow us to observe trends and patterns that are not as easily discernible when looking at an equation. Graphs can also be used to help us solve equations and systems of equations.

Example 2

A person graduates with $12 thousand in student loan debt. After graduating, the person pays off debt at a rate of $3 thousand per year until the debt is paid off. In this situation, we will assume that this is a family loan and that no interest is being charged.

Words
A $12 thousand debt is paid off at a rate of $3 thousand per year.

Table
Let x represent the number of years after graduation, and let y represent the amount of debt, in thousands of dollars. See Figure 4.

When reading this table, remember that y represents the amount of debt in *thousands* of dollars. A y-value of 9 corresponds to a debt of $9,000. When interpreting tables, equations and graphs, we need to understand what each variable represents.

Let's use the table to determine how many years after graduation the debt will be $6,000. This amount of debt is represented by a y-value of 6. The table shows us that when the y-value is 6, the corresponding x-value is 2. This means that 2 years after graduation, the person will have $6,000 in student loan debt.

x	y
0	12
1	9
2	6
3	3
4	0

Figure 4.

Equation
If x represents the number of years after graduation and y represents the amount of debt, in thousands of dollars, then the equation is $y = 12 - 3x$. Equivalently, we could write $y = -3x + 12$. To use our equation, we replace y with 6 because a y value of 6 corresponds with $6,000 in debt. This gives us $6 = -3x + 12$.

After solving this equation, we have $x = 2$. Therefore, two years after graduation, the debt will be $6,000.

Graph
Figure 5 presents the graph of our equation. To use the graph to determine how many years after graduation the debt will be $6,000, we locate 6 on the y-axis and move horizontally to find the corresponding point on the graph. We then move down and locate the value on the x-axis, 2, which lines up with this point. Therefore, two years after graduation, the debt will be $6,000.

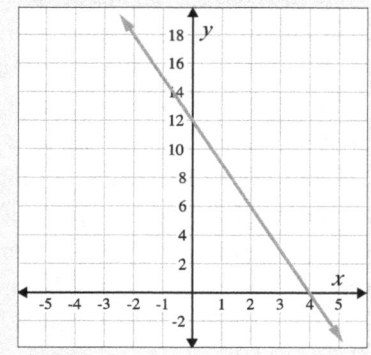

Figure 5.

As in Example 1, we can use the equation and graph in Example 2 to find results involving decimals. In this case, however, that isn't necessarily a problem. In theory, a person could be making little payments all the time so that after 1.48 years, for example, the debt could be 7.56 thousand dollars. This scenario is plausible in real life.

However, as you saw with Example 1, results involving negative numbers do not make sense in real life. According to the equation and graph in Example 2, if it was negative-two years after graduation — in other words, two years *before* graduation — the person would owe $18,000. In real life, however, student loan debt does not work like this. Also, according to the equation and graph in Example 2, the debt would be negative-three thousand dollars five years after graduation. In real life, a person stops payments once the debt is paid off. In this situation, the debt will be paid off in four years, at which time the person will stop making payments.

Practice A

Now it's your turn. You will be given a situation in words and an accompanying model in table, equation, or graph form. Use the model to answer the questions. When you are done, turn the page and check your solutions.

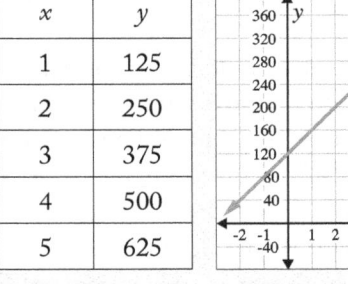

x	y
1	125
2	250
3	375
4	500
5	625

Figure 6.

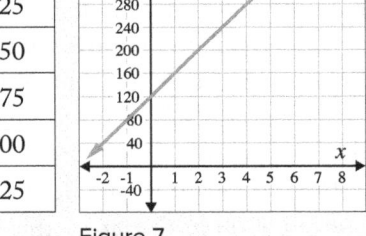

Figure 7.

1. An electronics store sells graphing calculators for $125 each. Let x represent the number of calculators purchased, and let y represent the total cost. See Figure 6.

    a. If you purchase three graphing calculators, what will the total cost be?

    b. How many graphing calculators can you purchase for $500?

2. A person who initially weighs 250 pounds loses 2 pounds per week. Let t represent the number of weeks and let W represent the person's weight, in pounds: $W = -2t + 250$

    a. What is the person's weight after twelve weeks?

    b. After how many weeks will the person weigh 234 pounds?

3. A mechanic charges a very reasonable $40 per hour to work on cars. For a particular repair, $120 worth of parts are required. Let x represent the number of hours that the mechanic works on the car, and let y represent the total cost of the repair. See Figure 7.

    a. If the mechanic works on the car for two hours, what is the total cost of the repair?

    b. If the repair costs a total of $320, how many hours of work did the mechanic put in?

B. Identifying Tables and Graphs that Correctly Model Relationships

When solving application problems, we often construct models that match given situations. It's important to be able to identify whether or not a model has been constructed correctly so that we can check our work for accuracy. For now, we will be analyzing tables and graphs. We'll hold off on analyzing equations until after we have learned about slope and intercepts.

Let's take another look at the problem from the beginning of this section.

Example 3

A person decides to save money in a jar at home. The person puts $200 in the jar, and every month $20 is added. Let x represent the number of months that pass and let y represent the amount of money in the jar. Select the table in Figure 8 that correctly models this situation.

x	y
0	0
1	20
2	40
3	60
4	80

x	y
200	0
220	1
240	2
260	3
280	4

x	y
0	200
1	180
2	160
3	140
4	120

x	y
0	200
1	220
2	240
3	260
4	280

Figure 8a.   Figure 8b.   Figure 8c.   Figure 8d.

Solution

In many cases, we can identify incorrect tables simply by looking at a single row. In this problem, we can eliminate Figure 8a and Figure 8b because the top row of each is incorrect. The top row of Figure 8a indicates that the person began by putting $0 in the jar. The top row of Figure 8b indicates that after 200 months, there is $0 in the jar. Neither of these scenarios match the situation that was described to us.

The top row of Figure 8c and the top row of Figure 8d both indicate that the person began by putting $200 in the jar — this is correct. Therefore, we need to look at another row. The second row of Figure 8c indicates that after one month, the jar will contain $180. This is not correct because the person is *adding* $20 to the jar each month. On the other hand, the second row of Figure 8d indicates that after one month, the jar will contain $220. This reflects the addition of money, so Figure 8d correctly models the situation.

Example 4

A 600-gallon tank is initially full of water. Water is drained from the tank at a rate of 25 gallons per minute. Let x represent the number of minutes that pass while the water is draining, and let y represent the number of gallons of water left in the tank. Select the graph that correctly models this situation.

Solution

We know that the 600-gallon tank is initially full of water and that the amount of water in the tank decreases by 25 gallons per minute. Therefore, after 2 minutes of draining, there will be 550 gallons of water left in the tank: $600 - 25 - 25 = 550$.

Figure 9a indicates that there will be 300 gallons of water in the tank after 2 minutes. Figure 9b indicates that there will be 50 gallons of water in the tank after 2 minutes. Figure 9c indicates that there will be 650 gallons of water in the tank after 2 minutes. Figure 9d indicates that there will be 550 gallons of water in the tank after 2 minutes. Therefore, Figure 9d is correct.

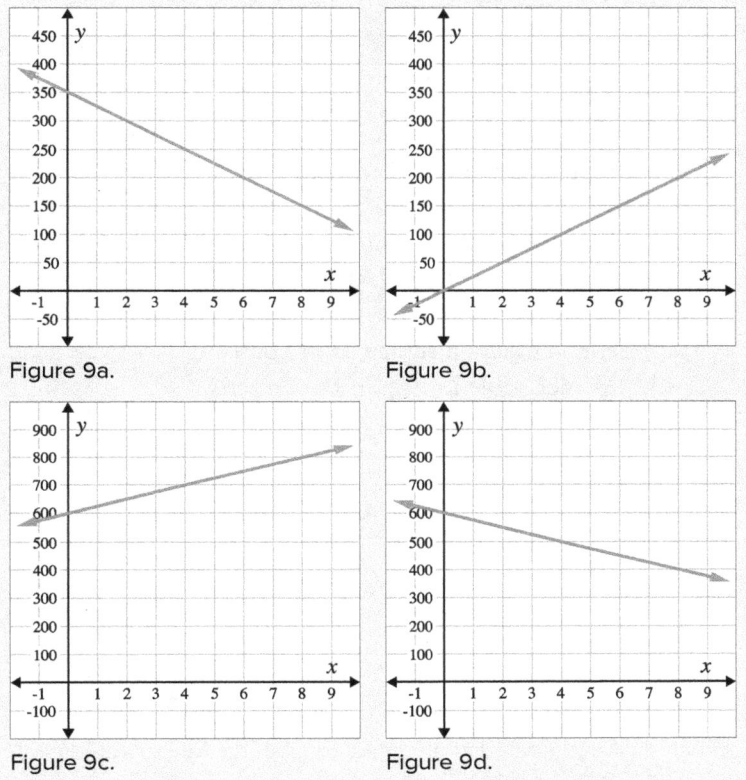

Figure 9a.

Figure 9b.

Figure 9c.

Figure 9d.

In Example 4, we could have begun by identifying which graphs indicate the correct starting amount of water. We know that the tank holds 600 gallons, and that it's initially full of water. In other words, after 0 minutes of draining, there are 600 gallons of water in the tank. Since Figures 9c and 9d both show a y-value of 600 corresponding with an x-value of 0, we know that one of those graphs must be correct. Figure 9c indicates that the amount of water in the tank is increasing as time passes, while Figure 9d indicates that the amount of water in the tank is decreasing as time passes. Therefore, Figure 9d is correct.

Later in this chapter, we'll look more closely at the concepts of increasing vs. decreasing and "starting amount" when the concepts of slope and y-intercept are introduced.

Practice A — Answers

1. a. $375;   b. 4 graphing calculators

2. a. 226 pounds; 8 weeks

3. a. $200;   b. 5 hours

Practice B

Select the table or graph that correctly models each situation. When you are done, turn the page and check your solutions.

4. Manuel spends the weekend with friends who live 385 miles from his home. After his visit, he drives back home at an average rate of 55 miles per hour. Let x represent the time spent driving, in hours, and let y represent Manuel's distance from home, in miles. Which table in Figure 10 correctly models the situation?

5. Lana is getting cable TV at her house. Installation will cost $125, and the cable TV service will cost $64 per month. Let x represent the number of months during which Lana is paying for cable TV, and let y represent the total cost, in dollars.

x	y
385	0
330	1
275	2
220	3
165	4

Figure 10a.

x	y
0	385
1	330
2	275
3	220
4	165

Figure 10b.

x	y
0	385
1	440
2	495
3	550
4	605

Figure 10c.

x	y
0	0
1	55
2	110
3	165
4	220

Figure 10d.

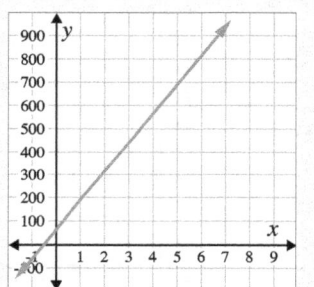

Figure 11a.

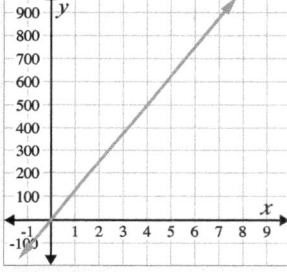

Figure 11b.

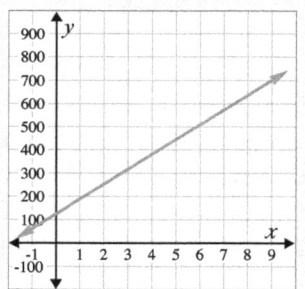

Figure 11c.

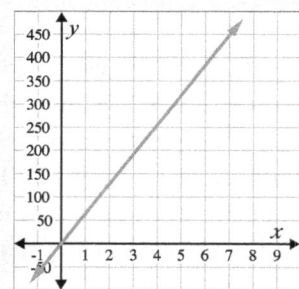

Figure 11d.

Exercises 2.1

For the following exercises, use the table, equation, or graph that accompanies each situation to help you answer the questions. If a problem involves an equation, round your answer off to the nearest hundredth if necessary.

1. Chick-O-Bell sells chicken burritos for $3.65 each. Let b represent the number of chicken burritos purchased, and let C represent the total cost, in dollars.

b	1	2	3	4	5
C	3.65	7.30	10.95	14.60	18.25

   a. You purchase two chicken burritos. What is the total cost?

   b. Your friends then pool their money together, and come up with a total of $17.50. How many burritos can they afford to purchase?

2. Chick-O-Bell is struggling because of weak sales of chicken burritos. They decide to start a new advertising campaign. A marketing expert determines that for every $33 spent on the new ad campaign, 5 additional customers will be acquired. Let a represent the amount of money spent on the new advertising campaign, in dollars, and let C represent the total number of customers that the company has.

a	0	33	66	99	132	165	198
C	160	165	170	175	180	185	190

   a. How many customers does Chick-O-Bell have before any money is spent on the new advertising campaign?

   b. How many customers will Chick-O-Bell have if $165 is spent on the new advertising campaign?

   c. If the Chick-O-Bell's goal is to have a total of 180 customers, how much money should be spent on the new advertising campaign?

3. Janet sets up a budget for discretionary spending each month. Last month, she averaged $52 per week in discretionary spending. Let n represent the number of weeks gone by, and let D represent the amount of discretionary money that Janet had available to spend.

n	0	1	2	3	4
D	240	188	136	84	32

   a. How much discretionary money did Janet have available at the beginning of the month?

   b. How much discretionary money did Janet have available after three weeks?

4. At the end of the Grateful Dead tribute show, there are 1,328 people inside of the auditorium. People leave the auditorium at an average rate of 63 per minute. Let m represent the number of minutes after the show ends, and let P represent the number of people still in the auditorium.

m	1	2	3	4	5	6	7
P	1,265	1,202	1,139	1,076	1,013	950	887

   a. How many people are in the auditorium two minutes after the end of the show?

   b. After how many minutes will there be 950 people in the auditorium?

5. Packy Manniao, a professional boxer, weighing 160 pounds, gains weight at a rate of 1.5 pounds per week. Let t represent the elapsed time in weeks, and let W represent Manniao's weight in pounds.

t	2	3	4	5	6	7	8
W	163	164.5	166	167.5	169	170.5	172

   a. How many weeks does it take him to reach a weight of 170.5 pounds?

   b. How much does he weigh after three weeks?

   c. In order to be in the super middleweight class, Manniao must weigh between 168 and 175 pounds. How many weeks until he reaches the super middleweight class?

6. The corner store sells bananas for $0.59 per pound. Let w represent the weight of bananas purchased, in pounds, and let C represent the total cost in dollars.

   $C = 0.59w$

   a. What is the cost of 3 pounds of bananas?

   b. How many pounds of bananas can I buy for $5?

7. Gus's Gas Station charges $2.75 per gallon for unleaded gasoline. Let g represent the number of gallons of gasoline purchased, and let C represent the total cost, in dollars.

   $C = 2.75g$

   a. What is the cost of 5.6 gallons of gasoline?

   b. How much gasoline can be purchased for $15?

8. Your uncle's 32,000-gallon swimming pool only has 13,500 gallons of water in it, so he uses a hose to add water. The hose delivers water at a flow rate of 24 gallons per minute. Let t represent the amount of time elapsed, in minutes, and let V represent the volume of water in the swimming pool, in gallons.

   $V = 24t + 13,500$

   a. How much water will be in the pool after 39 minutes?

   b. How much water will be in the pool after 4 hours?

   c. How long until the pool has 25,000 gallons of water in it so that we can go for a swim?

9. A 1,400-gallon water tank is initially full of water. The tank is drained at a rate of 32 gallons per minute. Let x represent the elapsed time, in minutes, and let y represent the number of gallons of water still in the tank.

   $y = -32x + 1,400$

   a. How much water will be in the tank after 14 minutes?

   b. After how many minutes will there be 600 gallons of water left in the tank?

   c. After how many minutes will the tank be empty?

10. In 2010, the population of Salem, Oregon, was 154,637. Since then, the population has grown at an average rate of 1,982 people per year. Let n represent the number of years after 2010 and let P represent the population of Salem.

    $P = 1,982n + 154,637$

    a. What was the population of Salem in 2012?

    b. If the population continues to grow at the same rate, what will the population of Salem be in the year 2029?

    c. If the population continues to grow at the same rate, in what year would you expect the population of Salem to reach 225,000?

11. Andre is riding a bicycle across the country. He intends to cover a total distance of 2,860 miles and will ride an average of 110 miles per day. Let x represent the number of days Andre rides his bike, and let y represent the distance, in miles, that Andre still needs to cover before reaching his destination.

    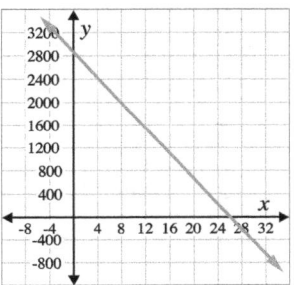

    a. How far will Andre be from his destination after four days of riding?

    b. After how many days of riding will Andre be 1,540 miles away from his destination?

    c. After how many days of riding will Andre reach his destination?

12. Got Dirt, Inc. charges a one-time fee of $180 plus $90 per cubic yard of dirt moved. Let x represent the number of cubic yards of dirt moved, and let y represent the total cost, in dollars.

    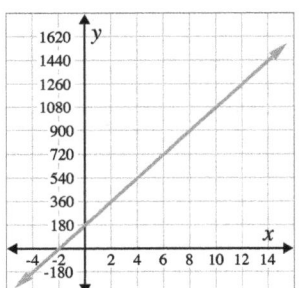

    a. What is the total cost of moving 4 cubic yards of dirt?

    b. If the total cost is $1,260, how many cubic yards of dirt did Got Dirt, Inc., remove?

13. It is estimated that there are 18,000 fish in Loch Lake. The fish population is expected to grow by 500 per year. Let x represent the elapsed time in years, and let y represent the fish population in thousands.

    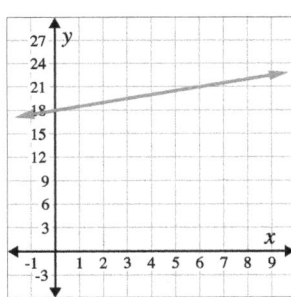

    a. How many fish will be in Loch Lake after two years?

    b. After how many years will there be 22,000 fish in Loch Lake?

Practice B — Answers

4. Figure 10b is correct.   5. Figure 11c is correct.

14. On a particular job, a cabinet-maker uses $150 worth of parts and charges $30 per hour for labor. Let x represent the number of hours of labor, and let y represent the total cost of the job in dollars.

    a. If the cabinet-maker puts in eleven hours of labor, how much will the job cost?

    b. How many hours of labor go into the job if the total cost is $360?

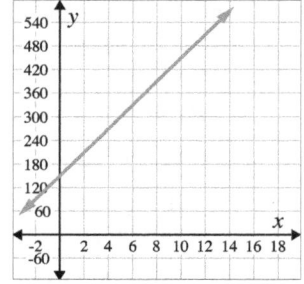

For the following exercises, match each situation with the correct table. In each table, let x represent the number of months elapsed.

x	y
0	20
1	320
2	620
3	920
4	1220

Table 1.

x	y
300	0
320	1
340	2
360	3
380	4

Table 2.

x	y
0	300
1	320
2	340
3	360
4	380

Table 3.

x	y
0	300
1	280
2	260
3	240
4	220

Table 4.

15. Sadie has saved $300. She saves an additional $20 each month.

16. Conchita owes her friend $300. She pays off $20 each month.

17. Nikolay has 20 followers on Twitter. Each month, he gains 300 followers.

For the following exercises, match each situation with the correct table. In each table, let x represent the number of years elapsed.

x	y
67	0
78	1
89	2
100	3
111	4

Table 5.

x	y
0	67
1	56
2	45
3	34
4	23

Table 6.

x	y
0	11
1	78
2	145
3	212
4	279

Table 7.

x	y
0	67
1	78
2	89
3	100
4	111

Table 8.

18. A 67-inch oak tree grows at a rate of 11 inches per year.

19. A family who has owned a large plot of land for 67 years decides to subdivide the land and build houses. The county allows the family to split off one small parcel and build one house every 11 years.

20. A coin collector has 67 rare coins. Every year, the collector auctions off 11 coins to raise money for charity.

For the following exercises, match each situation with the correct graph. For each graph, let x represent the number of days elapsed.

21. A restaurant serves an average of 240 customers per day.

22. A rock collector who owns 240 rocks finds 20 new rocks every day.

23. An elementary school teacher has 240 stickers and gives out 20 stickers every day.

24. An author writes 20 pages per day.

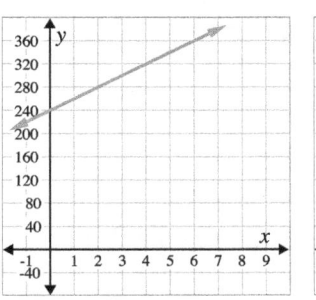

Graph 1.

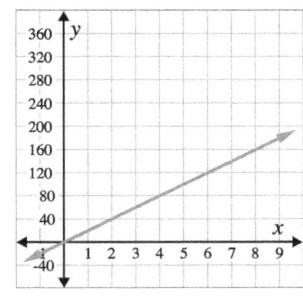

Graph 2.

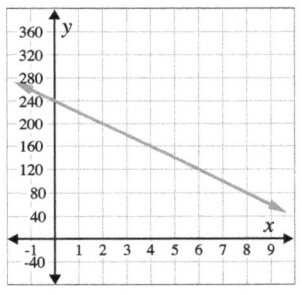

Graph 3.

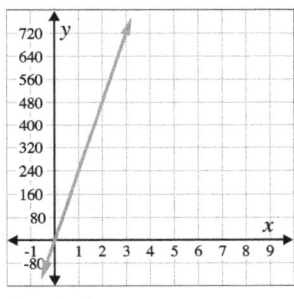

Graph 4.

For the following exercises, match each situation with the correct graph. For each graph, let x represent the number of minutes elapsed.

25. At a car factory, a new car comes off the production line every 42 minutes.

26. The doors to an auditorium are opened before a show. Every minute, 42 people enter the auditorium.

27. A couple picnicking near an anthill notices 35 ants on their picnic blanket. Every minute, 42 more ants show up.

28. A Facebook post has 42 "likes". Every minute, the post gets 35 more "likes".

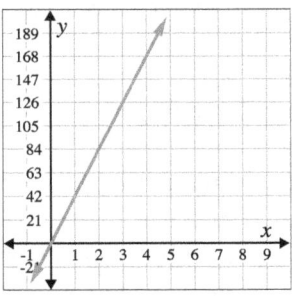

Graph 5.

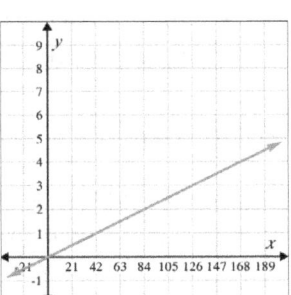

Graph 6.

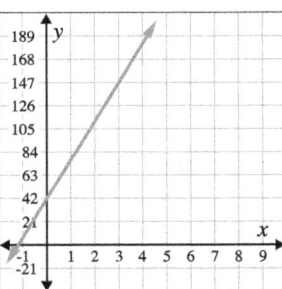

Graph 7.

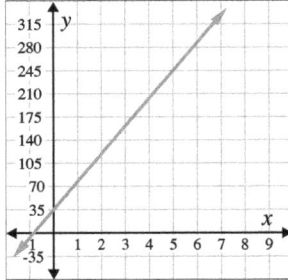

Graph 8.

2.2 Tables and Graphs

Overview

Now that you are familiar with linear models and the various forms they can take, it's time for you to begin learning how to convert a model from one form into another.

Here's a problem:

Graph the equation $x - 3y = -12$.

We are given an equation and instructed to create a graph from it. Before we can do this, we need to cover some of the basics of graphing.

In this section, you will learn how to:

◆ Plot ordered pairs on a coordinate plane and identify which quadrant (if any) a given point is in

◆ Create a scatter plot and determine if it represents a relationship that is exactly linear, approximately linear, or not linear

◆ Find ordered pair solutions of a given equation, and use the ordered pairs to help you graph the equation

A. Introduction to Graphing

A **plane** is a flat surface that extends forever in all directions. We can think of a plane as a never-ending piece of paper. A coordinate-system has been developed for designating position on a plane. To use this system, we place a horizontal number line and a vertical number line on the plane. When this is done, we refer to the plane as a **coordinate plane**. The mathematician and philosopher René Descartes developed the coordinate plane, so some people refer to it as a Cartesian plane.

On a coordinate plane, the horizontal number line is typically called the **x-axis**, and the vertical number line is typically called the **y-axis**. The point where these number lines intersect is called the **origin**. Notice that the origin corresponds to the number 0 on both lines. On the x-axis, the numbers to the right of the origin are positive and the numbers to the left of the origin are negative. On the y-axis, the numbers above the origin are positive and the numbers below the origin are negative.

The axes ("axe – EES") divide the coordinate plane into four regions called **quadrants**. The upper-right quadrant is Quadrant I,

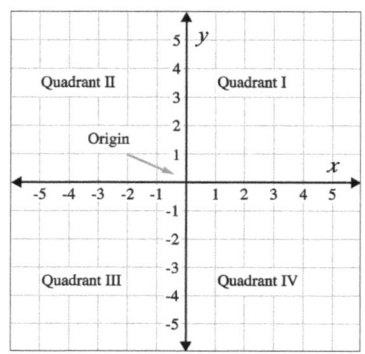

Figure 1.

the upper-left quadrant is Quadrant II, the lower-left quadrant is Quadrant III, and the lower-right quadrant is Quadrant IV. Figure 1 shows what a coordinate plane looks like.

The position of every point on the coordinate plane has a horizontal component and a vertical component. The horizontal component is called the *x*-coordinate, and the vertical component is called the *y*-coordinate. These coordinates correspond with numbers on the axes. We write these coordinates as an *ordered pair*: (*x*-coordinate, *y*-coordinate). The most common strategy for plotting a point is to:

1. Locate the number on the *x*-axis that corresponds with the *x*-coordinate.

2. From this number, move up (or down) the number of spaces indicated by the *y*-coordinate. If the *y*-coordinate is positive move up, if the *y*-coordinate is negative, move down.

Example 1

Plot each point on a coordinate plane.

1. $(3, 2)$

2. $(0, -3)$

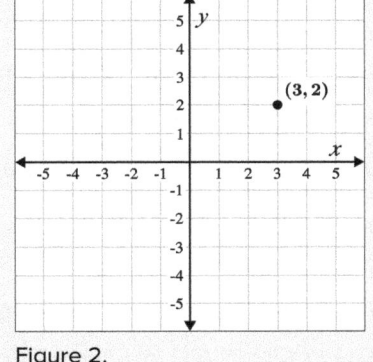

Figure 2.

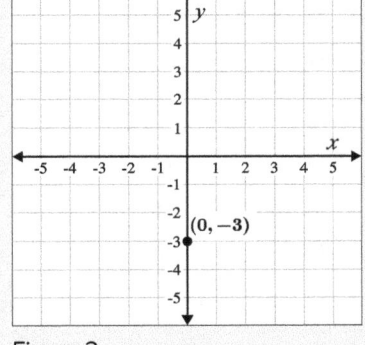

Figure 3.

Solutions

1. $(3, 2)$

We begin by locating the number 3 on the *x*-axis. From this number, we move *up* 2 spaces to locate the position of the point. Notice in Figure 2 that this point is vertically lined up with the number 3 on the *x*-axis and horizontally lined up with the number 2 on the *y*-axis.

2. $(0, -3)$

We begin by locating the number 0 on the *x*-axis — this occurs at the origin. From here, we move *down* 3 spaces to locate the position of the point in Figure 3.

Sometimes, we need to identify which quadrant a given point is in. If we understand how the sign of a coordinate affects the location of the point, then we can identify the quadrant without actually graphing the point. The table below provides a summary of the rules relating coordinates to quadrants.

Sign of x-coordinate	*Sign of y-coordinate*	*Location of Point*
positive	positive	quadrant I
negative	positive	quadrant II
negative	negative	quadrant III
positive	negative	quadrant IV

If a point is on an axis, then one of the coordinates will be 0. In that case, the point is not in a quadrant. If the x-coordinate is 0, then the point is on the y-axis. If the y-coordinate is 0, then the point is on the x-axis. The only point that is on both axes is $(0, 0)$. Remember that this point is called the origin.

Example 2

Where possible, identify the quadrant where each point is located. If a given point is not in a quadrant, tell which axis or axes the point is located on.

1. Point A, $(4, -7)$
2. Point B, $(-12, 15)$
3. Point C, $(9, 0)$
4. Point D, $(-5, -20)$
5. Point E, $(0, 0)$
6. Point F, $(21, 2)$
7. Point G, $(0, 6)$

Solutions

1. Point A has a positive x-coordinate and a negative y-coordinate, so it is in Quadrant IV.
2. Point B has a negative x-coordinate and a positive y-coordinate, so it is in Quadrant II.
3. Point C has a nonzero x-coordinate, but the y-coordinate is 0, so it is on the x-axis.
4. Point D has a negative x-coordinate and a negative y-coordinate, so it is in Quadrant III.
5. Both coordinates of Point E are 0, so it is on both axes. In other words, Point E is the origin.
6. Point F has a positive x-coordinate and a positive y-coordinate, so it is in Quadrant I.
7. Point G has an x-coordinate of 0 and a nonzero y-coordinate, so it is on the y-axis.

Practice A

The following exercises will let you practice the basics of plotting points. When you are done, turn the page and check your solutions.

1. Plot and label each point on a coordinate plane.
   a. $(-2, 5)$
   b. $(3, -1)$
   c. $(-5, 0)$
   d. $(1, 4)$
   e. $(-4, -3)$

2. Tell which quadrant each point is located in. If a given point is not in a quadrant, indicate the axis on which the point is located.
   a. $(27, 49)$
   b. $(0, -115)$
   c. $(68, -706)$
   d. $(-32.9, -19.6)$
   e. $(0, 0)$

B. Scatter Plots

We've discussed four ways in which mathematical models can be constructed — words, tables, equations, and graphs. In addition to interpreting models, we need to be able to convert a model from one form to another. We will begin by converting from table-to-graph.

When we graph multiple points on the same coordinate plane, we create a *scatter plot*. Scatter plots are useful in determining the relationship between two quantities. If two quantities are related, we use the word *correlation* to describe the strength of the relationship.

Example 3

Create a scatter plot from the information in this table.

x	1	2	3	4	5	6	7	8	9
y	8	3	7	1	10	1	6	4	9

Solution

The Figure 4 table gives us parings of x-coordinates and y-coordinates. We can write the information given in this table as a list of ordered pairs: $(1, 8), (2, 3), (3, 7), (4, 1), (5, 10), (6, 1), (7, 6),$ $(8, 4),$ and $(9, 9)$. We can now plot each point on a coordinate plane in order to create a scatter plot. See Figure 5.

The points on the Figure 5 scatter plot don't appear to exhibit any sort of pattern. In other words, they don't show any *correlation* between x and y. If a scatter plot shows correlation, then we can identify the type of relationship that exists between x and y. In this chapter, we are particularly interested in quantities that show a *linear* relationship.

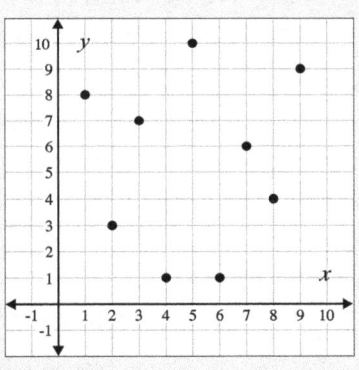

Figure 5.

Example 4

Create a scatter plot from the information in the following table.

x	1	3	5	7	9
y	2	3	4	5	6

Solution

To create the scatter plot, graph the points $(1, 2)$, $(3, 3)$, $(5, 4)$, $(7, 5)$, and $(9, 6)$.

The points on the Figure 6a scatter plot show strong correlation between x and y. In fact, the points appear to line up perfectly. We can verify this by drawing a line. See Figure 6b.

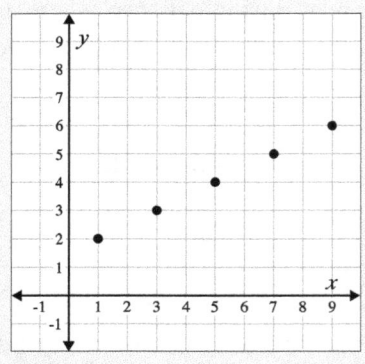

Figure 6a.

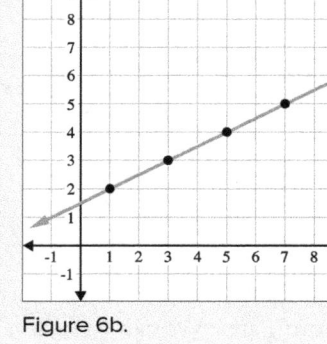

Figure 6b.

When every point on a scatter plot lies on a line, as in Example 4, we say that the points show an **exact linear relationship**. In many cases, the points on a scatter plot do not perfectly line up. However, the scatter plot may still show correlation between x and y.

Example 5

Create a scatter plot from the information below.

x	1	3	4	6	7	9	10
y	10	8	8	5	6	4	2

Solution

To create the scatter plot, graph the points $(1, 10)$, $(3, 8)$, $(4, 8)$, $(6, 5)$, $(7, 6)$, $(9, 4)$ and $(10, 2)$. See Figure 7a.

The points on the scatter plot show strong correlation between x and y. Unlike the scatter plot in Example 4, these points do not perfectly line up. but they appear to exhibit a linear relationship. We can verify this by drawing a line. See Figure 7b.

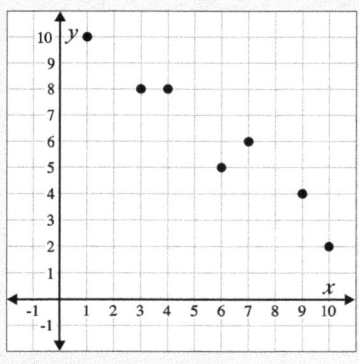

Figure 7a.

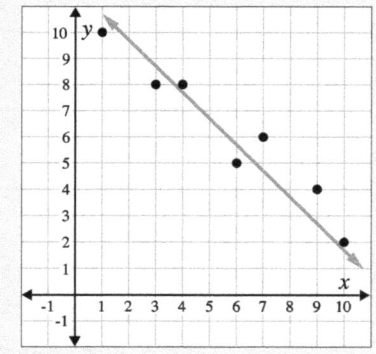

Figure 7b.

When the points on a scatter plot do not perfectly line up, but still exhibit a linear relationship, then we say that the points show an **approximate linear relationship**.

In some cases, the points show strong correlation, but the relationship exhibited is not linear.

Example 6

Create a scatter plot from the information in this table.

x	2	3	4	5	6	7	8	9	10
y	9	5.5	3	1.5	1	1.5	3	5.5	9

Solution

To create the scatter plot, graph the points $(2, 9)$, $(3, 5.5)$, $(4, 3)$, $(5, 1.5)$, $(6, 1)$, $(7, 1.5)$, $(8, 3)$, $(9, 5.5)$ and $(10, 9)$.

The points in Figure 8a show strong correlation, but in this case the relationship between x and y is not linear. Figure 8b shows that the graph modeling this relationship is actually a parabola.

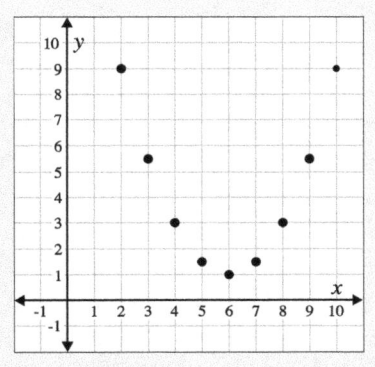

Figure 8a.

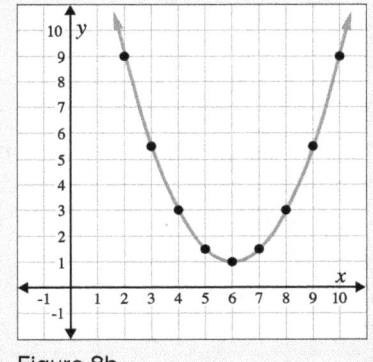

Figure 8b.

We will look at parabolas more in Chapter 6. For now, we will simply state that these points show a relationship that is *not linear*.

Practice A — Answers

1.

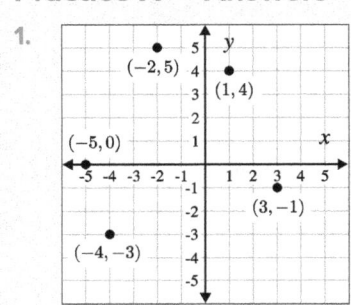

2. a. Quadrant I

b. y-axis

c. Quadrant IV

d. Quadrant III

e. origin (both axes)

Practice B

For each table, create a scatter plot. Then, state whether the points show an exact linear relationship, approximate linear relationship, or a relationship that is not linear. When you are done, turn the page and check your solutions.

1.

x	2	3	4	6	7	8	10
y	5	6	5	7	8	8	9

3.

x	1	2	3	4	5	6
y	11	9	7	5	3	1

2.

x	2	3	5	6	7	9	10
y	0	5	7	8	8	6	1

C. Graphing a Line

Recall from Chapter 1 that a number is a *solution* of an equation in one variable if the equation is true when the variable is replaced with the number.

Now we consider equations in *two* variables. An ordered pair represents the solution of an equation in two variables if the equation is true when each variable is replaced by the appropriate coordinate from the ordered pair. For example, $(5, 8)$ is a solution of the equation $3x - y = 7$ because if x is replaced with 5 and y is replaced with 8, then we get a true equation: $3(5) - 8 = 7$.

Example 7

Determine whether or not each point represents a solution of the equation $5x - 2y = 1$.

   1. $(5, -12)$ **2.** $(1, 2)$ **3.** $(7, 17)$

Solution

1. To determine whether or not $(5, -12)$ is a solution, we replace x with 5 and y with -12 and simplify the result.

$$5x - 2y = 1$$
$$5(5) - 2(-12) = 1$$
$$25 - (-24) = 1$$
$$49 = 1$$

By replacing x with 5 and y with -12, we have created a false equation. Because of that, $(5, -12)$ is not a solution of the original equation.

2. Next we check to see if $(1, 2)$ is a solution.

$$5x - 2y = 1$$
$$5(1) - 2(2) = 1$$
$$5 - 4 = 1$$
$$1 = 1 \checkmark$$

When x is replaced with 1 and y is replaced with 2, the resulting equation is true, so $(1, 2)$ is a solution of the original equation.

3. Finally, we check to see if $(7, 17)$ is a solution.

$$5x - 2y = 1$$
$$5(7) - 2(17) = 1$$
$$35 - 34 = 1$$
$$1 = 1 \checkmark$$

Replacing x with 7 and y with 17 also causes this to be a true equation, so $(7, 17)$ is a solution of the original equation.

In the previous example, we verified two different solutions for the equation $5x - 2y = 1$. In general, an equation in two variables has infinitely many solutions.

Because these solutions can be expressed as ordered pairs, they can be represented by points on a coordinate plane. The phrase "graph the equation" means to locate and graph *every* point that is a solution of the given equation on a coordinate plane.

Example 8

Graph the equation: $-2x + y = -3$

Solution

We'll graph six solutions of this equation (ordered pairs) on the coordinate system below. We'll find these solutions by choosing x-values from -1 to $+4$. Then we'll substitute those x-values into the equation $-2x + y = -3$ and solve the resulting equation to obtain the corresponding y-values. We'll use the table below to keep track of the ordered pairs.

If x =	the equation reads	and y =	Ordered Pair
−1	$-2(-1) + y = -3$	−5	$(-1, -5)$
0	$-2(0) + y = -3$	−3	$(0, -3)$
1	$-2(1) + y = -3$	−1	$(1, -1)$
2	$-2(2) + y = -3$	1	$(2, 1)$
3	$-2(3) + y = -3$	3	$(3, 3)$
4	$-2(4) + y = -3$	5	$(4, 5)$

We plot our six solutions of the equation $-2x + y = -3$ in Figure 9a.

However, an equation in two variables has infinitely many solutions. Let's take another look at the six points we've plotted. These points show an exact linear relationship. Therefore, we should be able to determine the location of all the other points that are solutions of the equation because all of the solution points will lie on the line that contains these six points.

From our discussion in previous sections, we know that this equation is linear because the highest power of each variable is 1. Now we have a second definition for linear equation: A **linear equation** is an equation whose graph is a straight line. In fact, the word "linear" comes from the word "line." We can graph *all* of the solutions of $-2x + y = -3$ by drawing the line that passes through the six solution points we've already plotted, as you see in Figure 9b.

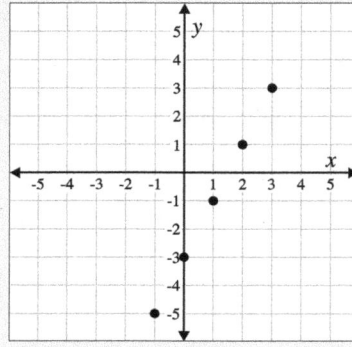

Figure 9a.

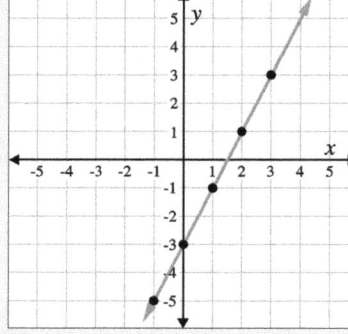

Figure 9b.

Standard Form of a Linear Equation

In Example 8, we graphed an equation that was written in standard form. Suppose that a, b, and c are real numbers and that a and b are not both zero at the same time. In that case, this is the **standard form** of a linear equation:

$$ax + by = c$$

Practice B — Answers

3.

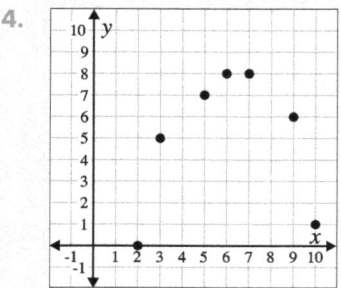

This is an approximate linear relationship.

4.

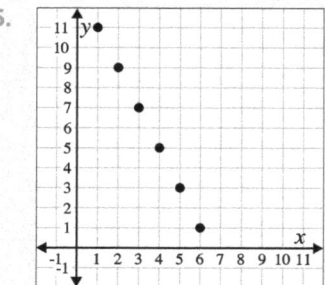

The relationship is not linear.

5.

This is an exact linear relationship.

Ideally, when a linear equation is written in standard form, a, b and c are integers. For example, if we have the equation $0.5x + 3y = 4.5$, then both sides of the equation can be multiplied by 2 to form the equivalent equation $x + 6y = 9$.

It's also best if a, b and c don't have any common factors. For example, if we have the equation $10x - 25y = 35$, both sides of the equation can be divided by 5 to form the equivalent equation $2x - 5y = 7$.

When defining the standard form of a linear equation, it's important to specify that a and b cannot both equal zero at the same time. If a and b were both equal to zero, then we would have $0x + 0y = c$, which simplifies to $0 = c$. This equation has no variables because c represents a real number and the variables x and y have disappeared. A linear equation must have at least one variable.

We can model many situations in real life with linear equations in standard form. However, when graphing a linear equation, it's often easier to do if the equation is first rewritten in slope-intercept form, $y = mx + b$.

Graphing a Line by Using Any Three Points

In Example 8, we found six solution points for a linear equation. In practice, we do not need to find this many points to help us graph a linear equation. Two points are sufficient to define a line. Therefore, we could graph a linear equation by finding just two solution points, plotting those points, and then drawing the line that passes through the points.

In the following example, we'll find three solution points for each equation. The third point serves as a check to make sure our graph is correct. If all three points do not form a straight line, then we've made a mistake that needs to be found and corrected.

We will make sure that each equation is written in slope-intercept form before finding solution points. When a linear equation is solved for y, we can strategically choose the x-coordinates for our ordered pairs and then calculate the corresponding y-coordinates relatively quickly.

Because we get to select the x-coordinates for our ordered pairs, we should *choose wisely*. We want numbers that will give us ordered pairs that are easy to plot. Plotting ordered pairs that involve fractions or decimals can be difficult, so try to choose integer x-coordinates that will lead to integer y-coordinates.

Example 9

Find three solution points for each equation. Then graph each equation on a coordinate plane.

1. $y = x - 4$ 2. $8x + 4y = 0$ 3. $x - 3y = -12$

Solutions

1. $y = x - 4$

This equation is already in slope-intercept form. The equation tells us that each y-coordinate is found by subtracting 4 from the corresponding x-coordinate. This means that, for any integer x-coordinate, the corresponding y-coordinate will also be an integer. We will choose -1, 0, and 1 for x-coordinates.

x	with x-coordinate plugged in	y	(x, y)
−1	$y = -1 - 4$	−5	(−1, −5)
0	$y = 0 - 4$	−4	(0, −4)
1	$y = 1 - 4$	−3	(1, −3)

Next, we plot the three points on a coordinate plane. All three points in Figure 10a form a straight line. Therefore, we can be reasonably sure that our calculations were correct. We finish by drawing a line through the points in Figure 10b.

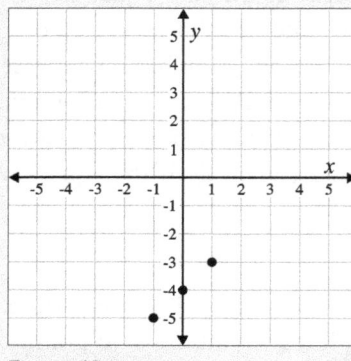

Figure 10a.

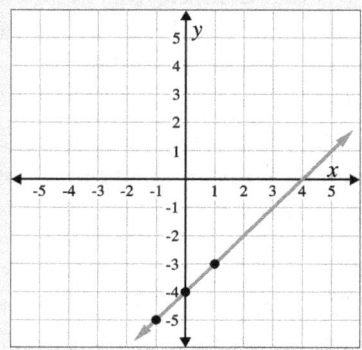

Figure 10b. This line is the graph of the equation $y = x - 4$.

2. $8x + 4y = 0$.

This equation is in standard form. We begin by solving for y.

$$8x + 4y = 0$$
$$\underline{-8x \qquad -8x}$$
$$4y = -8x$$
$$\frac{4y}{4} = \frac{-8x}{4}$$
$$y = -2x$$

This equation indicates that each y-coordinate is found by multiplying the corresponding x-coordinate by −2. Therefore, for any integer x-coordinate, the corresponding y-coordinate will also be an integer. We will choose −1, 0, and 1 for x-coordinates.

x	with x-coordinate plugged in	y	(x, y)
−1	$y = -2(-1)$	2	(−1, 2)
0	$y = -2(0)$	0	(0, 0)
1	$y = -2(1)$	−2	(1, −2)

Now, we can plot the three points from the table on a coordinate plane.

After verifying that the points form a straight line in Figure 11a, we finish by drawing the line in Figure 11b.

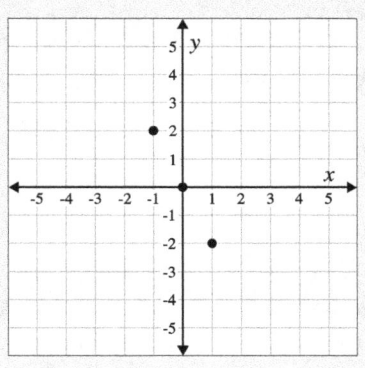

Figure 11a.

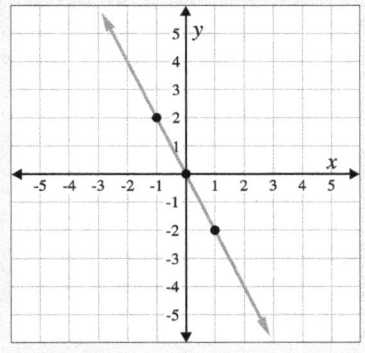

Figure 11b.

3. $x - 3y = -12$. This is the problem we saw at the beginning of the section. Solve for y.

$$x - 3y = -12$$
$$\underline{\quad -x \qquad\quad -x \quad}$$
$$-3y = -x - 12$$
$$\frac{-3y}{-3} = \frac{-x - 12}{-3}$$
$$y = \frac{-x}{-3} - \frac{12}{-3}$$
$$y = \frac{1}{3}x + 4$$

This equation tells us that each y-coordinate is found by multiplying the corresponding x-coordinate by $\frac{1}{3}$ and then adding 4. This means that only an x-coordinate that is a multiple of 3 will yield an integer y-coordinate. Because of this, we choose $-3, 0,$ and 3 for x-coordinates.

x	with x-coordinate plugged in	y	(x, y)
-3	$y = \frac{1}{3}(-3) + 4$	3	$(-3, 3)$
0	$y = \frac{1}{3}(0) + 4$	4	$(0, 4)$
3	$y = \frac{1}{3}(3) + 4$	5	$(3, 5)$

We can now plot the three points from table above, verify that they form a straight line, and draw the line. The points and line can be seen in Figure 12.

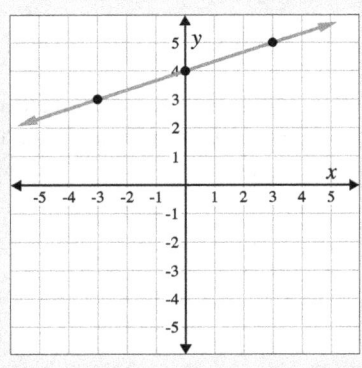

Figure 12.

Practice C

Graph the following equations. When you are done, turn the page and check your solutions.

6. $y=3x-1$

7. $2x+y=3$

8. $x-5y=5$

9. $x+2y=6$

Exercises 2.2

For the following exercises, plot and label each point on the same coordinate plane.

1. a. $(5, 2)$ b. $(1, 4)$ c. $(-7, -3)$ d. $(3, -5)$ e. $(0,1)$

2. a. $(-6, 4)$ b. $(-3, 0)$ c. $(1, -7)$ d. $(0, 5)$ e. $(-2, -8)$

3. a. $(-9, -2)$ b. $(4, 7)$ c. $(-5, 5)$ d. $(6, 0)$ e. $(3, -8)$

4. a. $(0, 0)$ b. $(7, -5)$ c. $(-1, 9)$ d. $(-6, -3)$ e. $(0, -7)$

5. a. $(4.5, 2.8)$ b. $(-5.2, 0)$ c. $(-1.3, -3.75)$ d. $(6, -7.25)$ e. $(-2.4, 4.9)$

6. a. $(9.1, -7.8)$ b. $(2.6, -0.9)$ c. $(0, 5.8)$ d. $(-3.2, -8.7)$ e. $(-6.5, 4.4)$

7. a. $\left(\frac{7}{2}, 4\right)$ b. $\left(-\frac{29}{7}, \frac{5}{3}\right)$ c. $\left(5, -\frac{11}{5}\right)$ d. $\left(-\frac{63}{10}, -\frac{17}{6}\right)$ e. $\left(\frac{8}{9}, 0\right)$

8. a. $\left(-5, \frac{22}{3}\right)$ b. $\left(\frac{51}{8}, -\frac{5}{11}\right)$ c. $\left(0, -\frac{12}{7}\right)$ d. $\left(\frac{57}{20}, \frac{59}{8}\right)$ e. $\left(-\frac{24}{5}, 0\right)$

For the following exercises, tell which quadrant the point is located in. If a given point is not in a quadrant, indicate the axis or axes on which the point is located.

9. $(38, -94)$

10. $(27, 106)$

11. $(0, -549)$

12. $(-712, -120)$

13. $(-95, 164)$

14. $(48, -169)$

15. $(-553, -71)$

16. $(138, 0)$

17. $(-187, 46)$

18. $(544, 329)$

19. $(-143, 0)$

20. $(-19, -28)$

21. $(0, 0)$

22. $(279, -133)$

23. $(59, 294)$

For the following exercises, create a scatter plot from the given table. Then, state whether the points show an exact linear relationship, approximate linear relationship, or a relationship that is not linear.

24.

x	0	1	3	4	6	8	9
y	9	7	3	1	-3	-7	-9

25.

x	-6	-4	-2	0	2	4	6
y	7	4	1	-2	-5	-8	-11

26.

x	-9	-7	-5	-3	-1	1	3
y	8	4	2	1	3	5	9

27.

x	−4	−3	−2	−1	0	1	2
y	−7	−4	−2	1	2	5	7

28.

x	−2	−1	0	1	2	3	4
y	8	4	1	−3	−5	−9	−11

29.

x	−9	−6	−3	0	3	6	9
y	−8	−3	0	1	0	−3	−8

30.

x	−5	−2	−1	3	5	6	8
y	5	2	1	−3	−5	−6	−8

31.

x	−7	−4	−3	−1	2	4	8
y	−10	−8	−8	−6	−5	−4	−1

32.

x	−3	−1	1	3	5	7	9
y	1	7	10	11	9	6	0

33.

x	−5.5	−4	−2.5	−1	0.5	2	3.5
y	8	5.5	3	0.5	−2	−4.5	−7

34.

x	0	1	4	5	7	9	10
y	5.5	4.5	3	3.5	1.5	1	0.5

35.

x	−6.5	−4.5	−3	−1	0.5	2	3.5
y	6.5	3.5	4.5	1	−1.5	−4.5	−4

36.

x	−9.5	−7.5	−6	−4	−1	1	3.5
y	11.5	7.5	4.5	0.5	−5.5	−9.5	−14.5

37.

x	0	1.5	3.5	5	7	8.5	10.5
y	−6	−0.5	1.5	2	1	−1.5	−9

For the following exercises, determine whether or not each ordered pair represents a solution of the given equation.

38. $y = 2x − 11$:   a. $(5, −1)$ b. $(7, 2)$ c. $(−3, −17)$

39. $y = −x + 7$:   a. $(3, 10)$ b. $(13, −20)$ c. $(9, −2)$

40. $3x + 2y = 14$:   a. $(8, −5)$ b. $(−2, 10)$ c. $(0, 7)$

41. $5x − y = 31$:   a. $(7, 4)$ b. $(3, −16)$ c. $(9, −14)$

42. $−2x + 6y = 28$:   a. $(4, −6)$ b. $(11, 1)$ c. $(−20, −2)$

43. $7x − 2y = 21$:   a. $(0, 3)$ b. $(7, 13)$ c. $(−1, 14)$

For the following exercises, find three solutions for each equation. Then graph each equation on a coordinate plane.

44. $y = 2x - 7$

45. $y = x + 3$

46. $y = \frac{1}{2}x - 4$

47. $y = -\frac{2}{3}x + 8$

48. $y = 3x - 1$

49. $y = -2x - 1$

50. $y = -\frac{1}{2}x + 5$

51. $y = \frac{5}{4}x - 3$

52. $x + y = -1$

53. $x - y = -6$

54. $4x + y = 9$

55. $6x + 2y = 4$

56. $-3x + 9y = -18$

57. $x - 4y = 20$

58. $3x - 2y = -10$

59. $12x - 4y = 24$

60. $2.4x + 1.2y = -3.6$

61. $18.8x - 4.7y = 37.6$

62. $-7.3x - 7.3y = 51.1$

63. $\frac{3}{8}x - \frac{3}{4}y = -\frac{3}{2}$

64. $\frac{39}{32}x - \frac{13}{8}y = \frac{13}{2}$

Practice C — Answers

6.

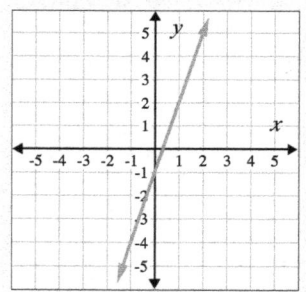

7.

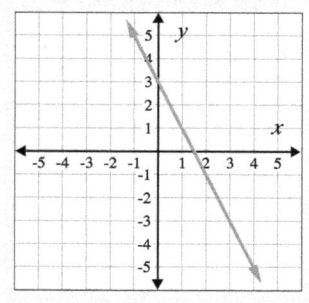

8.

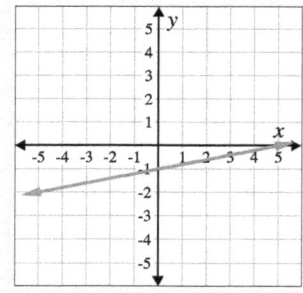

9.

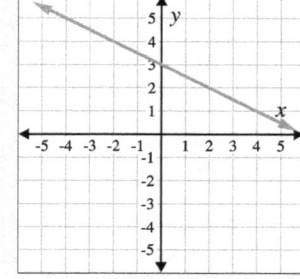

2.3 Slope

Overview

When analyzing the graph of a line, we need to be able to precisely describe where the line is headed and how steep the line is. We can do this if we fully understand the concept of slope.

Here's a problem:

> Belinda's Bodacious Burritos is putting in some new parking spaces outside of their building. If x represents the number of car parking spaces and y represents the number of motorcycle parking spaces, then the equation $10x + 5y = 70$ represents the number and type of new parking spaces that are possible. Find the slope of the line whose equation is $10x + 5y = 70$. Then interpret the result as it pertains to Belinda's situation.

Our goal is to determine the slope of a line and then interpret what that means within the context of this problem. Before we do that, however, we need to cover the basics of slope. As you study this section, you will learn:

- The definition of slope
- How to find the slope of a line, given the graph
- How to calculate the slope of a line, given two ordered pairs
- How to determine the slope of a line, given the equation

A. Introduction to Slope

Slope is a measure of the steepness of a line. In addition to indicating a line's steepness, slope also lets us know whether a line is rising or falling from left-to-right. Formally, we can define slope as the ratio of a line's vertical change to its horizontal change:

$$\text{Slope} = \frac{\text{vertical change}}{\text{horizontal change}}$$

We often use the word "rise" for vertical change and the word "run" for horizontal change:

$$\text{Slope} = \frac{\text{rise}}{\text{run}}$$

The farther a line's slope is from zero, the steeper the line. For example, a line whose slope is 10 is steeper than a line whose slope is 3 because 10 is farther from zero than 3. Similarly, a line whose slope is −7 is steeper than a line whose slope is 4 because −7 is farther from zero than 4.

The sign of a line's slope indicates whether a line is rising or falling from left-to-right. Every line with a positive slope *rises* from left-to-right and every line with a negative slope *falls* from left-to-right. A horizontal line moves from left-to-right without rising or falling. This means that the slope of a horizontal line is 0. A vertical line does not move from left-to-right. This means that the slope of a vertical line doesn't exist. In this case, we can say that the slope is unde-fined.

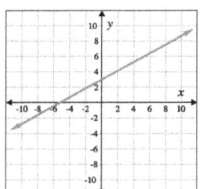

Figure 1a.
Positive slope.

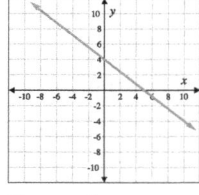

Figure 1b.
Negative slope.

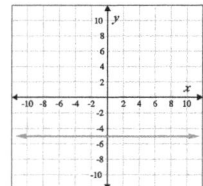

Figure 1c.
Slope = 0

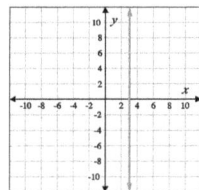

Figure 1d.
Slope is undefined.

B. Determining Slope from a Graph

We can often determine the slope of a line by analyzing the graph. When doing this, we need to understand the following properties of rise and run:

Properties of Rise	Properties of Run
When we move *up*, rise is positive.	When we move *right*, run is positive.
When we move *down*, rise is negative.	When we move *left*, run is negative.

We need to identify two points that the line passes through. We can then determine the vertical movement (rise) and horizontal movement (run) required to get from one point to the other.

Example 1

Determine the slope of the line in Figure 2a.

Solution
The line in Figure 2a is rising from left-to-right, so we know that the slope is positive. We begin by identifying two points that the line passes through, (2, 2) and (5, 4) (Figure 2b). We start at

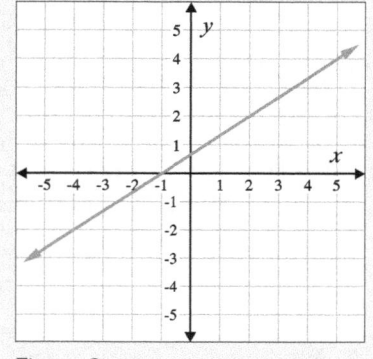

Figure 2a.

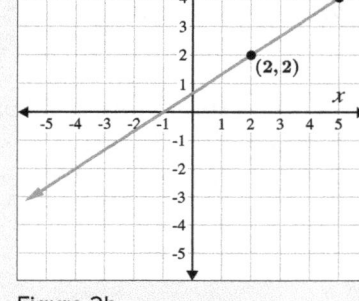

Figure 2b.

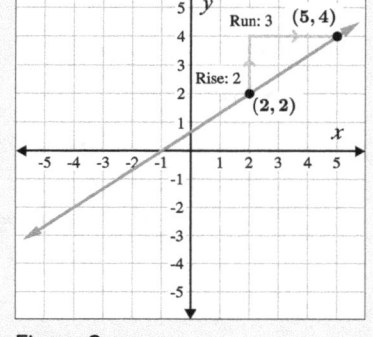

Figure 2c.

(2, 2), the point that is farthest left, and move to (5, 4). This requires us to move *up* 2 spaces and *right* 3 spaces (Figure 2c). The rise is 2, and the run is 3. Therefore, the slope of this line is $\frac{2}{3}$.

It is possible to move from right-to-left when determining slope. In Example 1, we could have started at (5, 4) and moved to (2, 2). In that case, we would have moved *down* 2 spaces and *left* 3 spaces. The rise would have been −2 and the run would have been −3, and the slope would have come out as: $\frac{-2}{-3}$, as in Figure 3.

Of course, when we simplify this fraction, we still get the correct result: $\frac{2}{3}$.

It is also important to note that the slope of a line is not affected by which points we choose to look at. In Example 1, we could also have chosen to look at the points (−4, −2) and (2, 2), as in Figure 4.

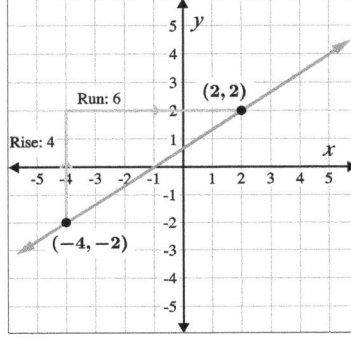

Figure 3.   Figure 4.

In this case, we would get a rise of 4 and a run of 6, and the slope would come out as $\frac{4}{6}$. When simplified, the slope is still $\frac{2}{3}$.

Example 2

Determine the slope of the line in Figure 5a.

Solution

The line in Figure 5a is falling from left-to-right, so we know that the slope is negative. In order to calculate the slope, we begin by identifying two points that the line passes through. we'll choose

(−3, 1) and (1, −4). We start at (−3, 1), the point that is farthest left, and move to (1, −4). This requires us to move *down* 5 spaces and *right* 4 spaces.

The rise is −5, and the run is 4. Therefore, the slope of this line is $\frac{-5}{4}$. It's common to write a fraction like this with the negative sign moved to the left side of the fraction: $-\frac{5}{4}$.

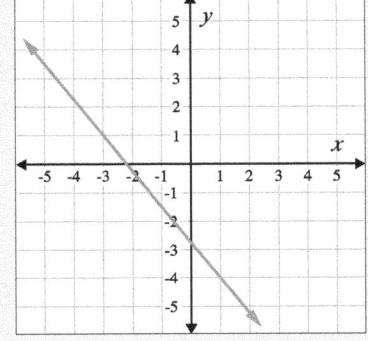

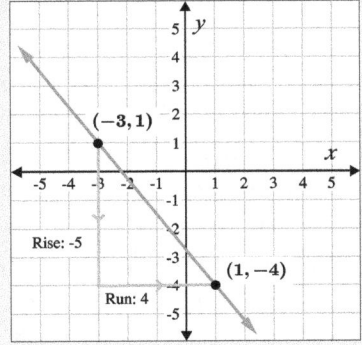

Figure 5a.   Figure 5b.

Practice B

Determine the slope of each line. When you are done, turn the page and check your solutions.

1.
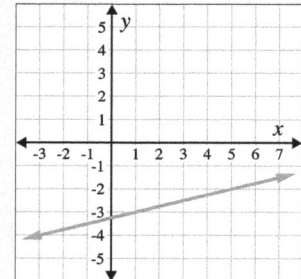

3.
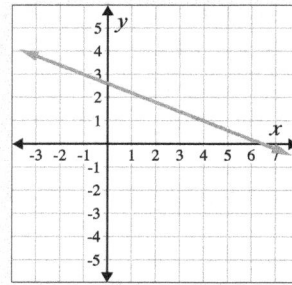

2.
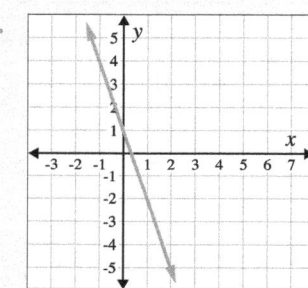

4.
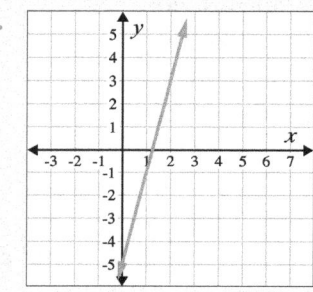

C. Determining Slope, Given Two Ordered Pairs

With a lot of situations, it's important to know the slope of a line. In many of these cases, we're not provided with a graph of the line. Instead, we're given other bits of information about the line. Drawing a coordinate plane, graphing a line, and counting spaces to determine rise and run is often cumbersome or impractical. Because of that, we need to explore some other methods for determining slope.

If we are given ordered pairs for two points that lie on a line, there is a formula we can use to calculate the slope. To help us understand this formula, let's consider a line passing through two given points. We'll call these points (x_1, y_1) and (x_2, y_2).

Note that the numbers next to each letter are subscripts and *not* exponents. The expression x_1 doesn't mean "x to the first power," and x_2 doesn't mean "x squared." Instead, the subscripts help us to identify the two different points on the line: (x_1, y_1) is the first point, and (x_2, y_2) is the second point. This means that x_1 is the x-coordinate of the first point and x_2 is the x-coordinate of the second point. As you see in Figure 6, the difference in x values $(x_2 - x_1)$ gives us the horizontal change (run), and the difference in y values $(y_2 - y_1)$ gives us the vertical change (rise).

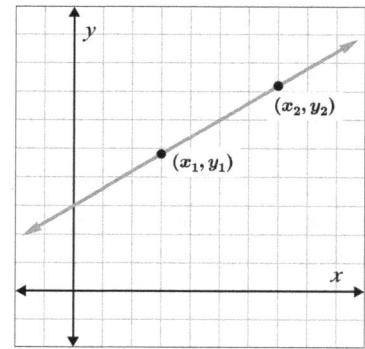

Figure 6.

At the beginning of this section, we defined slope as the ratio of a line's vertical change to its horizontal change. Since y-coordinates are associated with vertical change and x-coordinates are associated with horizontal change, we can think of this ratio as:

$$\text{Slope} = \frac{\text{change in } y}{\text{change in } x}$$

Mathematically, we can write the change in y as "$y_2 - y_1$" and the change in x as "$x_2 - x_1$." We've previously noted that the letter m is traditionally used to represent slope. With all that in mind, we have the *slope formula*:

Slope Formula

The slope, m, of a line passing through the points (x_1, y_1) and (x_2, y_2) is given by:

$$m = \frac{y_2 - y_1}{x_2 - x_1}$$

To help us remember that the y-coordinates are used to calculate the numerator of this fraction, we can think of the informal definition of slope we saw earlier:

$$\text{Slope} = \frac{\text{rise}}{\text{run}}$$

The word "rise" rhymes with "y's," which helps us remember that the y-values must go in the numerator.

In general, if a line is steep, then as we move from the first point to the second point, we'll see a greater vertical change than horizontal change as in Figure 7. This will cause the value $\frac{y_2 - y_1}{x_2 - x_1}$ to be farther away from 0. On the other hand, if a line is not very steep, then as we move from the first point to the second point, we'll see a vertical change that is smaller when compared to the horizontal change as in Figure 8. This will cause the value $\frac{y_2 - y_1}{x_2 - x_1}$ to be closer to 0.

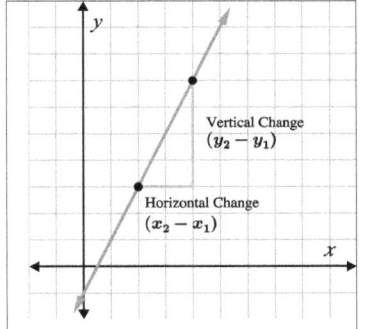

Figure 7.

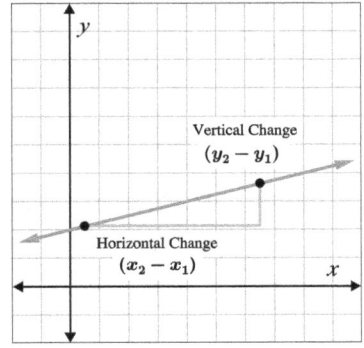

Figure 8.

Let's look at how the slope formula is used in the next example.

Example 3

Find the slope of the line passing through $(0, 1)$ and $(1, 3)$.

Solution

We choose (x_1, y_1) to be $(0, 1)$ and (x_2, y_2) to be $(1, 3)$. This means that $x_1 = 0$, $x_2 = 1$, $y_1 = 1$ and $y_2 = 3$.

$m = \dfrac{y_2 - y_1}{x_2 - x_1}$ Begin with the slope formula.

$m = \dfrac{3 - 1}{1 - 0}$ In the numerator, replace y_2 with 3 and replace y_1 with 1. In the denominator, replace x_2 with 1 and replace x_1 with 0.

$m = \dfrac{2}{1}$ Simplify the numerator and denominator.

$m = 2$ Finish by writing the simplified answer.

Figure 9 shows us what the graph looks like for Example 3. Notice that as we look from left-to-right, the line rises. It's not surprising then that the slope is positive. Also notice that when moving from $(0, 1)$ to $(1, 3)$, the rise is 2 and the run is 1. These values match the slope we calculated by using the formula.

Does the choice of (x_1, y_1) and (x_2, y_2) matter? What if we decided to choose (x_1, y_1) to be $(1, 3)$ and (x_2, y_2) to be $(0, 1)$? In that case, the problem would look like this:

$m = \dfrac{y_2 - y_1}{x_2 - x_1}$ Start with the slope formula.

$m = \dfrac{1 - 3}{0 - 1}$ In the numerator, replace y_2 with 1 and replace y_1 with 3. In the denominator, replace x_2 with 0 and replace x_1 with 1.

$m = \dfrac{-2}{-1}$ Simplify the numerator and denominator.

$m = 2$ Finish by writing a simplified answer.

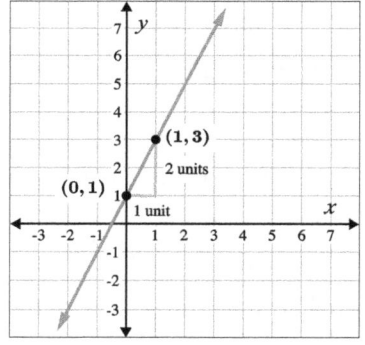

Figure 9.

Practice B — Answers

1. $\dfrac{1}{4}$ 2. -3 3. $-\dfrac{2}{5}$ 4. 4

As you can see, we get the same result regardless of our choice of (x_1, y_1) and (x_2, y_2).

Keep in mind that even though the choice of (x_1, y_1) and (x_2, y_2) doesn't matter, it is important to stick with our choice. The starting value in the numerator, y_2, and the starting value in the denominator, x_2, must come from the same ordered pair.

Example 4

Find the slope of the line passing through each pair of points.

1. $(2, 2)$ and $(4, 3)$

2. $(-2, 4)$ and $(1, 1)$

3. $(1, 3)$ and $(5, 3)$

4. $(4, 4)$ and $(4, 0)$

Solutions

1. $(2, 2)$ and $(4, 3)$

We choose (x_1, y_1) to be $(2, 2)$, and we choose (x_2, y_2) to be $(4, 3)$.

$m = \frac{y_2 - y_1}{x_2 - x_1}$ Begin with the slope formula.

$m = \frac{3 - 2}{4 - 2}$ In the numerator, substitute 3 for y_2 and substitute 2 for y_1. In the denominator, substitute 4 for x_2 and substitute 2 for x_1.

$m = \frac{1}{2}$ Simplify the numerator and denominator. The resulting fraction cannot be simplified any further.

2. $(-2, 4)$ and $(1, 1)$.

We choose (x_1, y_1) to be $(-2, 4)$, and we choose (x_2, y_2) to be $(1, 1)$.

$m = \frac{y_2 - y_1}{x_2 - x_1}$ Begin with the slope formula.

$m = \frac{1 - 4}{1 - (-2)}$ In the numerator, replace y_2 with 1 and replace y_1 with 4. In the denominator, replace x_2 with 1 and replace x_1 with -2.

$m = \frac{-3}{1 + 2}$ Finish by simplifying the resulting expression.

$m = \frac{-3}{3}$

$m = -1$

If we were to plot the ordered pairs $(-2, 4)$ and $(1, 1)$ and draw the line passing through those points, we would notice that the line falls from left-to-right — and so it is not surprising that the slope is negative.

3. $(1, 3)$ and $(5, 3)$.

We choose (x_1, y_1) to be $(1, 3)$, and we choose (x_2, y_2) to be $(5, 3)$. We will work through the steps of this example without explanation.

$$m = \frac{y_2 - y_1}{x_2 - x_1}$$

$$m = \frac{3 - 3}{5 - 1}$$

$$m = \frac{0}{4}$$

$$m = 0$$

Notice that we did not leave our answer as $\frac{0}{4}$. When 0 is divided by any nonzero number, the result is simply 0. Also notice that if we were to plot the ordered pairs and draw the line, we would see a horizontal line. We mustn't make the mistake of saying that this line has "no slope." Every line that moves from left-to-right has a slope. This line is no different. It has a slope, and the slope is equal to 0.

4. $(4, 4)$ and $(4, 0)$.

We choose (x_1, y_1) to be $(4, 4)$, and we choose (x_2, y_2) to be $(4, 0)$.

$$m = \frac{y_2 - y_1}{x_2 - x_1}$$

$$m = \frac{0 - 4}{4 - 4}$$

$$m = \frac{-4}{0}$$

$$m \text{ is undefined.}$$

It is impossible to divide by 0. Therefore, if a fraction has a denominator of 0, that fraction does not represent a number. In math, we call an expression like this *undefined*. It would also be acceptable to say that this line has "no slope." If we were to plot the ordered pairs, we would see a vertical line.

Practice C

Find the slope of the line passing through each pair of points. Be sure to simplify your answer. When you are done, turn the page and check your solutions.

5. $(2, 1)$ and $(6, 3)$

6. $(-7, 8)$ and $(5, 5)$

7. $(-2, 9)$ and $(-2, 16)$

8. $(8, 4)$ and $(-1, 4)$

D. Determining Slope, Given the Equation of the Line

Let's explore how the equation of a line is related to the line's slope.

Example 5

Find the slope of the line whose equation is $3x + 6y = 24$.

Solution

$$3x + 6y = 24$$

We begin by rewriting the equation in slope-intercept form. To write a linear equation in slope-intercept form, we must solve for y.

$$\underline{-3x \qquad\qquad -3x}$$

$$6y = -3x + 24$$

$$\frac{6y}{6} = \frac{-3x + 24}{6}$$

$$y = \frac{-3x}{6} + \frac{24}{6}$$

$$y = -\frac{1}{2}x + 4$$

Now, we find two points that lie on the graph of the line.

If $x =$	then the equation reads	and $y =$	Ordered Pair
0	$y = -\frac{1}{2}(0) + 4$	4	$(0, 4)$
2	$y = -\frac{1}{2}(2) + 4$	3	$(2, 3)$

Finally, we use the slope formula to calculate the slope.

$$m = \frac{y_2 - y_1}{x_2 - x_1}$$

$$m = \frac{3 - 4}{2 - 0}$$

$$m = \frac{-1}{2}$$

This line has a slope of $-\frac{1}{2}$.

Let's take another look at the equation in slope-intercept form, $y = -\frac{1}{2}x + 4$.

Notice that the constant factor, or **coefficient**, of the term containing x is $-\frac{1}{2}$. The fact that this number is equal to the slope is not unique to this equation. If any linear equation is written in slope-intercept form, the coefficient of the term containing x will be equal to the slope.

Remember, m is used to represent slope, and slope-intercept form is generally denoted as $y = mx + b$. We'll discuss the meaning of b in the next section.

Example 6

Find the slope of each line below.

1. $5x - 3y = 18$

2. $6y = 45$

3. $-5x = 65$

Solutions

1. $5x - 3y = 18$

$$5x - 3y = 18 \qquad \text{Rewrite the equation in slope-intercept form.}$$

$$\underline{-5x} \qquad \qquad \underline{-5x}$$

$$-3y = -5x + 18$$

$$\frac{-3y}{-3} = \frac{-5x + 18}{-3}$$

$$y = \frac{-5x}{-3} + \frac{18}{-3}$$

$$y = \frac{5}{3}x - 6$$

The coefficient of the term containing x is $\frac{5}{3}$, so the slope of this line is $\frac{5}{3}$.

2. $6y = 45$

$$6y = 45 \qquad \qquad \text{Begin by solving for } y.$$

$$\frac{6y}{6} = \frac{45}{6}$$

$$y = \frac{15}{2} \qquad \qquad \text{The right side of the equation doesn't contain the letter } x.$$

$$y = 0x + \frac{15}{2} \qquad \text{We can rewrite the right side with a ``}0x\text{'' term.}$$

The slope of this line is 0.

3. $-5x = 65$

This equation does not contain the letter y, so it's impossible to solve for y. This means that it's also impossible to write the equation in slope-intercept form. Because of this, the line has no slope. We can say that the slope is *undefined*.

Remember that a horizontal line has a slope of 0 and a vertical line has an undefined slope. Parts 2 and 3 of the previous example illustrate that the equation of a horizontal line only contains the variable y, while the equation of a vertical line only contains the variable x.

We'll wrap things up by finding the slope within the context of an application problem and then interpreting our result. Let's take another look at the problem from the beginning of the section.

Example 7

Belinda's Bodacious Burritos is putting in some new parking spaces outside of their building. If x represents the number of car parking spaces and y represents the number of motorcycle parking spaces, then the equation $10x + 5y = 70$ represents the number and type of new parking spaces that are possible. Find the slope of the line whose equation is $10x + 5y = 70$. Then interpret the result as it pertains to this situation.

Solution

As we've done before, we will rewrite the equation in slope-intercept form so that we can easily identify the slope.

$$10x + 5y = 70$$
$$\underline{-10x \qquad\qquad -10x}$$
$$5y = -10x + 70$$
$$\frac{5y}{5} = \frac{-10x + 70}{5}$$
$$y = \frac{-10x}{5} + \frac{70}{5}$$
$$y = -2x + 14$$

The slope of this line is -2.

Now, let's interpret the result. We will write the slope in fraction form so that we can determine the change in y and the corresponding change in x.

$$-2 = \frac{-2}{1}$$
$$\frac{\text{change in } y}{\text{change in } x} = \frac{-2}{1}$$

When the value of y decreases by 2, the value of x increases by 1. Since y represents the number of motorcycle parking spaces and x represents the number of car parking spaces, we can say:

> If the number of motorcycle parking spaces is reduced by 2, then 1 additional car parking space can be made.

Note that -2 can also be written as $\frac{2}{-1}$. This leads to the following interpretation:

> If 2 additional motorcycle parking spaces are wanted, then the number of car parking spaces will have to be reduced by 1.

Both of these interpretation statements are equivalent.

Practice Set C — Answers

5. $\frac{1}{2}$ 6. $-\frac{1}{4}$ 7. undefined   8. 0

Practice D

Now it's your turn. For problems 9 – 12, find the slope of the lines whose equations are given. For problem 13, find the slope and interpret the result within the context of the problem. When you are done, turn the page and check your solutions.

9. $2x + 5y = 15$ 10. $12x - 4y = 3$ 11. $-11y = 15$ 12. $12x = 108$

13. When the outside temperature is at least seventy degrees, then the number of customers at Francisco's Frozen Yogurt can be modeled by the equation $17x - 2y = -76$. In this equation, x represents the amount (in degrees) by which the temperature exceeds seventy degrees, and y represents the number of customers.

Exercises 2.3

Determine the slope of the line shown in each graph below.

1.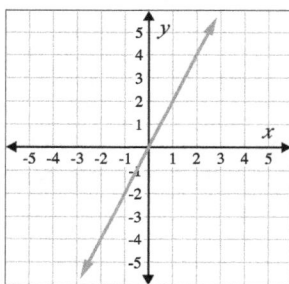

3.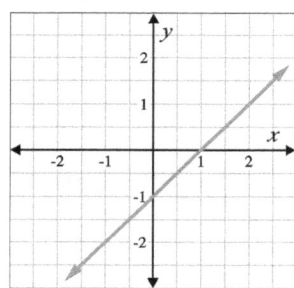

5.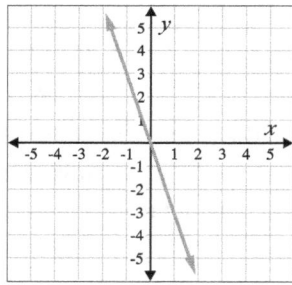

2.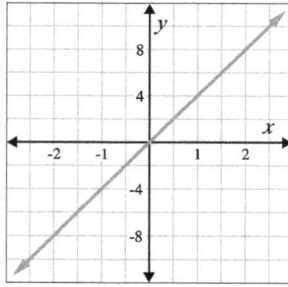

4.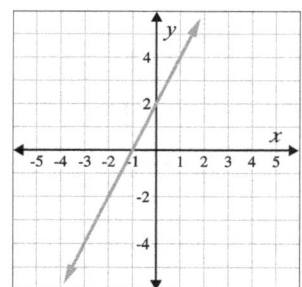

6.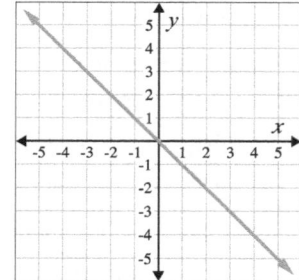

7.
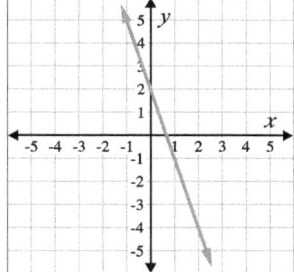

11.
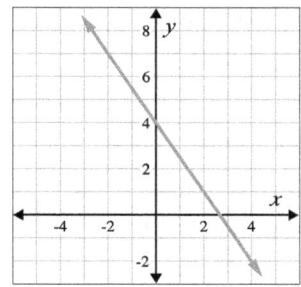

15.
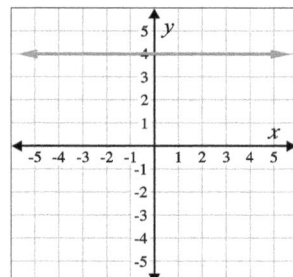

8.
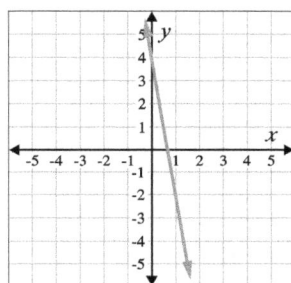

12.
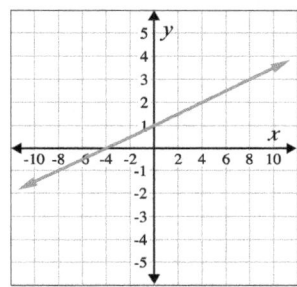

16.
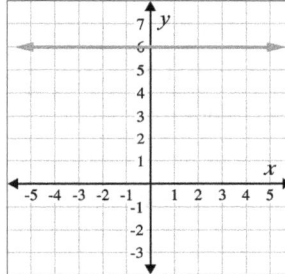

9.
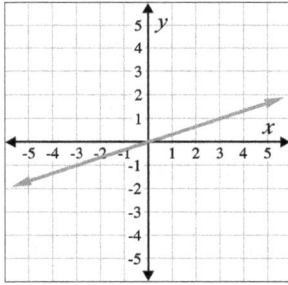

13.
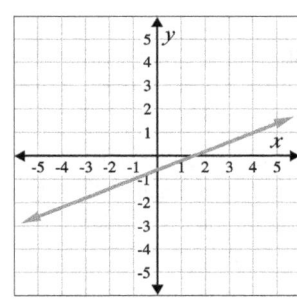

17.
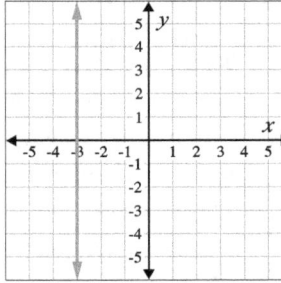

10.
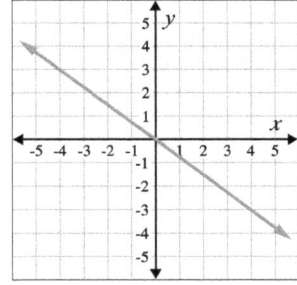

14.
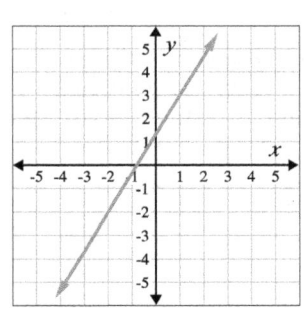

18.
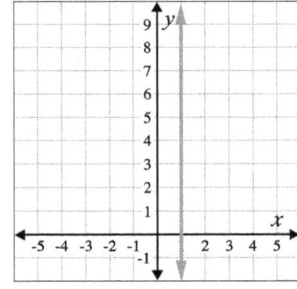

Find the slope of the line passing through the two given points.

19. $(1, 6), (4, 9)$

20. $(2, 3), (4, 7)$

21. $(5, 1), (1, 9)$

22. $(2, 8), (3, 2)$

23. $(0, 2), (3, 4)$

24. $(8, 4), (6, 11)$

25. $(0, 4), (2, -6)$

26. $(-2, 1), (0, 5)$

27. $(2, 2), (-12, 4)$

28. $(1, -9), (5, 1)$

29. $(2, -7), (9, -7)$

30. $(8, -9), (8, -1)$

31. $(-3, -8), (4, 6)$

32. $(0, 0), (6, 6)$

33. $(-6, 9), (-6, 4)$

34. $(-6, -3), (-14, -3)$

35. $(-4, -10), (5, -4)$

36. $(-5, 0), (0, -8)$

Find the slope of each line whose equation is given.

37. $2y = 4x + 8$

38. $-3y = 12x - 27$

39. $-4y = 5x + 8$

40. $10y = -12x + 1$

41. $-y = x + 8$

42. $-y = -5x + 3$

43. $12x + 4y = 24$

44. $-6x + 2y = -16$

45. $7x - 7y = 21$

46. $3x + 8y = 24$

47. $4x - y = 7$

48. $5x + 3y = 6$

49. $5y = 35$

50. $-4y = 22$

51. $13x = 52$

52. $20 = -6x$

53. $-7y - 6x = -12$

54. $4y - x = -1$

For each situation, find the slope of the line whose equation is given. Then interpret your result within the context of the situation.

55. Micah parks his car in the parking lot at the mall on a sunny day. If x represents the number of minutes the car is parked and y represents the temperature inside the car, in degrees Fahrenheit, then the equation $15x - 12y = -960$ represents the relationship between elapsed time and temperature inside the car.

56. Pat's Pets is installing new dog and cat enclosures. If x represents the number of new dog enclosures and y represents the number of new cat enclosures, then the equation $18x + 6y = 126$ represents the number and type of new enclosures that are possible.

57. Liljana is hiking in the mountains. If x represents elevation at which Liljana is hiking, in thousands of feet, and y represents the temperature, in degrees Fahrenheit, then the equation $7x + 2y = 126$ represents the relationship between elevation and temperature.

58. Professor Epsilon is thinking about adding questions to an Algebra test. If x represents the number of questions added and y represents the average additional time, in minutes, that a student will need to complete the test then the equation $45x - 10y = -300$ represents the relationship between additional questions and completion time.

Practice D — Answers

9. $-\frac{2}{5}$

10. 3

11. 0

12. undefined

13. Slope $= \frac{17}{2}$; The number of customers increases by 17 if the outside temperature increases by 2 degrees.

2.4 Slope-Intercept Form

Overview

Most mathematicians are lazy. They never want to take 5 steps to work through a process that can be completed in just 2 steps. As strange as it may sound, this laziness is one of the greatest motivating factors in putting in the time and effort necessary to learn and understand mathematical rules, patterns, and formulas. These rules, patterns, and formulas can significantly reduce the time and effort needed to complete mathematical processes.

We've been learning about the basics of graphing linear equations, and now it's time for us to use this knowledge to help us develop some shortcuts for graphing lines. Our main focus now is on quickly transforming an equation into a graph and vice versa, but we'll also take a look at some application problems. Here's a problem for you to consider:

> Chrissie's Custom T-shirts charges $5 per shirt to make t-shirts. There is also a $45 charge to set up the silkscreen and other equipment that is necessary for the job. Write and graph an equation that models this situation.

We saw some situations like this problem in Section 2.1. At that time, the equations and graphs were provided for us. Now it's time for us to generate equations and graphs on our own. Before we do, however, let's learn some more about slope-intercept form and see how it can allow us to quickly draw graphs of linear equations.

In this section, you will learn how to:

- Use slope-intercept form to graph a line whose equation is given
- Use a graphing calculator to graph a line whose equation is in slope-intercept form
- Use slope-intercept form to model real-life situations
- Write the equation of a line whose graph is given

A. Graphing Equations in Slope-Intercept Form

The point where a graph crosses the x-axis is called the x-intercept. The point where a graph crosses the y-axis is called the y-intercept. We are now ready to formally define slope-intercept form.

Slope-Intercept Form

The slope-intercept form of a linear equation is

$$y = mx + b$$

where m represents the slope of the line and b represents the y-intercept of the line.

When a linear equation is presented in slope-intercept form, we have different options for constructing a graph. One option is to choose two or three x-values, plug them into the equation and compute the corresponding y-values. This is the approach we took in Section 2.2. This approach is necessary whenever we want to graph an equation by hand but don't know any graphing shortcuts. Unfortunately, these computations can sometimes be difficult and time consuming. More difficult computations also tend to lead to more errors.

Now that we have a good understanding of slope, we can use a graphing shortcut that makes use of slope and y-intercept. The method is quick and simple, and it involves no computations.

Using *y*-intercept and Slope to Graph an Equation

1. Plot the y-intercept $(0, b)$.

2. Starting at the y-intercept, use the slope (m) to determine the location of another point.

3. Draw a line through the two points.

Recall that in the previous section, we learned that

$$\text{slope} = \frac{\text{rise}}{\text{run}}$$

We also learned that a line with a positive slope rises as it moves from left-to-right, while a line with a negative slope falls as it moves from left-to-right. For example, a line whose slope is $\frac{2}{3}$ will move *up* 2 spaces and *right* 3 spaces in order to get from one point to the next, while a line whose slope is $-\frac{5}{7}$ will move *down* 5 spaces and *right* 7 spaces in order to get from one point to the next. Understanding the properties of rise and run from Section 2.3 will be crucial as we use this graphing shortcut.

Let's look at how this shortcut works in a few examples.

Example 1

Graph $y = \frac{3}{4}x + 2$

Solution

Step 1: $b = 2$. This means that the y-intercept of the line is $(0, 2)$. We begin by plotting $(0, 2)$.

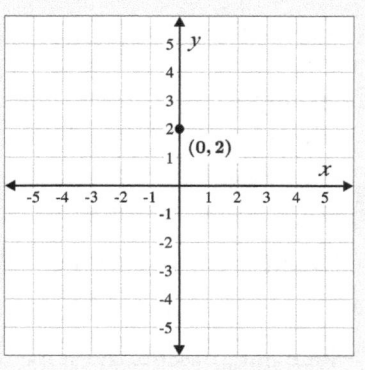

Step 2: The slope, m, is $\frac{3}{4}$. This means that the line will move *up* 3 spaces and *right* 4 spaces in order to get from one point to the next. We know that the line passes through the point $(0, 2)$, so we'll start there. From this point, we move up 3 spaces and right 4 spaces in order to find the location of another point on the line. We arrive at the point $(4, 5)$.

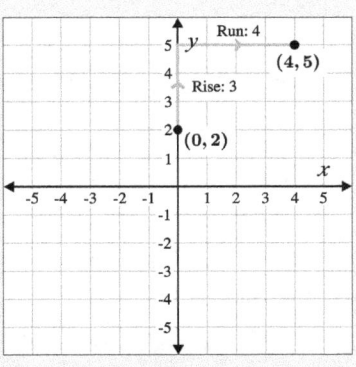

Step 3: The last step is to draw a line through both points.

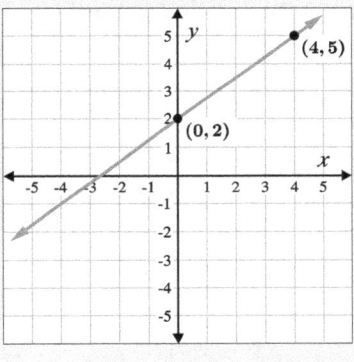

Although it's not generally recommended, we could move from right-to-left while using this graphing shortcut. However, to do so, we would need to make use of the fact that $\frac{3}{4} = \frac{-3}{-4}$. This means that if we start at the point $(0, 2)$ and move our pencil *down* 3 spaces and *left* 4 spaces, we will find another point on the line. See Figure 1.

It is also useful to note that $\frac{3}{4} = \frac{3/4}{1}$. This means that if we start at any point on the line and move one unit to the right, we have to move up $\frac{3}{4}$ of a unit to get back on the line. See Figure 2.

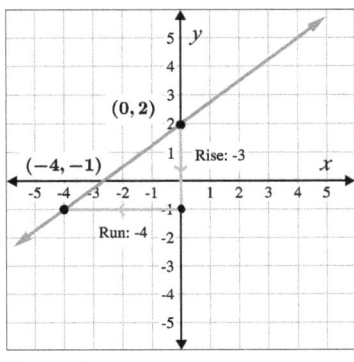

Figure 1.

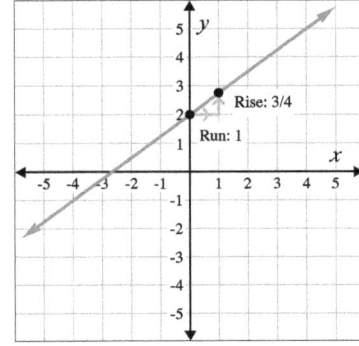

Figure 2.

Example 2

Use the y-intercept and slope to graph each equation by hand.

1. $y = -\frac{1}{2}x + \frac{7}{2}$ **2.** $y = \frac{2}{5}x$ **3.** $y = 2x - 4$

Solutions

1. $y = -\frac{1}{2}x + \frac{7}{2}$

Step 1: $b = \frac{7}{2}$. This means that the y-intercept of the line is $\left(0, \frac{7}{2}\right)$. Another way to write this ordered pair is $\left(0, 3\frac{1}{2}\right)$ or $(0, 3.5)$. We begin by plotting this point, as in Figure 3.

Step 2: The slope, m, is $-\frac{1}{2}$. We can write $-\frac{1}{2}$ as $\frac{-1}{2}$. We now know that we can start at the point $(0, 3.5)$ and then move *down* 1 space and *right* 2 spaces to get to another point on the line. In Figure 4, we see that after moving the appropriate number of spaces, we arrive at the point $(2, 2.5)$.

Step 3: In Figure 5, we draw a line through both points.

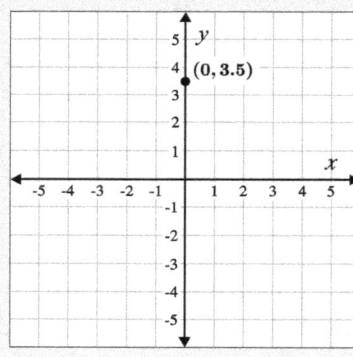
Figure 3.

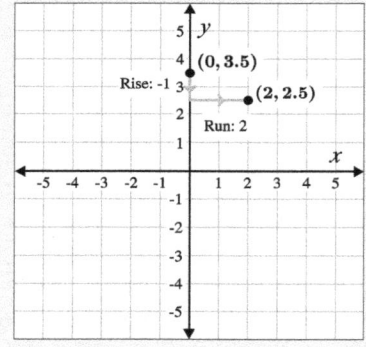
Figure 4.

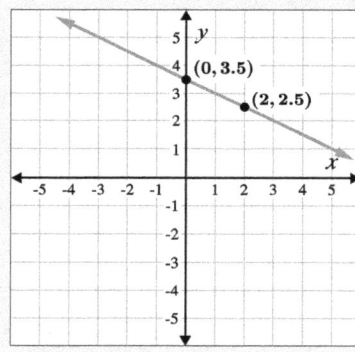
Figure 5.

2. Graph $y = \frac{2}{5}x$

Step 1: Technically, this equation is written in slope-intercept form. However, we can more clearly see the value of b if we re-write the equation as $y = \frac{2}{5}x + 0$. Now we can see that $b = 0$. This means that the y-intercept of the line is $(0, 0)$. This line goes right through the origin. See Figure 6.

Step 2: The slope, m, is $\frac{2}{5}$. Starting at the origin, we move up 2 spaces, then move to the right 5 spaces. We mark a point at this location.

Step 3: We finish by drawing a line through the two points. The line and both points are shown in Figure 7.

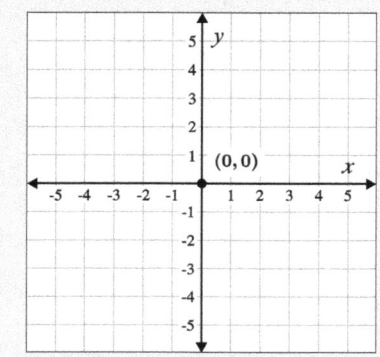
Figure 6.

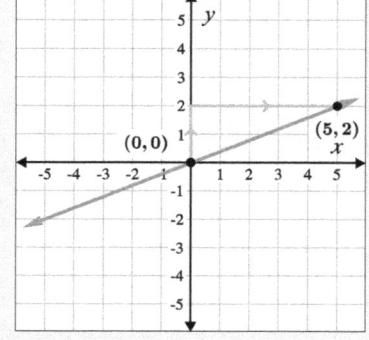
Figure 7.

3. Graph $y = 2x - 4$

Step 1: The y-intercept is the point $(0, -4)$.

Step 2: The slope, m, is 2. It's easier to see the rise and run if we write the slope as a fraction: $2 = \frac{2}{1}$. We start at the point $(0, -4)$ and move up 2 spaces. Then we move right 1 space and mark a point at this location.

Step 3: Then we draw a line through the two points in Figure 9.

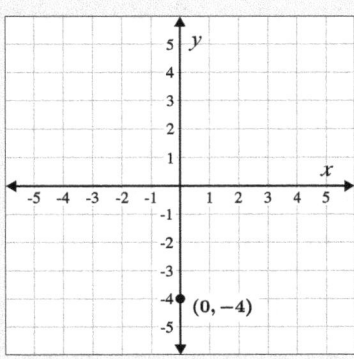

Figure 8.

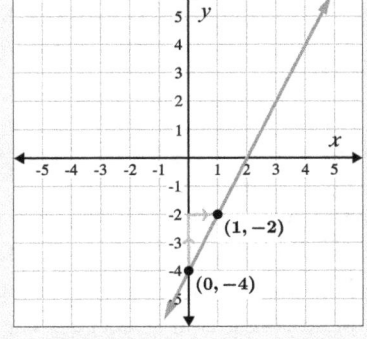

Figure 9.

Notice that in each of the previous examples, the equations were given to us in slope-intercept form. If an equation is *not* in slope-intercept form, then we need to do a bit of work to rewrite it in slope-intercept form before we can use the graphing shortcut.

Example 3

Use the y-intercept and slope to graph $-2x + 3y = -9$ by hand.

Solution

This equation is not written in slope-intercept form. We will begin by solving the equation for y.

$$-2x + 3y = -9$$
$$\underline{+2x \qquad +2x}$$
$$3y = 2x - 9$$
$$\frac{3y}{3} = \frac{2x - 9}{3}$$
$$y = \frac{2x}{3} - \frac{9}{3}$$
$$y = \frac{2}{3}x - 3$$

Now, we proceed with the graphing shortcut:

1. Graph the y-intercept, $(0, -3)$. From $(0, -3)$, move *up* 2 spaces and *right* 3 spaces to locate another point on the line.

2. Draw the line through these two points. Figure 10 shows the finished graph.

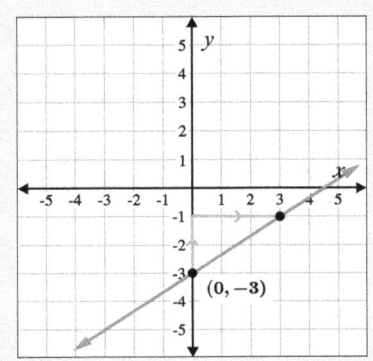

Figure 10.

Practice A

Now you can practice using the y-intercept and slope to graph each line. When you are done, turn the page and check your solutions.

1. $y = -\frac{2}{3}x + 4$ **2.** $y = \frac{3}{4}x$ **3.** $3x + 2y = 2$

B. Graphing Lines on a Graphing Calculator

Previously, we used a graphing calculator to create a table of values representing both sides of an equation we solved. Now, we're going to see how to use a graphing calculator to graph a linear equation.

When graphing a linear equation on a graphing calculator, the equation *must* be in slope-intercept form. That is because the equation we are graphing will be entered on the $\boxed{Y=}$ screen, which means that the equation must be solved for y.

Example 4

Graph the following equations on a graphing calculator.

   1. $y = 2x - 5$ **2.** $y = \frac{9}{5}x + 31.48$

Solutions

1. $y = 2x - 5$

Step 1: Begin by entering "$2x - 5$" in the Y_1 equation.

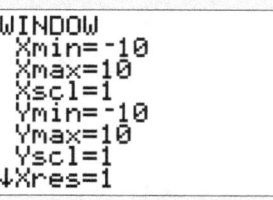

Step 2: You can check the settings of your viewing window by pressing the $\boxed{\text{WINDOW}}$ button. By setting Xmin to −10 and Xmax to 10, you will be able to view the x-axis between the values of −10 and 10. Similarly, by setting Ymin to −10 and Ymax to 10, you will be able to view the y-axis between the values of −10 and 10.

   If you press the $\boxed{\text{ZOOM}}$ button and select ZStandard, the viewing window settings will automatically be updated so that Xmin and Ymin are both −10, and Xmax and Ymax are both 10.

Step 3: To view the line, press the $\boxed{\text{GRAPH}}$ button.

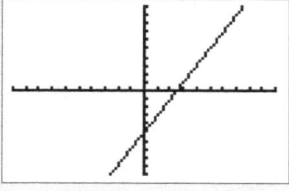

2. $y = \frac{9}{5}x + 31.48$

Step 1: Begin by entering "$\frac{9}{5}x + 31.48$" in the Y_1 equation. Use the ÷ button to put in the fraction bar. The parentheses are located above the numbers 8 and 9 on the calculator. They are optional, but they help set the fraction apart from the variable when viewing the equation.

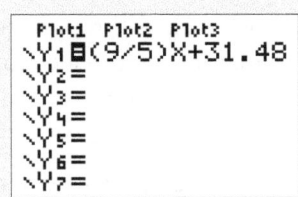

Step 2: The viewing window needs to be adjusted, because the y-intercept of this line is 31.48. Use a Ymax value of 40. For now, leave the other values alone.

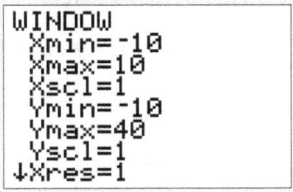

Step 3: Now press the GRAPH button to view the line.

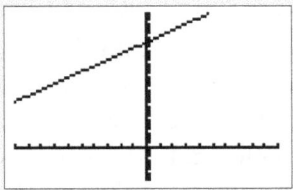

Step 4: Now try setting up the viewing window so that Xmin and Ymin are both −40, and Xmax and Ymax are both 40.

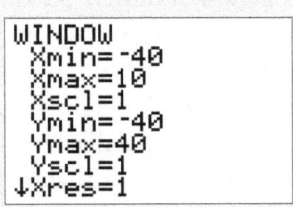

Step 5: Notice how that changes the appearance of the graph on the calculator screen.

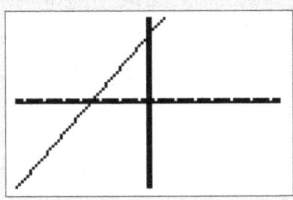

This example illustrates how adjusting the viewing window on your calculator can dramatically affect the appearance of a graph. You should keep this in mind when using a calculator to graph an equation.

Practice B

If you worked through the last example successfully, you should be ready to use your graphing calculator to graph a couple more equations on your own. When you are done, turn the page and check your solutions — but remember, if the viewing window on your calculator isn't set up to match the values given in these problems, your graph won't look like the answer graphs.

4. Use a graphing calculator to graph $y = \frac{4}{7}x + 2$ {Viewing window: $-10 < x < 10$ and $-10 < y < 10$}

5. Use a graphing calculator to graph $y = 3.762x - 27.95$. {Viewing window: $-10 < x < 10$ and $-35 < y < 10$}

C. Slope-intercept Form and Real-Life Situations

Many real life situations involving two variable quantities can be modeled using linear equations in slope-intercept form. These scenarios typically involve a quantity or value that depends on another quantity or value.

At the beginning of Chapter 2, for example, we considered a store that sells 1-gallon jugs of milk for $4 each. In this situation, the total cost of the milk depends on the number of gallons purchased. If 1 gallon of milk is purchased, then the total cost is $4. If 2 gallons of milk are purchased, then the total cost is $8. And so on. We modeled this situation with the equation $y = 4x$. In this equation, the variable y is used to represent the quantity (total cost, in dollars) that depends on the other quantity (gallons of milk purchased).

That example illustrates that whenever a literal equation with two variables is solved for a particular variable, that variable is the **dependent** variable. That's because its value *depends* on the value of the other variable in the equation. The other variable is known as the **independent** variable. When we're dealing with a linear equation in slope-intercept form, the equation is solved for y. This means that y is the dependent variable and x is the independent variable.

Practice A — Answers

1.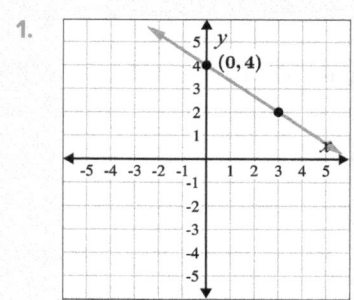

2.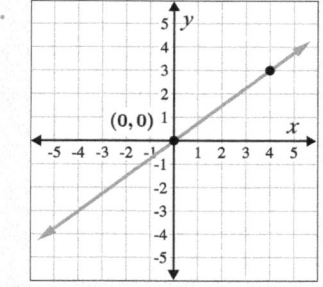

3.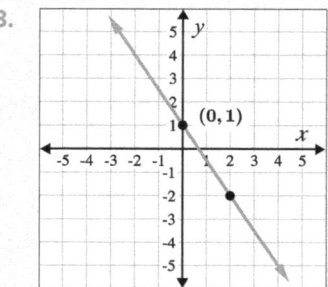

We can see this more clearly by considering a few simple equations:

$A = 3B + 7$ This equation indicates that the value of A can be found by assigning a value to B, multiplying this value by 3, and adding 7 to the result. In other words, the value of A depends on the value assigned to B. A is the dependent variable, and B is the independent variable. An easy way to identify the dependent variable is to determine which variable is by itself. Because the equation is solved for A, A is the dependent variable.

$y = -2x + 10$ This equation is solved for y, so y is the dependent variable, and x is the independent variable.

$2n + 5p = 30$ This equation has not been solved for a variable, so we can't identify which variable is dependent and which variable is independent.

When using a linear equation in slope-intercept form to model a real-life situation, we must identify which variable quantity is dependent and which is independent. We then let y represent the dependent quantity and let x represent the independent quantity.

There are a couple things to notice with the example of the store selling milk for \$4 per gallon:

1. The total cost goes up by \$4 when the number of gallons purchased increases by 1. This corresponds to the fact that the slope of the line modeling the situation is $\frac{4}{1}$, or when simplified, 4. See Figure 11.

2. Before any gallons of milk are purchased, the initial cost is \$0. This corresponds to the fact that the y-intercept of the line modeling the situation is 0.

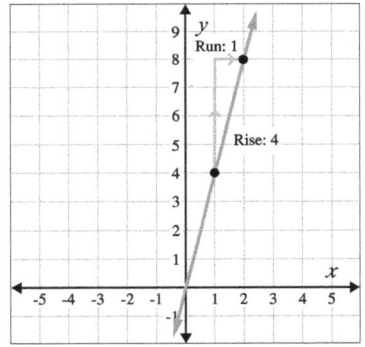

Figure 11.

In general, when we write linear equations in slope-intercept form to model real-life situations, the *rate of change* will correspond with the slope and the *initial quantity* will correspond with the y-intercept. In the milk problem, the phrase "dollars-per-gallon" indicates a rate of change. You'll see how this works in the next examples.

Example 5

Chrissie's Custom T-shirts charges \$5 per shirt to make t-shirts. They also charge \$45 to set up the silkscreen and other equipment needed for the job.

a. Identify what the variables x and y will represent in this situation, using the proper units.

b. Then write and graph an equation that models this situation.

c. Determine the cost to produce 30 t-shirts.

Solution

a. In this situation, the variable quantities are the number of t-shirts produced and the total cost. The units are t-shirts, and dollars. The total cost *depends* on the number of t-shirts produced. Therefore:

x = the number of t-shirts produced

y = the total cost, in dollars.

The rate of change in this problem is indicated by the phrase "$5 per shirt." The word "per" almost always indicates a rate of change. This means that the slope is 5. Before any t-shirts are produced, the initial cost is $45. This means that the y-intercept is 45.

b. We can now write the equation for this situation: $y = 5x + 45$. To draw the graph, we'll set up the coordinate plane so that both axes have hash marks every 5 spaces. See Figure 12.

c. Finally, just as we did in Section 2.1, we can use either the equation or the graph to determine the total cost of producing 30 t-shirts. We'll use the equation:

$$y = 5x + 45$$
$$y = 5(30) + 45$$
$$y = 195$$

It will cost $195 to produce 30 t-shirts.

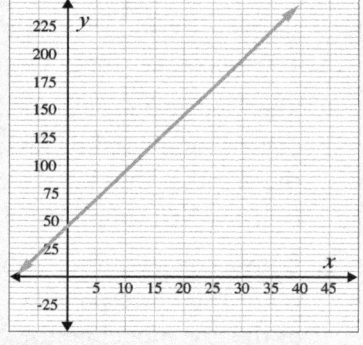

Figure 12.

Look again at the graph for Example 5. By scaling both axes the same — in this case, the distance between hash marks is 5 — we're able to plot our points correctly by starting at the y intercept and then moving *up* 5 spaces and *right* 1 space.

If the axes are scaled differently, then a slope of 5 will appear differently. We saw something similar happen when we adjusted the viewing window of our graphing calculator in Example 7. Figures 13 and 14 illustrate how changes in the scales will change the graphing for a line with a slope of 5.

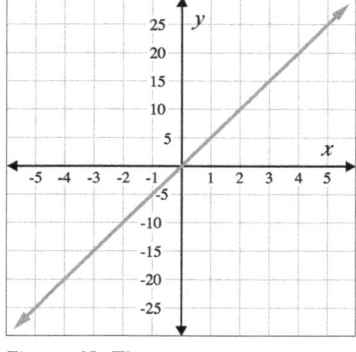

Figure 13. The x-axis goes by 1s, and the y-axis goes by 5s.

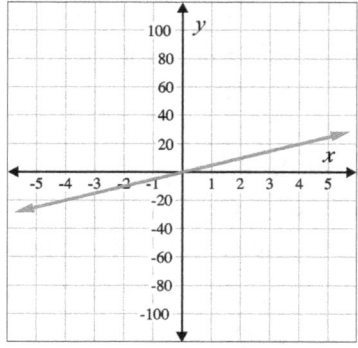

Figure 14. The x-axis goes by 1s, and the y-axis goes by 20s.

Example 6

An 84-gallon tank is full of water. It is then drained at a rate of 6 gallons per minute.

a.   Define variables for this situation and include the proper units.

b.   Write and graph an equation in slope-intercept form that models this situation.

c.   Determine how much water will be in the tank after 8 minutes.

Solution

a. In this problem, the variable quantities are the amount of time elapsed while the tank drains and the amount of water in the tank. The units, respectively, are minutes and gallons. The amount of water left in the tank *depends* on the amount of time elapsed. Therefore:

x = the time elapsed, in minutes

y = the amount of water in the tank, in gallons

The rate of change in this problem is indicated by the phrase "6 gallons per minute". However, the tank is *losing* 6 gallons of water per minute. The amount of water in the tank is going down, not up, so the slope is negative. In order for our equation to be correct, our slope must be −6.

The initial amount of water in the tank is 84 gallons. This means that the y-intercept is 84.

b. We can now write the equation: $y = -6x + 84$.

When we draw the graph, both axes will have hash marks every 4 spaces, as in Figure 15.

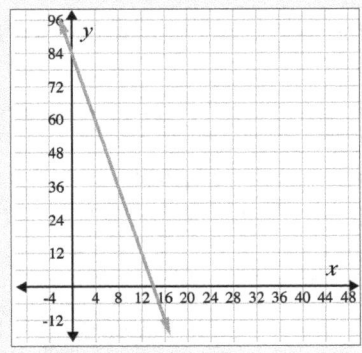

Figure 15.

c. Finally, we can use either the equation or the graph to determine the amount of water left in the tank after 8 minutes. This time, we'll use the graph. By looking at our graph, we can see that the point (8, 36) is on the line. This means that after 8 minutes, there are 36 gallons of water left in the tank.

Practice B — Answers

4.

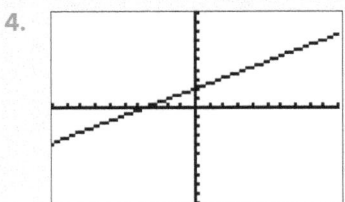

5.

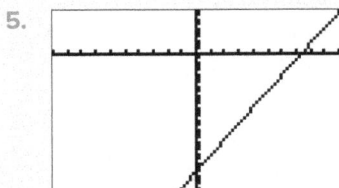

Practice C

Now it's your turn analyze some real-life situations. You'll define variables, write equations, draw graphs, and answer questions just like we did in the examples. When you are done, turn the page and check your solutions.

6. Maurice signs up for an online coffee buying service. He pays a membership fee of $12, and then he can buy a variety of premium coffees for $6 per pound.

    a. Define variables for this situation and include the proper units.

    b. Write and graph an equation in slope-intercept form that models this situation.

    c. Determine the total amount of money Maurice will pay if he purchases 9 pounds of coffee.

7. My uncle's barn has 24 rats in it. Every day, his ancient cat and kills two rats.

    a. Define variables for this situation and include the proper units.

    b. Write and graph an equation in slope-intercept form that models this situation.

    c. Determine the total number of rats that will be in my uncle's barn after 8 days.

D. Writing the Equation of a Line

Now, it is time to use the information we've learned about slope-intercept form and graphs of linear equations to help us write the equations of lines whose graphs are given to us.

Example 7

Write the equation of the line in Figure 16 using slope-intercept form.

Solution

In order to write our equation, we must determine the slope, m, and the y-intercept, b. The y-intercept of this line is $(0, 1)$, so $b = 1$. From this point, we must move *up* 4 spaces and *right* 3 spaces to reach the next point whose coordinates are integers: $(3, 5)$, as in Figure 17.

Because of this, we know that $m = \frac{4}{3}$. We can now write the equation of the line: $y = \frac{4}{3} x + 1$.

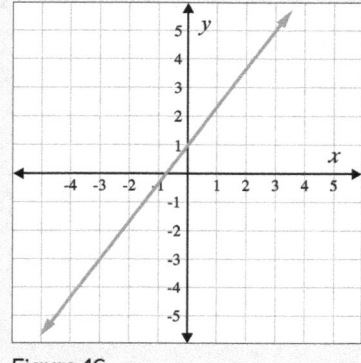

Figure 16.

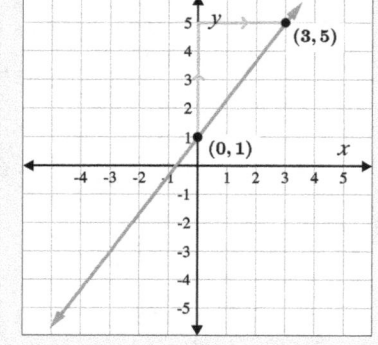

Figure 17.

It's important to note that when we were figuring out the slope of the line in the previous example, we used points on the line whose coordinates are integers, (0, 1) and (3, 5). The points on the line that are between (0, 1) and (3, 5) do not have integer coordinates, so they are very hard to identify from the graph. For the record, here are a few of those points: $\left(\frac{3}{4}, 2\right), \left(1, 2\frac{1}{3}\right), \left(1\frac{1}{2}, 3\right), \left(2, 3\frac{2}{3}\right),$ and $\left(2\frac{1}{4}, 4\right)$.

Example 8

Write the equation of the line shown in Figure 18 in slope-intercept form.

Solution

The y-intercept of the line is (0, –5), so $b = -5$. However, when we start at (0, –5), the graph in Figure 21 doesn't allow us to move from left-to-right in order to locate another point with integer coordinates. Therefore, we'll have to move from right-to-left. We start at (0, –5) and move *up* 2 spaces and *left* 3 spaces to locate the point (–3, –3). See Figure 19.

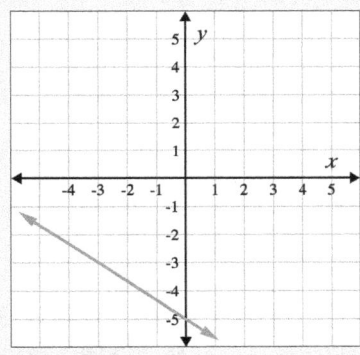

Figure 18.

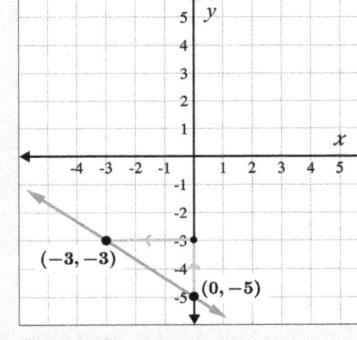

Figure 19.

Because of this, $m = \frac{2}{-3}$. We should re-write the slope as $m = -\frac{2}{3}$. We can now write the equation of the line: $y = -\frac{2}{3}x - 5$.

It was a little more difficult to find the slope of the line in the previous example. However, we can verify that our result makes sense by noting that the line falls from left-to-right. Because of this, we know that the slope is negative.

Practice D

Using slope-intercept form, write the equation of the line shown in each graph. When you are done, turn the page and check your solutions.

8.

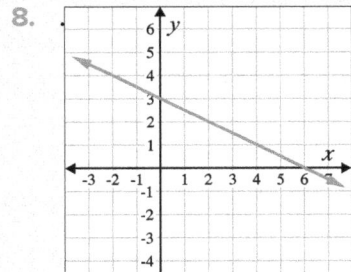

9.

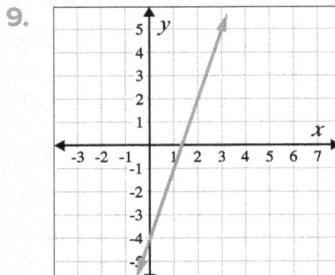

Exercises 2.4

For the following problems, use the y-intercept and slope to help you graph each equation by hand.

1. $y = \frac{2}{3}x + 1$

2. $y = \frac{1}{4}x - 2$

3. $y = -\frac{6}{5}x - 3$

4. $y = -\frac{8}{3}x + 4$

5. $y = \frac{1}{5}x + 2$

6. $y = \frac{3}{2}x - 5$

7. $y = -5x + 2$

8. $y = 3x - 1$

9. $y = \frac{3}{5}x$

10. $y = -\frac{4}{3}x$

11. $y = -2x + \frac{3}{4}$

12. $y = 4x - \frac{8}{3}$

13. $y = x + 2$

14. $y = x$

15. $4x + y = -7$

16. $x + y = 0$

17. $3y - 2x = -3$

18. $6x + 8y = 32$

19. $-8x + 4y = -20$

20. $6x - 15y = -45$

21. $7x + 42y = 168$

22. $16x + 2y = 12$

23. $14x - 12y + 31 = 31$

24. $2x + 7y = 6x + 4y - 15$

For the following six exercises, graph the equation with your calculator, then sketch your graph by hand.

25. $y = \frac{5}{3}x - 5$

26. $y = -\frac{1}{8}x + 3$

27. $y = 2.185x - 7.61$

28. $y = -6.31x + 12.95$

29. $8x - 3y = 17$

30. $-5.6x + 4.2y = -18.4$

For the following exercises, answer all three questions about each situation.

31. Brittany is going on a Caribbean cruise in three weeks and decides to start a new workout routine to get in shape. Before starting, she can do 15 sit-ups without tiring. Each day, she pushes herself to do 3 more sit-ups than her previous best.

    a. Define variables for this situation and include the proper units.

    b. Write and graph an equation in slope-intercept form that models this situation.

    c. Determine how many sit-ups Brittany will be able to do on the 21st day, right before leaving on her cruise.

32. In 2013, someone in Salem, Oregon began leaving money in cereal boxes, blankets, and other items in stores for people to find. One day, this person, whom some call "Benny," leaves home with $2300 and hides a $100 bill in each of many stores.

    a. Define variables for this situation and include the proper units.

    b. Write and graph an equation in slope-intercept form that models this situation.

    c. Using your equation, determine how much money "Benny" will have left after visiting 9 stores.

33. An average-sized bathtub holds about 42 gallons of water. After pulling the drain plug, water flows out at a rate of about 12 gallons per minute.

    a. Define variables for this situation and include the proper units.

    b. Write and graph an equation in slope-intercept form that models this situation.

    c. How much water is left after 3 minutes?

34. When Makayla plugs in her cell phone, it has 23% battery remaining. Using the charger that came with the phone, it charges at a rate of 12% per hour.

    a. Define variables for this situation and include the proper units.

    b. Write and graph an equation in slope-intercept form that models this situation.

    c. If Makayla charges the phone for 6 hours, what percent battery will it have?

35. To host a birthday party for her twins, Mrs. Michiko wants to rent a room at a local pizza place. The restaurant charges $44.95 for the room rental for the afternoon, plus $7.80 per guest for food.

    a. Define variables for this situation and include the proper units.

    b. Write and graph an equation in slope-intercept form that models this situation. Consider using your calculator to help you graph.

    c. How much will the party cost Mrs. Michiko if there are 18 guests?

36. William begins a rafting trip down the Rogue River, in southern Oregon, at Whitehorse Park and plans to take out at the Grave Creek boat ramp, 26.3 miles away. He can cover 3.6 miles every hour he spends on the water.

    a. Define variables for this situation and include the proper units.

    b. Write and graph an equation in slope-intercept form that models this situation. Consider using your calculator to help you graph.

    c. How far will William still have to go after 4 hours?

Practice Set C — Answers

6. a. Variables: x = amount of coffee purchased, in pounds; y = total cost, in dollars.

    b. Equation: $y = 6x + 12$

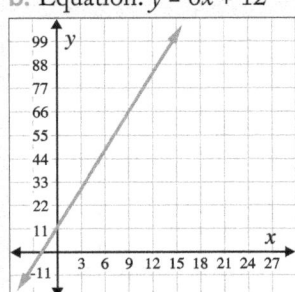

    c. Maurice will pay $66 for 9 pounds of coffee.

7. a. Variables: x = number of days elapsed; y = total number of rats in the barn

    b. Equation: $y = -2x + 24$

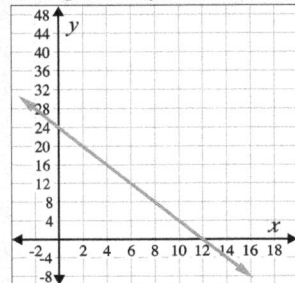

    c. After 8 days, there will be 8 rats in my uncle's barn.

Practice D — Answers

8. $y = -\frac{1}{2}x + 3$

9. $y = 3x - 4$

Write the equation, in slope-intercept form, of the line shown in each graph.

37.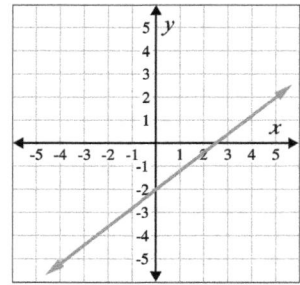

41.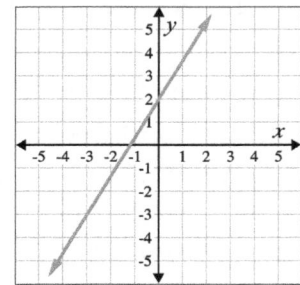

45.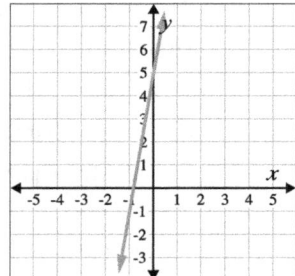

38.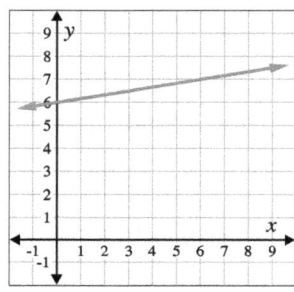

42.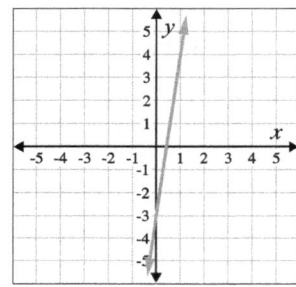

46.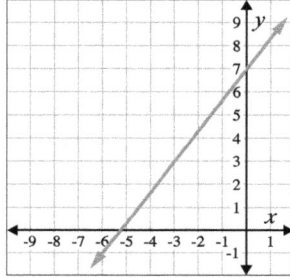

39.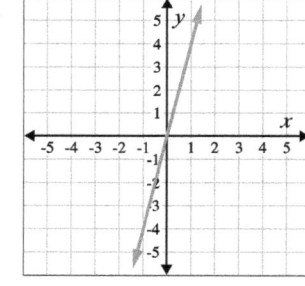

43.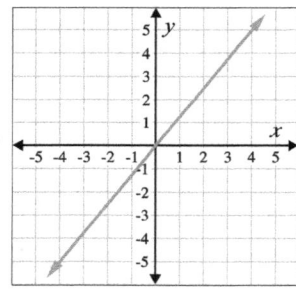

47.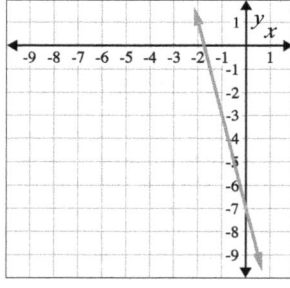

40.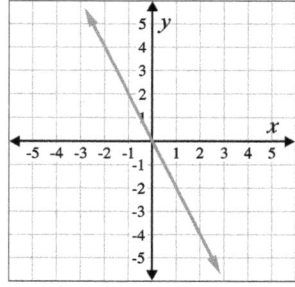

44.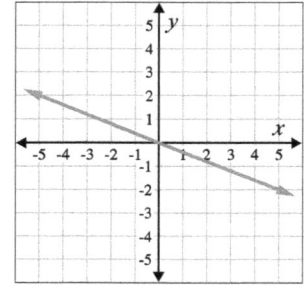

48.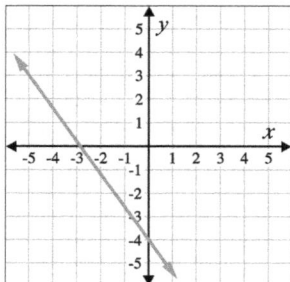

2.5 Standard Form

Overview

In the previous section, we learned how to use slope-intercept form to transition between equations and graphs. We also looked at applications in which one variable quantity depended upon another variable quantity and used equations written in slope-intercept form to model these situations. However, there are some linear relationships in which *both* variable quantities seem to be independent.

Here's a question for you:

> Joscelyn and Aiyana are playing basketball and decide that the first person to score 30 points wins. Points are scored with 2- and 3-point baskets. There are no free throws. Write an equation and draw a graph that illustrates all the different ways a player can score 30 points.

Unlike the problems in the last section, this sort of problem makes it difficult to state with certainty which variable quantity is dependent upon the other. Does the number of 2-point baskets depend on the number of 3-point baskets, or vice versa? Are both variable quantities independent, or are they both dependent? Situations like this are best modeled by writing equations in standard form. We will return to this problem after learning about standard form.

In this section, you will learn how to:

- Use intercepts to graph a line whose equation is given in standard form
- Graph horizontal and vertical lines
- Use standard form to model real-life situations

A. The Intercept Method of Graphing

In section 2.2, we saw that the standard form of a linear equation is: $ax + by = c$. When a linear equation in two variables is given to us in standard form, it is often convenient to algebraically find both of the intercepts of the line. These are the points at which the line crosses the two axes, as you see in Figure 1.

In order to algebraically find both intercepts, we need to understand the following facts:

- Every point on the x-axis has a y-coordinate of 0
- Every point on the y-axis has an x-coordinate of 0.

Those facts allow us to use the following method for finding intercepts and graphing a line.

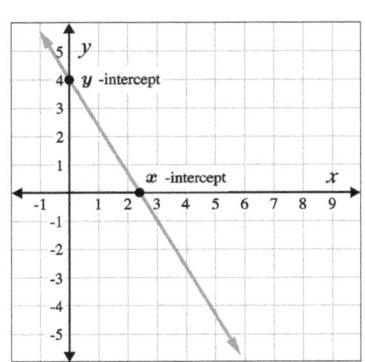

Figure 1.

Using the Intercept Method to Graph an Equation

1. Algebraically find the x-intercept by replacing y with 0 and solving for x.

2. Algebraically find the y-intercept by replacing x with 0 and solving for y.

3. Plot both intercepts.

4. Draw a line through the two points.

Because we need at least two points in order to determine where the line is located, the intercept method won't work if both the x- and y-intercept turn out to be 0. This situation arises when an equation in standard form has a constant term of 0, such as $2x + 3y = 0$.

Let's see how the intercept method works with a few examples.

Example 1

Use the intercept method to graph the line whose equation is $15x - 12y = 60$.

Solution

$15x - 12(0) = 60$ In order to find the x-intercept, replace y with 0 and simplify.

$15x - 0 = 60$

$15x = 60$

$\dfrac{15x}{15} = \dfrac{60}{15}$ Now solve for x.

$x = 4$ The x-intercept is the point $(4, 0)$.

Next, we'll find the y-intercept.

$15(0) - 12y = 60$ Replace x with 0 and simplify.

$0 - 12y = 60$

$-12y = 60$

$\dfrac{-12y}{-12} = \dfrac{60}{-12}$ Now solve for y.

$y = -5$ The y-intercept is the point $(0, -5)$.

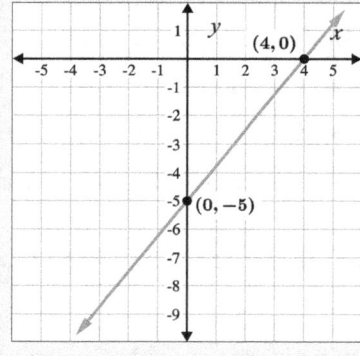

Figure 2.

We can now plot both of the intercepts and draw the line, as you see in Figure 2.

The intercept method gives us some insight into why, when an equation is written in slope-intercept form, b represents the y-intercept. When we replace x with 0, we have the following:

$$y = mx + b$$
$$= m(0) + b$$
$$= b$$

The y-intercept is $(0, b)$. Let's look at another example.

Example 2

Use the intercept method to help you graph the lines whose equations are given below.

1. $y - 2x = -3$ **2.** $-2x + 3y = 3$

Solutions

1. $y - 2x = -3$

This equation isn't written in standard form, but we can still use the intercept method.

$0 - 2x = -3$ Replace y with 0 and simplify.

$-2x = -3$

$\dfrac{-2x}{-2} = \dfrac{-3}{-2}$ Finish by solving for x.

$x = \dfrac{3}{2}$ The x-intercept is the point $\left(\dfrac{3}{2}, 0\right)$.

When plotting points, it is often helpful to rewrite improper fractions as mixed numbers or decimals. In this case, we can rewrite the x-intercept as $\left(1\dfrac{1}{2}, 0\right)$ or $(1.5, 0)$.

$y - 2(0) = -3$ Replace x with 0 and simplify.

$y - 0 = -3$

$y = -3$ The y-intercept is the point $(0, -3)$.

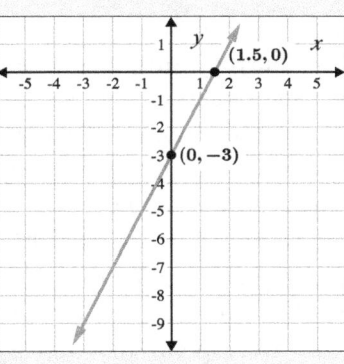

Figure 3.

Now that we know both of the intercepts, we can plot them and graph the line in Figure 3.

2. Graph $-2x + 3y = 3$

This time we'll go through the process without explaining each step. See if you can following along. Start by finding the x-intercept.

$-2x + 3(0) = 3$

$-2x = 3$

$\dfrac{-2x}{-2} = \dfrac{-3}{-2}$

$x = -\dfrac{3}{2}$

Then find the y-intercept.

$-2(0) + 3y = 3$

$3y = 3$

$\dfrac{3y}{3} = \dfrac{3}{3}$

$y = 1$

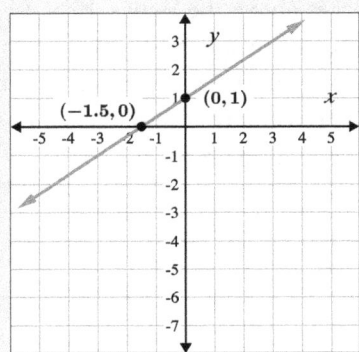

Figure 4.

When written in decimal form, the x-intercept of this line is $(-1.5, 0)$. The y-intercept is $(0, 1)$. We plot the points and graph the line in Figure 4.

Practice A

Now you give it a try. Use the intercept method to help you graph the lines whose equations are given below. When you are done, turn the page and check your solutions.

1. $3x + y = 3$ **2.** $7x - 6y = 21$ **3.** $4x = 2y - 10$

B. Slanted, Horizontal, and Vertical Lines

In all the graphs we've worked with so far in this section, the lines have been slanted. This will always be the case when both variables appear in the equation. If only one variable appears in the equation, then the line will be either horizontal or vertical.

To illustrate this, let's consider the equation $5y = 15$. This equation only contains the variable y. When we solve the equation for y, we get $y = 3$. This equation tells us that the only acceptable value for y is 3. In other words, the line will contain every point that has a y-coordinate of 3. Here are a few points that will be on the line: (1, 3), (2, 3), (3, 3), and

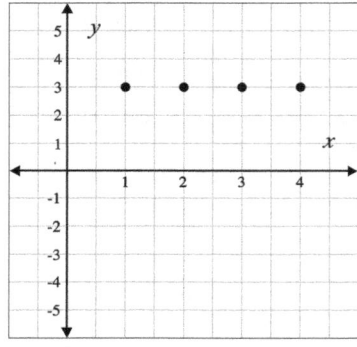

Figure 5.

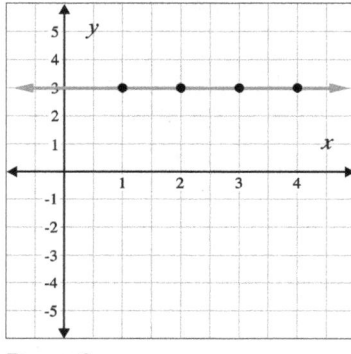

Figure 6.

(4, 3). When we plot these points in Figure 5, we notice that they line up horizontally. They move from left to right without rising or falling.

This shouldn't be surprising. If every y-coordinate is 3, then every point on this line will be exactly 3 spaces above the x-axis. Because of this, the line will be parallel to the x-axis. In Figure 6, we finish graphing the equation by drawing the line that passes through the points we've plotted.

We will see a similar result when the only variable that appears is x. In that case, every point on the line will have the same x-coordinate, and so the line will be parallel to the y-axis. In other words, the line will be vertical.

Example 3

Graph the following equations.

1. $y = 4$

2. $x = -2$

Solutions

1. $y = 4$

The only variable in this equation is y. No matter which x-value we choose, the y-value will always be 4, as we can see in this table.

x	y	(x, y)
-2	4	$(-2, 4)$
-1	4	$(-1, 4)$
0	4	$(0, 4)$
1	4	$(1, 4)$
2	4	$(2, 4)$

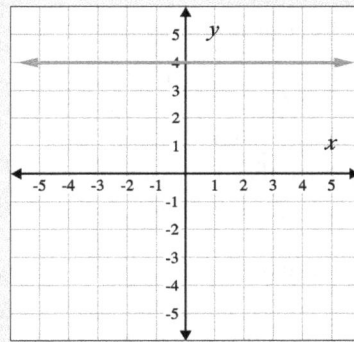

Figure 7.

All points with a y-value of 4 work in this equation, so in Figure 7, we have a horizontal line that iss 4 units above the x-axis.

2. $x = -2$

In this equation, the only variable that appears is x. No matter which y-value we choose, the x-value will always be -2, as you see in this table.

x	y	(x, y)
-2	-2	$(-2, -2)$
-2	-1	$(-2, -1)$
-2	0	$(-2, 0)$
-2	1	$(-2, 1)$
-2	2	$(-2, 0)$

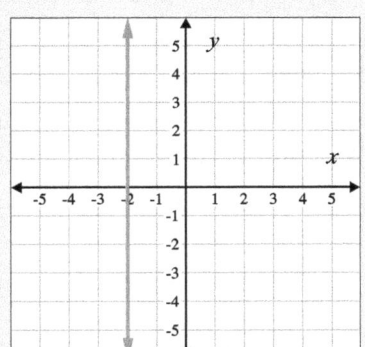

Figure 8.

The resulting graph in Figure 8 is a vertical line two units to the left of the y-axis.

☠ Warning! Incorrect Approach! ☠

When attempting to graph the equation $y = 4$, some people make the mistake of simply plotting the point $(0, 4)$. When attempting to graph $x = -2$, some people mistakenly plot $(-2, 0)$ and leave it at that.

It's important to remember that when we graph an equation on a coordinate plane, we must graph *every* point

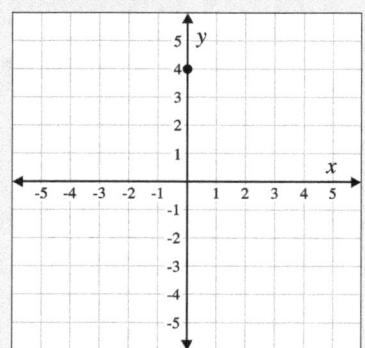

Incorrect graph of $y = 4$.

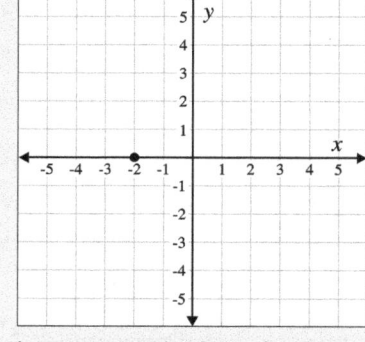

Incorrect graph of $x = -2$.

that satisfies the equation. For any linear equation, there are infinitely many points that satisfy the equation, and when graphed, those points form a straight line.

If you're having trouble remembering which type of equation gives a horizontal line and which gives a vertical line, this may be helpful:

> If a linear equation contains a variable, the graph must pass through the axis indicated by that variable. If an equation only contains the variable y, the graph will *only* pass through the y-axis and thus be horizontal. Similarly, if an equation only contains the variable x, then the graph will *only* pass through the x-axis and thus be vertical.

Practice B

Now it's your turn to graph equations that only contain one variable. When you are done, turn the page and check your solutions.

4. Graph $y = -3$.

5. Graph $x = 4$.

Practice A — Answers

1.

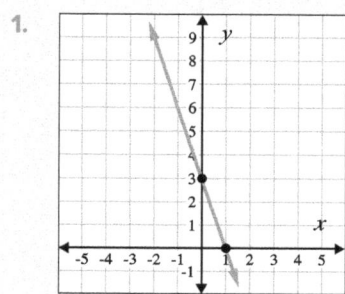

2.

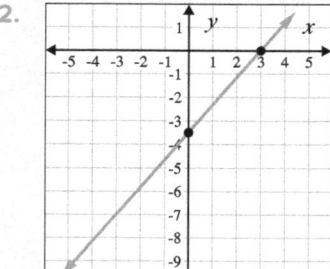

3.

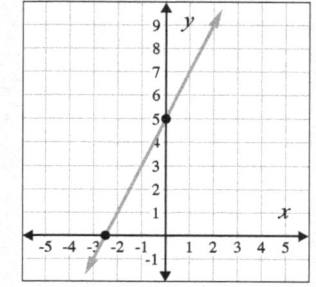

C. Standard Form and Real-Life Situations

As we saw in the introduction to this section, some real-life situations involving two variable quantities do not present one variable quantity as being dependent upon the other. Many of these situations, however, still involve value of some sort. Typically, equations that represent these scenarios take the following form:

$$\begin{array}{c}\text{value represented by}\\ \text{one variable quantity}\end{array} \; + \; \begin{array}{c}\text{value represented by}\\ \text{the other variable quantity}\end{array} \; = \; \begin{array}{c}\text{total}\\ \text{value}\end{array}$$

We often think of value as a monetary measurement, but value isn't always about money. In a basketball game, for example, value is measured in points. With that in mind, let's take another look at the problem from the beginning of this section.

> ## Example 4
>
> Joscelyn and Aiyana are playing basketball and decide that the first person to score 30 points wins. Points are scored with 2- and 3-point baskets. There are no free throws. Write an equation and draw a graph that illustrates all the different ways a player can score 30 points.
>
> ### Solution
>
> In this situation, the variable quantities are the number of 2-point baskets made and the number of 3-point baskets made. We start by defining our variables:
>
> x = the number of 2-point baskets made
> y = the number of 3-point baskets made
>
> Next, we must account for the *value* represented by each type of basket. Each 2-point basket is worth 2 points, and each 3-point basket is worth 3 points. Therefore:
>
> $2x$ = the value of the 2-point baskets made
> $3y$ = the value of the 3-point baskets made
>
> Finally, we note that the *total value* specified by this problem is 30 points. We can now write an equation that correctly models the situation: $2x + 3y = 30$.
>
> Because this equation is in standard form, we will use the intercept method to create the graph. We'll start by finding the x-intercept.
>
> $$2x + 3(0) = 30$$
> $$2x = 30$$
> $$\frac{2x}{2} = \frac{30}{2}$$
> $$x = 15$$

Now we determine the y-intercept.

$$2(0) + 3y = 30$$

$$3y = 30$$

$$\frac{3y}{3} = \frac{30}{3}$$

$$y = 10$$

The intercepts are $(15, 0)$ and $(0, 10)$.

Keep in mind that model breakdown can occur when a situation is modeled by a graph. As you see in Figure 9, the only values that make sense in real life are represented by points that have whole-number coordinates.

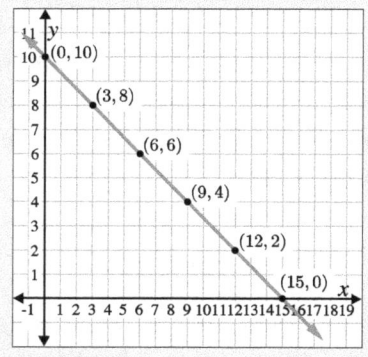

Figure 9.

Example 5

Monique has $40 to spend on beach balls and Frisbees for an upcoming beach party. Each beach ball costs $4.00, and each Frisbee costs $2.50. Write and graph an equation that illustrates all of the different ways that Monique can buy beach balls and Frisbees and spend exactly $40.

Solution

The variable quantities are the number of beach balls pruchances and the number of Frisbees purchased:

x = the number of beach balls purchased
y = the number of Frisbees purchased

Next, we must account for the *value* represented by each type of toy. Each beach ball is worth $4.00, and each Frisbee is worth $2.50. Therefore:

$4x$ = the value of the beach balls purchased
$2.5y$ = the value of the Frisbees purchased

Practice B — Answers

4.

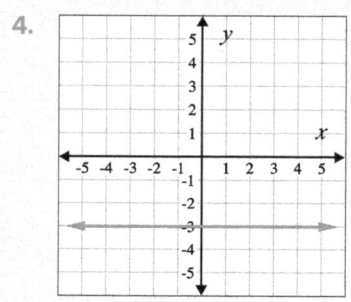

5.

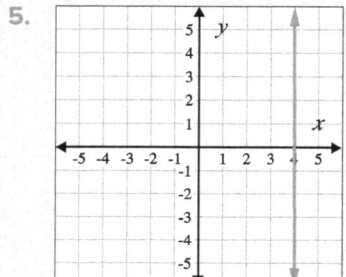

Using a *total value* of $40, we can write an equation: $4x + 2.5y = 40$. Now it's time to graph the equation. Once again, we'll use the intercept method, solving for x first and then for y.

$$4x + 2.5(0) = 40$$

$$4x = 40$$

$$\frac{4x}{4} = \frac{40}{4}$$

$$x = 10$$

$$4(0) + 2.5y = 40$$

$$2.5y = 40$$

$$\frac{2.5y}{2.5} = \frac{40}{2.5}$$

$$y = 16$$

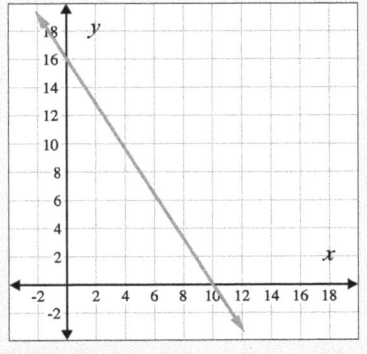

Figure 10.

The intercepts are $(10, 0)$ and $(0, 16)$. Figure 10 is our graph.

Practice C

It's time for you to create mathematical models — equations and graphs — representing some real-life situations. When you are done, turn the page and check your solutions.

6. Apple trees cost $30 each, and cherry trees cost $40 each. Rohan has $600 to spend on fruit trees.

   a. Define variables for this situation.

   b. Write an equation that illustrates the different ways Rohan can purchase apple trees and cherry trees and spend exactly $600.

   c. Graph the equation.

7. Phillipe and Tabatha are playing a trivia game in which two-part questions are asked. If a person is able to answer just one part of a question correctly, then they are partially-correct and are awarded 6 points. If a person answers both parts of a question correctly, then they are totally correct and are awarded 8 points. In order to win the game, a person must earn 72 points.

   a. Define variables for this situation.

   b. Write an equation that illustrates the different ways a person can earn exactly 72 points.

   c. Graph the equation.

Exercises 2.5

For the following problems, use the intercept method to help you graph the line whose equation is given.

1. $2x + 3y = 6$
2. $8x + 4y = 24$
3. $4x + 2y = -4$
4. $3x + 9y = -9$
5. $y = 7$
6. $y = -4$
7. $6x - 4y = 12$
8. $2x - 9y = 18$
9. $x - 3y = 6$
10. $5x - y = 5$

11. $x + y = 8$
12. $x - y = 4$
13. $x = -5$
14. $x = 1$
15. $-3x - 4y = 12$
16. $-2x - 6y = 18$
17. $4x = -12$
18. $6y = 6$
19. $6x - 8y = -2$
20. $4x - 7y = -14$

21. $-3x + 2y = 3$
22. $-6x + 4y = 4$
23. $5y = 4$
24. $2x = -7$
25. $-3x + 6y = -20$
26. $-9x + 8y = -60$
27. $6 = 8x$
28. $22 = 4x$

For the following problems, follow the instructions to model the situation mathematically.

29. To pay the band that's playing at her restaurant, Kyrsten needs to sell $1200 worth of tickets. She plans to sell general admission seats for $15 and preferred seats for $25. Define variables for this situation. Then write and graph an equation that illustrates all of the different ways Kyrsten can sell enough tickets to pay the band.

30. Pie is the most important part of the Thanksgiving meal at Fabian's house, even though his family doesn't like pumpkin! It's so important that he has to make 6 pies in order to feed everyone. Fabian needs a total of 36 cups of mixed berries to fill the pies. He can buy strawberries in 4-cup cartons and raspberries in 3-cup cartons. Define variables for this situation. Then write and graph an equation that illustrates all the ways Fabian can buy enough berries to fill the pies.

Practice C — Answers

6. a. Variables: x = number of apple trees purchased, and y = number of cherry trees purchased.
   b. Equation: $30x + 40y = 600$
   c.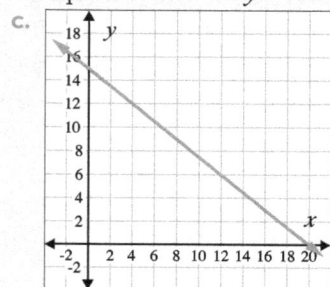

7. a. Variables: x = number of partially-correct answers, and y = number of totally-correct answers.
   b. Equation: $6x + 8y = 72$
   c.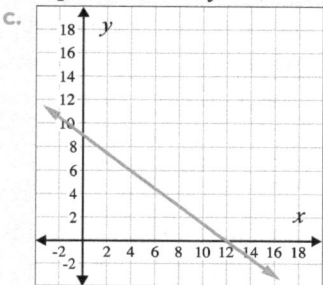

31. David is buying pizza for a group of 45 friends he hasn't seen in a while. A large pizza feeds 3 people and a family-size pizza will feed 5. Define variables for this situation. Then, write and graph an equation that illustrates all of the different ways that David can buy large and family-size pizzas to feed exactly 45 people.

32. At work, Isidro earns vacation leave hours each month. He's been saving for quite a while, and has 90 hours of time off he can use. Because he works shorter shifts on Fridays, taking a 3-day weekend costs him 5 hours of leave, while a full week off uses 40 hours. Define variables for this situation. Then, write and graph an equation that shows the different ways Isidro can use his 90 hours of vacation.

33. Kellie loves to read books, especially from her favorite used bookstore's clearance shelf. Clearance books are priced at $1.50 for paperbacks and $2.50 for hardbacks. Kellie has $30 to spend on books. Define variables for this situation. Then, write and graph an equation that illustrates all the different ways Kellie can buy books, spending only $30.

34. Kate works at a bakery, making dough for French bread and croissants. It takes her 25 minutes to make a batch of French bread dough and 40 minutes to make a batch of croissant dough. Kate's usual workday is 8 hours long. Define variables for this situation. Then, write and graph an equation that illustrates all the different ways Kate can spend her workday making different types of dough.

2.6 Ordered Pairs and Equations

Overview

So far in this chapter, we have constructed linear models in a number of different situations:

Given Information	Linear Model
a table (ordered pairs)	a graph (scatter plot)
an equation	a table and then a graph (line)
an equation	a graph (line)
a graph (line)	an equation

Now it's time for us to learn how to take information from a table (ordered pairs) and write the equation of the corresponding line. We need to be able to do this algebraically, without having to draw a graph. So here's a problem for you:

> There is a linear relationship between the number of credits a community college student is enrolled for and the total registration cost. A student taking 11 credits pays $1,091 to register. A student taking 14 credits pays $1,379 to register. Write a linear equation in slope-intercept form that models this situation.

The registration cost depends on the number of credits for which a student is enrolled, so we let x represent the number of credits that a student is enrolled for, and y represents the total registration cost. The problem gives us the ordered pairs (11, 1091) and (14, 1379). Graphing these points, drawing a line, and figuring out the y-intercept accurately would be difficult and time-consuming. That's why we need to determine the equation algebraically. We'll learn how to use algebra to find the equation of a line, given two points, and then revisit this problem.

As you work through this section, you will learn how to:

- Write linear equations in slope-intercept form and in point-slope form.
- Use slope-intercept form to help generate the equation of a line when given either the ordered pair for one point and the slope or the ordered pairs for two points.
- Use point-slope form to help generate the equation of a line when given either the ordered pair for one point and the slope or the ordered pairs for two points.
- Determine two ordered pairs from a story problem, and use these ordered pairs to write a linear equation.

A. The Slope-Intercept and Point-Slope Forms

There are two algebraic methods for finding the equation of a line when given two points. One method makes use of slope-intercept form, and the other method makes use of a form we haven't seen yet, point-slope form. Both of these forms are derived from the slope formula, $m = \frac{y_2 - y_1}{x_2 - x_1}$.

Slope-Intercept Form of a Linear Equation

First, we'll use the slope and y-intercept to derive a linear equation. We let the y-intercept $(0, b)$ represent one point on the line, and we let (x, y) represent any other point on the line. We can substitute this information into the formula for slope and go from there.

$m = \frac{y_2 - y_1}{x_2 - x_1}$	Begin with the slope formula.
$m = \frac{y - b}{x - 0}$	Rewrite the formula using $(0, b)$ in place of (x_1, y_1) and using (x, y) in place of (x_2, y_2).
$m = \frac{y - b}{x}$	Simplify the denominator.
$mx = x\left(\frac{y - b}{x}\right)$	Multiply both sides of the equation by x. This cancels out the denominator on the right side of the equation.
$mx = y - b$	
$\underline{ + b \qquad + b}$	Add b to both sides of the equation.
$mx + b = y$	
$y = mx + b$	Rewrite the equation so that y appears on the left side.

Because this equation was derived using the slope and the y-intercept, it is called the slope-intercept form of a linear equation. When using this form to write the equation of a line, we write numbers in place of m and b — for example, $y = 5x + 9$.

Example 1

Using the information given, write the equation of each line in slope-intercept form.

1. A line has a slope of 3 and a y-intercept of $(0, -7)$.

2. A line has a slope of $-\frac{4}{9}$ and a y-intercept of $(0, 11)$.

Solutions

1. A line has a slope of 3 and a y-intercept of $(0, -7)$.
   We are given the values of m and b: $m = 3$ and $b = -7$. Therefore, the equation of the line is $y = 3x - 7$.

2. A line has a slope of $-\frac{4}{9}$ and a y-intercept of $(0, 11)$.
   We are given the values of m and b: $m = -\frac{4}{9}$ and $b = 11$. Therefore, the equation of the line is $y = -\frac{4}{9}x + 11$.

Point-slope form of a linear equation

Now we'll use the slope and a given point on the line, (x_1, y_1), to derive a linear equation. We will let (x, y) represent any other point on the line.

$$m = \frac{y_2 - y_1}{x_2 - x_1}$$ Begin with the slope formula.

$$m = \frac{y - y_1}{x - x_1}$$ Leave (x_1, y_1) alone but use (x, y) in place of (x_2, y_2).

$$m(x - x_1) = (x - x_1) \cdot \frac{y - y_1}{x - x_1}$$ Multiply both sides of the equation by $(x - x_1)$ and simplify the right side of the equation.

$$m(x - x_1) = y - y_1$$

$$y - y_1 = m(x - x_1)$$ Rewrite the equation so that the terms containing y appear on the left side. This gives us the point-slope form of a linear equation

This equation was derived using a given point and the slope, so we call it the **point-slope form** of a linear equation. When using this form to write the equation of a line, we write numbers in place of m, x_1, and y_1.

Point-Slope Form

The point-slope form of a linear equation is

$$y - y_1 = m(x - x_1)$$

where m represents the slope of the line, and (x_1, y_1) represents a point on the line.

Example 2

Using the information given, write the equation of each line in point-slope form.

1.   A line has a slope of -5 and passes through the point $(6, 8)$.
2.   A line has a slope of $\frac{7}{4}$ and passes through point $(3, -2)$.

Solutions

1.   A line has a slope of -5 and passes through the point $(6, 8)$.

$$y - y_1 = m(x - x_1)$$ We start with the point-slope form of a linear equation.

$$y - 8 = -5(x - 6)$$ The problem gives us these values: $m = -5$, $x_1 = 6$ and $y_1 = 8$. We use them to rewrite the equation.

The equation of this line, in point-slope form, is $y - 8 = -5(x - 6)$.

2. A line has a slope of $\frac{7}{4}$ and passes through point $(3, -2)$.

A line has a slope of $\frac{7}{4}$ and passes through point $(3, -2)$.

$y - y_1 = m(x - x_1)$ We start with the point-slope form of a linear equation.

$y - (-2) = \frac{7}{4}(x - 3)$ The problem gives us these values: $m = \frac{7}{4}$, $x_1 = 3$ and $y_1 = -2$. We use them to rewrite the equation.

$y + 2 = \frac{7}{4}(x - 3)$ We finish by simplifying the left side of the equation.

▶ The equation of this line, in point-slope form, is $y + 2 = 7\_\_4(x - 3)$.

Practice A

Now it's your turn to practice writing linear equations. Use the given information to write the equation of the line in the specified form. When you are done, turn the page and check your solutions.

1. $m = -9$, y-intercept $(0, 3)$
   Write the equation in slope-intercept form.

2. $m = \frac{5}{8}$, y-intercept $(0, -11)$
   Write the equation in slope-intercept form.

3. $m = 4$, passes through $(-1, 6)$
   Write the equation in point-slope form.

4. $m = -2$, passes through $(7, 5)$
   Write the equation in point-slope form.

B. Writing the Equation of a Line (Given Slope and One Point)

Did you notice how both the slope-intercept and point-slope forms of a linear equation make use of the slope? This means that we must know the slope of a line before we can use either form to help us generate a linear equation.

When we're asked to write a linear equation and a form is not specified, it's traditional to write our final answer in slope-intercept form. We will now explore two methods for finding the equation of a line in slope-intercept form when we are given the slope and one point that the line passes through.

Method 1: Use Slope-Intercept Form

When given one point that lies on a line as well as the line's slope, we can make use of slope-intercept form to generate the line's equation. The steps of this approach are:

1. Begin with slope-intercept form: $y = mx + b$.

2. Replace x and y with coordinates from the ordered pair, and replace m with the slope.

3. Solve the equation for b.

4. Use values of m and b to write the final equation.

Example 3

Using the given information, write the equation of each line.

1. The line has a slope of 2, and passes through the point $(4, 3)$.
2. The line has a slope of $-\frac{6}{7}$ and passes through the point $(-14, 8)$.

In these problems, the form of the linear equation isn't specified, so we're expected to write our final answers in slope-intercept form.

Solutions

1. The line has a slope of 2, and passes through the point $(4, 3)$.

$y = mx + b$	Start with the slope-intercept form of a linear equation.
$3 = 2(4) + b$	Replace x with 4, y with 3, and m with 2.
$3 = 8 + b$	Then simplify the right side of the equation.
$\underline{-8 \qquad -8}$	Solve for b by subtracting 8 from both sides of the equation.
$-5 = b$	

We can now use the value of m that was given and the value of b that we just found to write our final equation, $y = 2x - 5$.

2. The line has a slope of $-\frac{6}{7}$ and passes through the point $(-14, 8)$.

$y = mx + b$	Begin with the slope-intercept form of a linear equation.
$8 = -\frac{6}{7}(-14) + b$	Using the given information, replace x, y and m.
$8 = 12 + b$	Simplify the right side of the equation.
$\underline{-12 \qquad -12}$	Solve for b by subtracting 12 from both sides.
$-4 = b$	

Now that we've found the value of b, we can write our equation, $y = -\frac{6}{7}x - 4$.

Method 2: Use Point-Slope Form

When we're given the line's slope and one point that a line passes through, as you saw in Example 3, point-slope form can also help us find the equation of the line. Here are the steps to follow with this approach:

1. Begin with point-slope form: $y - y_1 = m(x - x_1)$.

2. Replace x_1 and y_1 with coordinates from the ordered pair, and replace m with the slope.

3. Solve the equation for y.

We'll now revisit the problems from Example 3 to see how Method 2 works.

Example 4

Using the given information, write the equation of each line.

1. The line has a slope of 2, and passes through $(4, 3)$.
2. The line has a slope of $-\frac{6}{7}$ and passes through $(-14, 8)$.

Solutions

1. The line has a slope of 2, and passes through $(4, 3)$.

$y - y_1 = m(x - x_1)$	Begin with the point-slope form of a linear equation.
$y - 3 = 2(x - 4)$	Replace m with 2, x_1 with 4, and y_1 with 3.
$y - 3 = 2x - 8$	Use the distributive property on the right to remove the parentheses.
$\underline{+3 \qquad +3}$	Finish by adding 3 to both sides.
$y = 2x - 5$	

2. The line has a slope of $-\frac{6}{7}$ and passes through $(-14, 8)$.

$y - y_1 = m(x - x_1)$	Start with the point-slope form of a linear equation.
$y - 8 = -\frac{6}{7}(x - (-14))$	Use the given values of m, x_1, and y_1 to rewrite the equation.
$y - 8 = -\frac{6}{7}(x + 14)$	Simplify the expression inside the parentheses.
$y - 8 = -\frac{6}{7}x - 12$	Use the distributive property on the right and simplify the result.
$\underline{+8 \qquad +8}$	Finish the problem by adding 8 to both sides.
$y = -\frac{6}{7}x - 4$	

As you can see, Method 1 (Example 3) and Method 2 (Example 4) both give the same final answer. This illustrates the fact that a line's equation is not affected by the method we use to find it.

If a problem makes reference to a horizontal or vertical line, then we will not be given the slope directly. This is fine if we remember that the equation of a horizontal line will only contain the letter y, and the equation of a vertical line will only contain the letter x.

Practice A — Answers

1. $y = -9x + 3$
2. $y = \frac{5}{8}x - 11$
3. $y - 6 = 4(x + 1)$
4. $y - 5 = -2(x - 7)$

Example 5

Using the given information, write the equation of each line.

1. The line is horizontal and passes through $(7, 8)$.

2. The line is vertical and passes through $(5, 2)$.

Solutions

1. The line is horizontal and passes through $(7, 8)$.

Because this line is horizontal, the equation can only contain the letter y. The y-coordinate of the given point is 8, so the equation of this line is $y = 8$.

2. The line is vertical and passes through $(5, 2)$.

Because this line is vertical, the equation can only contain the letter x. The x-coordinate of the given point is 5, so the equation of this line is $x = 5$.

Practice B

Now it's your turn. For each problem, find the equation of the line whose information is given. You will be prompted to use both methods. In each case, write your final answer in slope-intercept form. When you are done, turn the page and check your solutions.

5. A line passes through $(7, -1)$ and has a slope of -2.

    a. Use slope-intercept form (method 1) to help you find the equation of the line.

    b. Use point-slope form (method 2) to help you find the equation of the line.

6. A line passes through $(8, 9)$ and has a slope of $\frac{3}{4}$.

    a. Use slope-intercept form (method 1) to help you find the equation of the line.

    b. Use point-slope form (method 2) to help you find the equation of the line.

7. Find the equation for the horizontal line and the vertical line passing through $(-10, 7)$

C. Writing the Equation of a Line Given Two Points

If we need to find the equation of a line, but are *not* given the slope, then we will need to know more than one point that is on the line. If we know two points that lie on a line, then we can use the slope formula to calculate the slope. Once we know the slope, we can then choose either of the given points and use it to help us generate the equation of the line.

Example 6

Find the equation of the line passing through the points $(6, -28)$ and $(24, -13)$.

Solution

The form of the linear equation is not specified, so we're expected to write our answer in slope-intercept form. Because we're only given two points, the first thing we have to do is find the slope.

$$m = \frac{y_2 - y_1}{x_2 - x_1} = \frac{-13 - (-28)}{24 - 6} = \frac{15}{18} = \frac{5}{6}$$

Now we can pick one of the given points and proceed. We'll choose the point $(6, -28)$ and use method 1 to find the equation.

$$y = mx + b \qquad \text{Begin with the slope-intercept form of a linear equation.}$$

$$-28 = \tfrac{5}{6}(6) + b \qquad \text{Replace } x \text{ with 6, } y \text{ with } -28, \text{ and } m \text{ with } \tfrac{5}{6}.$$

$$-28 = 5 + b \qquad \text{Simplify the right side of the equation.}$$

$$\underline{-5 \qquad -5} \qquad \text{Solve for } b.$$

$$-33 = b$$

$$y = \tfrac{5}{6}x - 33 \qquad \text{Finish by writing the equation of the line.}$$

Notice that if we had used the point $(24, -13)$ instead of $(6, -28)$ in Example 6, our final answer would still would have been $y = \tfrac{5}{6}x - 33$.

Example 7

Find the equation of the line that passes through the points $(4, 1)$ and $(3, 5)$.

Solution

Once again, we need to start by calculating the slope.

$$m = \frac{y_2 - y_1}{x_2 - x_1} = \frac{5 - 1}{3 - 4} = \frac{4}{-1} = -4$$

Now that we know the slope, we can pick one of the given points and proceed. We'll choose the point $(4, 1)$ and use method 2 from above.

$$y - y_1 = m(x - x_1) \qquad \text{Start with the point-slope form of a linear equation.}$$

$$y - 1 = -4(x - 4) \qquad \text{Replace } m \text{ with } -4, x_1 \text{ with 4, and } y_1 \text{ with 1.}$$

$$y - 1 = -4x + 16 \qquad \text{Use the distributive property to remove parentheses on the right.}$$

$$\underline{+1 \qquad\qquad +1} \qquad \text{Add 1 to both sides to solve for } y.$$

$$y = -4x + 17$$

If we had used the point $(3, 5)$ instead of $(4, 1)$, the final result would have been the same.

Practice C

Find the equation of the line passing through the given pair of points. Remember to write your final answer in slope-intercept form. When you are done, turn the page and check your solutions.

8. The line passes through $(4, 1)$ and $(6, 5)$ **10.** The line passes through $(8, -3)$ and $(20, 6)$

9. The line passes through $(-7, -1)$ and $(-4, 8)$

D. Application Problems

Now that we know how to find the equation of a line when we are given two points, we can try some application problems. Let's start with the problem from the beginning of this section.

Example 8

There is a linear relationship between the number of credits a community college student is enrolled for and the total registration cost. A student taking 11 credits pays \$1,091 to register. A student taking 14 credits pays \$1,379 to register. Write a linear equation, in slope-intercept form, that models this situation.

Solution

Because x represents the number of credits that a student is enrolled for, and y represents the total registration cost, we have the ordered pairs $(11, 1091)$ and $(14, 1379)$.

We start by finding the slope.

$$m = \frac{y_2 - y_1}{x_2 - x_1} = \frac{1379 - 1091}{14 - 11} = \frac{288}{3} = 96$$

Next, we use one of the given points — we'll choose $(11, 1091)$ — and the slope and find the equation of the line. Using method 1, the work looks like this:

$$y = mx + b$$
$$1091 = 96(11) + b$$
$$1091 = 1056 + b$$
$$\underline{-1056 \qquad -1056}$$
$$35 = b$$
$$y = 96x + 35$$

Based on what we learned previously in this chapter, our final equation tells us that the initial cost (probably a registration fee) before any credits are paid for is \$35. The equation also tells us that the cost-per-credit is \$96.

Example 9

Water drains out of a tank at a constant rate. This means that there is a linear relationship between the elapsed time and the amount of water in the tank. After 28 minutes, there are 888 gallons of water in the tank. After 37 minutes, there are 627 gallons of water in the tank. Write a linear equation that models this situation.

Solution

The amount of water in the tank depends on the amount of time that has elapsed. Therefore, we let x represent the elapsed time in minutes, and we let y represent the gallons of water left in the tank. Based on this, we have the ordered pairs (28,888) and (37,627). Because the form of the linear equation was not specified, our final answer will need to be in slope-intercept form.

The first step is to calculate the slope.

$$m = \frac{y_2 - y_1}{x_2 - x_1} = \frac{627 - 888}{37 - 28} = \frac{-261}{9} = -29$$

Next, we'll choose one of the given points, $(28, 888)$, and the slope to find the equation of the line. Using method 2, the work looks like this:

$$y - y_1 = m(x - x_1)$$
$$y - 888 = -29(x - 28)$$
$$y - 888 = -29x + 812$$
$$\underline{+ 888 \qquad\qquad + 888}$$
$$y = -29x + 1700$$

Based on previous knowledge, we can interpret our final equation. The starting volume of water is 1,700 gallons. The water drains away at a rate of 29 gallons per minute.

Practice B — Answers

5. a. $-1 = -2(7) + b$; $b = 13$; $y = -2x + 13$; b. $y + 1 = -2(x - 7)$; $y = -2x + 13$

6. a. $9 = \frac{3}{4}(8) + b$; $b = 3$; $y = \frac{3}{4}x + 3$; b. $y - 9 = \frac{3}{4}(x - 8)$; $y = \frac{3}{4}x + 3$

7. horizontal: $y = 7$; vertical: $x = -10$

Practice Set D

For each of these application problems, you'll have to define variables and then find an equation in slope-intercept form that correctly models the situation. When you are done, turn the page and check your solutions.

11. A misguided eighth-grader pulls the fire alarm at a middle school. There's a linear relationship between the amount of time after the alarm is pulled and the number of students in the building. After 13 seconds, there are 198 students in the building. After 21 seconds, there are 166 students in the building. Define the variables, including the correct units, and then find an equation that correctly models this situation.

12. There's a linear relationship between the number of credits a community college student is enrolled for and the total registration cost. A student taking 9 credits pays $983 to register. A student taking 13 credits pays $1,411 to register. Define the variables, including the correct units, and then find an equation that correctly models this situation.

Exercises 2.6

For the following problems, use the given information to write the equation of the line in slope-intercept form.

1. $m = 3$, y-intercept $(0, 4)$

2. $m = 2$, y-intercept $(0, 5)$

3. $m = 8$, y-intercept $(0, 1)$

4. $m = 5$, y-intercept $(0, -3)$

5. $m = -6$, y-intercept $(0, -1)$

6. $m = -4$, y-intercept $(0, 0)$

7. $m = -\frac{3}{2}$, y-intercept $(0, 0)$

For the following problems, use the given information to write the equation of the line in point-slope form.

8. $m = 7$, passes through $(9, 20)$

9. $m = -2$, passes through $(5, 1)$

10. $m = 3$, passes through $(11, -2)$

11. $m = 5$, passes through $(-10, 3)$

12. $m = -\frac{7}{8}$, passes through $(-16, 9)$

13. $m = \frac{3}{4}$, passes through $(-12, -6)$

14. $m = \frac{9}{2}$, passes through $(-8, -5)$

15. $m = -\frac{8}{3}$, passes through $(3, -10)$

For the following problems, use the given information to write the equation of the line.

16. $m = 3$, passes through $(1, 4)$

17. $m = 1$, passes through $(3, 8)$

18. $m = 2$, passes through $(1, 4)$

19. $m = 8$, passes through $(4, 0)$

20. $m = -3$, passes through $(3, 0)$

21. $m = -1$, passes through $(6, 0)$

22. $m = -6$, passes through $(0, 0)$

23. $m = -2$, passes through $(0, 1)$

24. $m = \frac{2}{3}$, passes through $(-9, 2)$

25. $m = -\frac{6}{7}$, passes through $(7, -3)$

26. $m = -\frac{1}{5}$, passes through $(20, 6)$

27. $m = \frac{5}{2}$, passes through $(12, 17)$

28. $m = \frac{11}{3}$, passes through $(6, -1)$

29. $m = -\frac{1}{9}$, passes through $(-27, 7)$

30. horizontal, passes through $(7, 12)$

31. horizontal, passes through $(-9, 5)$

32. vertical, passes through $(-1, 18)$

33. horizontal, passes through $(95, 321)$

34. vertical, passes through $(3, -14)$

35. vertical, passes through $(-57, -118)$

36. passes through $(0, 0)$ and $(3, 2)$

37. passes through $(0, 0)$ and $(5, 8)$

38. passes through $(4, 1)$ and $(6, 3)$

39. passes through $(2, 5)$ and $(1, 4)$

40. passes through $(5, -3)$ and $(6, 2)$

41. passes through $(-6, -5)$ and $(-2, -17)$

42. passes through $(-10, 24)$ and $(-7, 9)$

43. passes through $(2, 3)$ and $(5, 3)$

44. passes through $(-1, 5)$ and $(4, 5)$

45. passes through $(4, 1)$ and $(4, 2)$

46. passes through $(2, 7)$ and $(2, 8)$

47. passes through $(3, 13)$ and $(5, 5)$

48. passes through $(0, 0)$ and $(1, 1)$

49. passes through $(-2, 4)$ and $(3, -5)$

50. passes through $(1, 6)$ and $(-1, -6)$

51. passes through $(14, 12)$ and $(-7, -15)$

52. passes through $(0, -4)$ and $(5, 0)$

For the following problems, define the variables (include the correct units when applicable) and then find an equation that correctly models the situation.

53. Thaddeus is baking cookies for a party. There is a linear relationship between the number of cookies he bakes and the number of party guests. If there are 12 party guests, Thaddeus will bake 40 cookies. If there are 25 party guests, he will bake 79 cookies.

54. Professor Wagner has found that there is a linear relationship between the number of mathematicians attending a math conference and the number of cheesy jokes that are told. When there are 18 mathematicians present, 41 cheesy jokes are told. On the other hand, when there are 84 mathematicians present, 173 cheesy jokes are told.

55. There is a linear relationship between the number of police officers parked on the side of the highway and the number of passing drivers each minute who are speeding. When 2 police officers are parked on the side of the highway, 57 passing drivers per minute are speeding. When 5 officers are parked on the side of the highway, only 33 passing drivers per minute are speeding.

56. There is a linear relationship between the number of minutes elapsed after a movie ends and the number of people in the theater. There are 159 people in the theater one minute after the movie ends. There are 96 people in the theater four minutes after the movie ends.

57. A track-and-field coach has found that there is a linear relationship between the temperature and the number of athletes who complain at practice. If it is 75 degrees, then 7 athletes complain. However, if it is 87 degrees, then 11 athletes complain.

58. Kirstie is typing a lengthy research paper. She finds that there is a linear relationship between the amount of time she has been typing and her typing speed. After thirty minutes of typing, her typing speed is 41 words per minute. After eighty minutes of typing, her typing speed is only 26 words per minute.

Practice D — Answers

11. Variables: x = elapsed time in seconds; y = number of students in the building.
     Equation: $y = -4x + 250$

12. Variables: x = the number of credits a student enrolls for; y = total cost, in dollars.
     Equation: $y = 107x + 20$

2.7 Line of Best Fit

Overview

In the previous section, we used algebra to find the equation of a line when we were given ordered pairs for two points on the line. If two variable quantities exhibit an *exact* linear relationship, then two ordered pairs are all we need to find the equation for the linear model.

Now, let's think back to when we were constructing scatter plots earlier in the chapter. We saw some situations in which points exhibited an *approximate* linear relationship. In cases like these, we can still find an equation for a linear model. However, when doing this by hand, there is a bit of guesswork involved.

So here's a problem:

> The table below gives the net income of The Vegan Truck Stop, a small dining establishment in Mill City, from 2007 to 2013. Create a scatter plot from the information on the table. Then draw the line of best fit. Finally, use algebra to find the equation of the line of best fit.
>
> | Year | 2007 | 2008 | 2009 | 2010 | 2011 | 2012 | 2013 |
> |---|---|---|---|---|---|---|---|
> | Net Income | $4,000 | $8,000 | $9,000 | $7,000 | $10,000 | $11,000 | $14,000 |

We will work through this problem soon, but before we get to that, we need to know about the *line of best fit* and look at strategies for creating one by hand. Later on, we'll discover how to use technology to help us in this endeavor.

As you work through this section, you will learn how to:

- ◆ Find the equation of a line of best fit by hand.
- ◆ Create a scatter plot and find the equation of the line of best fit on a graphing calculator.
- ◆ Use the equation of the line of best fit to make predictions.

A. Scatter Plots and Linear Modeling by Hand

If a scatter plot exhibits an approximate linear relationship, then the line of best fit is the line that most closely matches the location and direction of the points. In general, the line of best fit will not hit most of the points in the scatter plot — and in fact, it may not hit *any* of the points. However, it should pass through the approximate center of the points and match the overall direction exhibited by the points.

When drawing a line of fit by hand, it's virtually impossible to draw the actual line of *best* fit. However, we should be able to draw a pretty close approximation. As we work through the process, we first need to identify two points — not necessarily from the scatter plot — that define a line that passes through the center of the scatter plot. This usually means that there will be about as many points above the line as there are below it. The two points must also define a line that matches the overall direction of points in the scatter plot. Once we have identified two acceptable points, we can draw our line of fit.

Example 1

Determine which diagram shows a line of best fit.

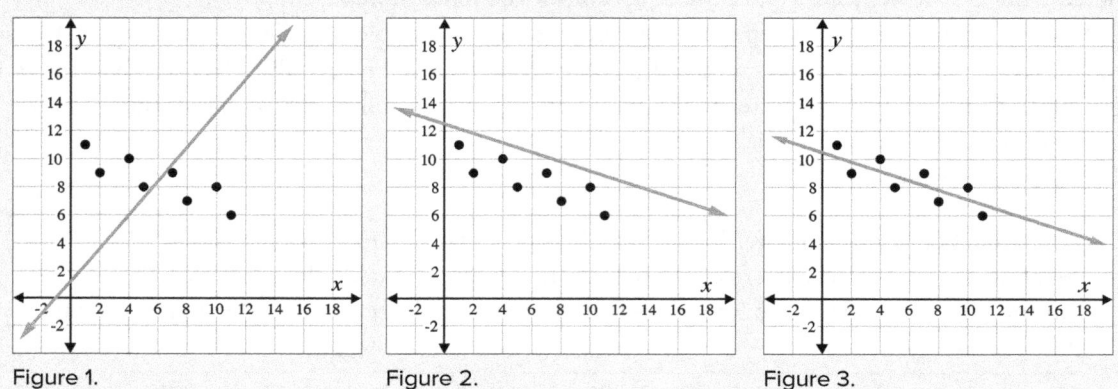

Figure 1.   Figure 2.   Figure 3.

The line in Figure 1 has four points above it and four points below it. However, it does *not* match the overall direction exhibited by the points. The line in Figure 2 matches the overall direction exhibited by the points. However, all eight points are below the line. The line in Figure 3 has four points above it and four points below it, *and* it matches the overall direction exhibited by the points. Therefore, Figure 3 shows a line of best fit.

Now, let's work through the process of constructing a line of best fit by hand and finding the equation for our line.

Now, let's work through the process of constructing a line of best fit by hand and finding the equation for our line.

Example 2

Create a scatter plot from the information on the table. Then, draw the line of best fit. Finally, use algebra to find the equation of the line of best fit.

x	1	3	4	6	7	9	10
y	2	5	4	9	8	11	11

Solution

We start by creating the scatter plot in Figure 4.

This is where the guesswork comes in. Now we need to find two points that will define a line that passes through the approximate center of the scatter plot and follows the overall direction exhibited by the scatter plot. We'll choose the points $(1, 3)$ and $(9, 10)$. See Figure 5.

These two points are not part of the scatter plot, but they do give us a pretty good line. See Figure 6.

There are four points above the line and three points below it. The line also follows the overall direction exhibited by the scatter plot. We have a reasonable line of best fit. Now it's time to find the equation of our line. Because the form was not specified, we need to make sure our final answer is in slope-intercept form.

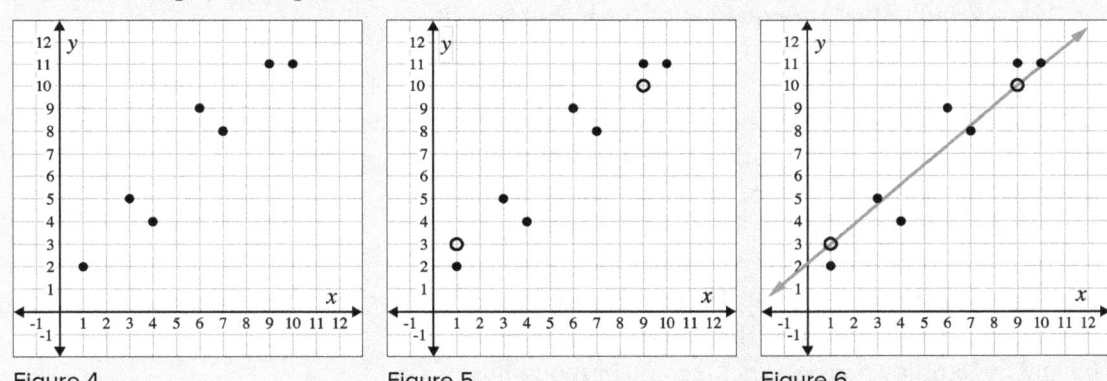

Figure 4.   Figure 5.   Figure 6.

$$m = \frac{y_2 - y_1}{x_2 - x_1} = \frac{10 - 3}{9 - 1} = \frac{7}{8}$$

For this example, we use slope-intercept form to help us generate the equation of our line.

$$y = mx + b$$

$$3 = \frac{7}{8}(1) + b \quad \text{Either point can be used here. We'll use } (1, 3).$$

$$3 = -\frac{7}{8} + b$$

$$\underline{-\frac{7}{8} \qquad -\frac{7}{8}}$$

$$2\frac{1}{8} = b$$

$$y = \frac{7}{8}x + 2\frac{1}{8} \quad \text{This is the equation of our line of best fit.}$$

Sometimes it's helpful to change how data is reported before creating a scatter plot. To illustrate this, let's look again at the problem stated at the beginning of this section.

Example 3

The table below gives the net income of The Vegan Truck Stop, a small dining establishment in Mill City, from 2007 to 2013.

Year	2007	2008	2009	2010	2011	2012	2013
Net Income	$4,000	$8,000	$9,000	$7,000	$10,000	$11,000	$14,000

Use this information to:

a. Create a scatter plot from the information on the table. Let x represent the number of years since 2005 and let y represent net income, in thousands of dollars.

b. Draw the line of best fit.

c. Use algebra to find the equation of the line of best fit.

Solution

a. It would be difficult to plot the numbers given in this table — particularly the years. In order to do so, our x-axis would need to go by ones and extend 2013 spaces from the origin, which isn't very practical. The y-axis wouldn't be as big of a problem because we could go by thousands, which only require us to extend the y-axis up 14 spaces from the origin. However, by letting x represent the number of years since 2005 and by letting y represent net income in thousands of dollars, the values become easier to graph:

x (years since 2005)	2	3	4	5	6	7	8
y (net income, in thousands)	4	8	9	7	10	11	14

Now, we create the scatter plot in Figure 7.

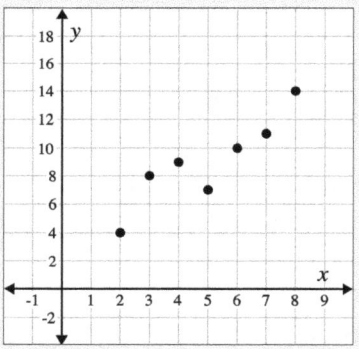

Figure 7.

b. In Figure 8, two of the points in the scatter plot, (2, 4) and (6, 10), appear to define a good line of best fit. We draw the line in Figure 9 and verify that it passes through the approximate center of the scatter plot and follows the overall direction exhibited by the points.

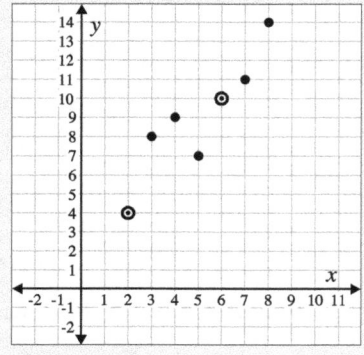

Figure 8.

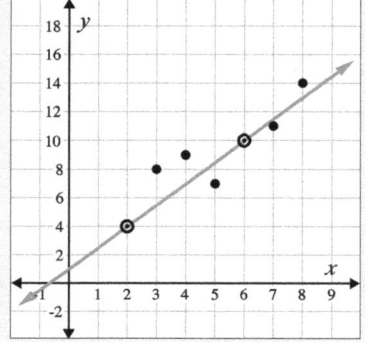

Figure 9.

c. Because the line in Figure 9 is reasonable, we proceed to find the equation. We'll start by calulating the slope.

$$m = \frac{y_2 - y_1}{x_2 - x_1} = \frac{10 - 4}{6 - 2} = \frac{3}{2}$$

For this example, we'll use point-slope form to help us find the equation of our line.

$$y - y_1 = m(x - x_1)$$

$$y - 4 = \frac{3}{2}(x - 2) \qquad \text{We can use either point. Let's use } (2, 4).$$

$$y - 4 = \frac{3}{2}x - 3$$

$$\underline{+ 4 \qquad \qquad + 4}$$

$$y = \frac{3}{2}x + 1 \qquad \text{This is the equation of our line of best fit.}$$

Don't forget that in this equation, x represents the number of years *since 2005* and y represents the net income *in thousands of dollars*. Later in this section, we'll use our equation to make predictions, and when we do that, it will be important to remember what each variable stands for.

Practice A

Now, you get to try your hand at creating a couple of lines of best fit. Follow the same procedure we used in the previous examples. When you are done, turn the page and check your solutions. Remember that there is a little bit of guesswork involved in creating a line of best fit by hand. Your slope and y-intercept may not perfectly match the slope and y-intercept of the equation given in the answer. As long as the values are fairly close, that's okay.

1. Use the following table to:

   a. Create a scatter plot and draw the line of best fit.

   b. Use algebra to find the equation of the line of best fit.

x	1	3	5	7	9	11	13
y	6	7	10	9	11	11	14

2. The table below gives the number of students enrolled at a college. Use this information to:

   a. Create a scatter plot and draw the line of best fit. Let x represent the number of years since 2000 and let y represent the number of students enrolled, in thousands.

   b. Use algebra to find the equation of the line of best fit.

Year	2003	2004	2005	2006	2007	2008	2009
Enrollment	9,000	11,000	8,000	6,000	8,000	7,000	5,000

B. Scatter Plots and Linear Regression with a Graphing Calculator

Now that you are familiar with finding a line of best fit by hand, it's time to get a bit more precise. As mentioned earlier, when working through these problems by hand, we can get a good approximation of the line of best fit. However, the likelihood of actually stumbling upon the *actual* line of best fit is slim to none. In other words, there is almost certainly a line that fits the data points better than the line we came up with.

There are multiple mathematical techniques for finding the *actual* line of best fit for a scatter plot, and the process of doing this is commonly referred to as regression. These techniques are quite complicated. It would be *very* time consuming to attempt to do them with pencil and paper. Fortunately, a graphing calculator can perform linear regression quickly, as we will see in the following examples.

▶ Example 4

Use a graphing calculator to create a scatter plot from the information in the table below. Then use the regression capabilities of the calculator to find the equation of the line of best fit.

x	1	3	4	6	7	9	10
y	2	5	4	9	8	11	11

This is the same set of points we used in Example 2, so it should be interesting to compare the equation our calculator generates with the equation we found by hand.

Solution

We must begin by entering the ordered pairs from the table into our graphing calculator and then viewing the scatter plot. With a TI-83 or TI-84, here are the steps to follow.

Practice A — Answers
Here are sample graphs and equations. Your answers may vary slightly.

1. a.
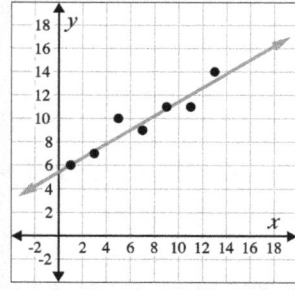

   b. $y = \frac{3}{5}x + 5\frac{2}{5}$ or $y = 0.6x + 5.4$

2. a.
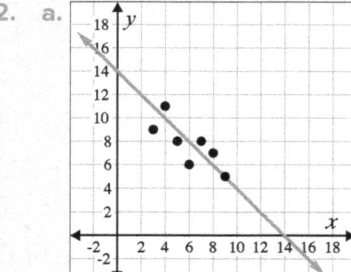

   b. $y = -x + 14$

Step 1: Put all of the *x*-coordinates into one list, and put the corresponding *y*-coordinates into another list. Access the lists by pressing the [STAT] key, located in the third row of buttons from the top. Once you press the [STAT] key, the screen should look like this.

Step 2: Select the first option, Edit. This will let you view the lists. The first list on your screen should be List 1 (L1).

If you're unable to see L1, press the [STAT] key and select the fifth option, SetUpEditor. The SetUpEditor command will now be on your home screen. If you press [ENTER], the Stat Editor should be reset, and you should be able to view L1.

If there are already numbers in L1, you can clear the list by using the [▲] arrow to highlight the label at the top of L1. Now press [CLEAR] and [ENTER] to clear the list.

Step 3: Enter the *x*-coordinates from the table into L1 and the corresponding *y*-coordinates into L2.

Step 4: In order to view the scatter plot, the plotting feature of your calculator must be turned on. You can do this by accessing the [STAT PLOT] menu - do this by pressing [2nd] and then [Y=]. To turn Plot 1 on, press [ENTER] and then toggle Plot 1 to On.

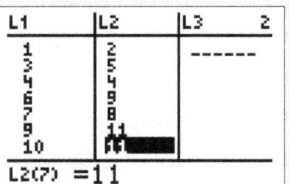

Step 5: Now you can press [GRAPH] to see the scatter plot. If you can't see the scatter plot, or can't see it very well, you may want to adjust your Window settings.

Step 6: Now it's time to have your calculator perform the linear regression. Press the [STAT] key. Then use your arrow key to highlight [CALC]. Finally, scroll down to the fourth option, LinReg.

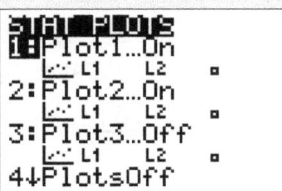

Step 7: When you press [CALC] one of two things will happen:

1. If you have an older calculator, the LinReg command will appear on your home screen. Press [ENTER] and the calculator will perform the regression.

2. If you have a newer calculator, you will see a screen verifying that you used L1 and L2 for your coordinates. At the bottom of this screen is a Calculate command. Highlighting this and pressing [ENTER] will cause the calculator to perform the regression.

When the calculator is done, you should see the screen on your right.

If your calculator has diagnostic tools turned on, there may be additional values (r^2 and r) showing. Disregard these values for now.

Your calculator gives you you a generic linear equation, $y = ax + b$, and the values that can be substituted for a and b. If we round the values off to the nearest thousandth, the equation of our line of best fit is $y = 1.045x + 1.171$.

Turning Off the Plotting Feature

If the plotting feature of your calculator is turned on but there aren't any numbers in your lists, the calculator will give you an error message any time you try to graph *anything*. For this reason, it's a good idea to turn the plotting feature off when you're finished creating scatterplots.

To turn the plotting feature off:

1. Access the [STAT PLOT] , menu by pressing y and then o.

2. Press [ENTER] and then toggle Plot 1 to Off.

Let's compare this line with the line we found by hand in Example 2. See Figure 10.

The line we created by hand is pretty good, but the line we found with a graphing calculator, Figure 11, is a slightly better fit.

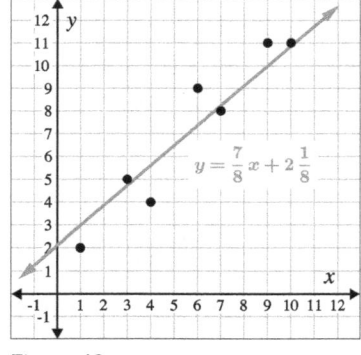

Figure 10.

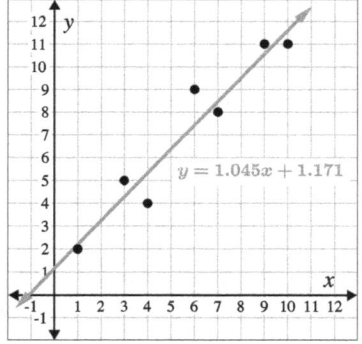

Figure 11.

Example 5

Use a graphing calculator to create a scatter plot from the information in the Figure 15 table. Then use the regression capabilities of the calculator to find the equation of the line of best fit. Round values off to the nearest thousandth.

x	2	4	6	8	10	12	14
y	13	9	11	8	7	8	4

Solution

Step 1: Begin by entering the *x*-coordinates into List 1, and the corresponding *y*-coordinates into List 2.

Step 2: Adjust the viewing window so that all of the points will be visible when graphing. The Xmax value needs to be greater than 14, and the Ymax value needs to be greater than 13.

Step : Now the calculator displays the whole scatterplot.

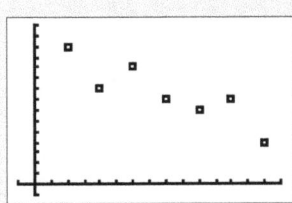

Step 4: Finally, perform the linear regression.

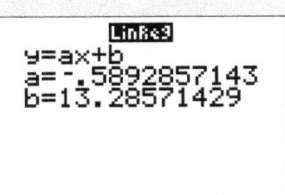

Rounding to the nearest thousandth, the equation of the line of best fit is $y = -0.589x + 13.286$.

Practice B

Now it's time for you to try a couple of problems on your own. For each table:

a. Use a graphing calculator to create a scatterplot.

b. Use the regression capabilities of your calculator to find the equation of the line of best fit. Round values off to the nearest thousandth. When you are done, turn the page and check your solutions.

3.

x	1	2	3	4	5	6	7
y	2	6	4	7	9	8	9

4.

x	5	10	15	20	25	30	35
y	48	37	35	41	39	25	28

C. Using Equations to Make Predictions

When points in a data set show strong correlation, we can find an equation that closely matches that data set and then use the equation to make predictions. Modeling and predicting is one of the most common uses for mathematics in everyday life.

In this chapter, we are focusing on linear models. Because of that, we are looking at data sets that are linear, or approximately linear. In the future, as you become more mathematically savvy, you will encounter situations that involve more complicated, non-linear models.

Example 6

In Example 3, the equation $y = \frac{3}{2}x + 1$ modeled the net income of a small restaurant, The Vegan Truck Stop. In this equation, x represents the number of years since 2005, and y represents net income in thousands of dollars. Use this equation to predict the company's net income in 2025.

Solution

$y = \frac{3}{2}x + 1$ Begin with the equation.

$y = \frac{3}{2}(20) + 1$ Because the year 2025 is 20 years after 2005, we replace x with 20 and simplify the result.

$y = 30 + 1$

$y = 31$

Because y represents net profit, in thousands of dollars, we need to adjust our result accordingly when stating the answer. In 2025, the Vegan Truck Stop's net income will be $31,000.

Example 7

The table gives the percentage of young adults in the United States who own an MP3 player.

Year	2010	2011	2012	2013	2014	2015
Percent of U.S. young adults who own an MP3 player	75	71	67	62	56	51

Let x represent the number of years since 2010. Then do this:

a. Use the regression capabilities of a graphing calculator to find the equation of the line of best fit, rounding values off to the nearest thousandth.

b. Use your equation to predict the percentage of young adults in the United States that will own an MP3 player in the year 2023, rounding to the nearest tenth of a percent.

Solution

a. Start by re-writing the table so that x represents the number of years since 2010.

x (years since 2010)	0	1	2	3	4	5
y (percent of U.S. young adults)	75	71	67	62	56	51

Next, enter these values into the lists on a graphing calculator and have the calculator perform a linear regression. The equation of the line of best fit is $y = -4.857x + 75.810$.

b. Because x represents the number of years since 2010, we can predict the percentage of U.S. young adults that will own an MP3 player in 2023 by replacing x with 13, and simplifying the result.

$$y = -4.857x + 75.810$$
$$y = -4.857(13) + 75.810$$
$$y = -63.141 + 75.810$$
$$y = 12.669$$

We round the result off to the nearest tenth when stating our prediction: In 2023, 12.7 percent of young adults in the U.S. will own an MP3 player.

Practice C

Now you can give it a try. Use the equation that models a given situation to make predictions. If an equation is not provided, then you will need to use a graphing calculator to find the equation. When you are done, turn the page and check your solutions.

5. A group of entomologists has determined that the population of ladybugs at a local park can be modeled by the equation $y = -1.437x + 197.686$, where x represents the number of years since 2010 and y represents the number of ladybugs, in thousands.

   a. Predict the ladybug population at the park in 2024

   b. Predict the ladybug population at the park in 2060

6. The following table gives the number of people following the Seattle Seahawks on Facebook.

Year	2010	2011	2012	2013	2014	2015
Number of Seattle Seahawks followers on Facebook	2,900,000	2,980,000	3,110,000	3,360,000	3,490,000	3,610,000

Letting x represent the number of years since 2010 and letting y represent the number of followers on Facebook, in millions, use the regression capabilities of a graphing calculator to find the equation of the line of best fit. Round values off to the nearest thousandth. Then, use your equation to predict the following:

   a. The number of people following the Seahawks on Facebook in 2021

   b. The number of people following the Seahawks on Facebook in 2044

Practice B — Answers

3. a.

(Viewing window: $-1 < x < 10$ and $-1 < y < 10$)

b. $y = 1.071x + 2.143$

4. a.

(Viewing window: $-1 < x < 40$ and $-1 < y < 50$)

b. $y = -0.571x + 47.571$

Exercises 2.7

For the following problems, create a scatter plot and draw a line of best fit by hand for the information given in the table. Then use algebra to find the equation of the line of best fit.

1.

x	2	4	6	8	10	12	14
y	3	4	9	7	10	13	12

2.

x	1	2	4	6	7	8	10
y	3	5	6	5	8	7	9

3.

x	1	3	4	5	7	9	10
y	1	5	4	7	12	10	15

4.

x	1	3	5	7	9	11	13
y	17	16	11	12	8	6	6

5.

x	1	2	3	4	5	6	7
y	12	8	10	9	4	1	3

6.

x	5	6	7	8	9	10	11
y	14	14	15	13	10	8	9

7.

x	9	10	11	12	13	14	15
y	6	7	5	4	6	2	3

8.

x	1	2	4	5	7	8	9
y	3	6	9	7	12	10	16

9. Let x represent the number of dogs, in hundreds, and let y represent the number of fleas, in thousands.

Dogs	100	200	300	400	500	600	700
Fleas	9,000	8,000	12,000	16,000	15,000	19,000	20,000

10. Let x represent the amount by which a person's IQ exceeds 100.

IQ	100	105	110	115	120	125	130
Average age at which a college degree is obtained	25	23	24	24	21	22	20

11. Let x represent the amount by which a person's IQ exceeds 100, and let y represent the average number of Facebook friends, in tens.

IQ	100	105	110	115	120	125	130
Average number of friends	250	260	210	160	180	150	100

12. Let x represent the number of years since 2005, and let y represent the number of *American Idol* viewers, in millions.

Year	2006	2007	2008	2009	2010	2011	2012
American Idol Viewers	30,000,000	31,000,000	28,000,000	26,000,000	23,000,000	26,000,000	20,000,000

13. Let x represent the number of years since 2000.

Year	2002	2004	2006	2008	2010	2012	2014
Percentage of people who buy fuel-efficient cars	10	16	19	17	19	28	26

For the following problems, use the regression capabilities of a graphing calculator to find the equation of the line of best fit. Round values off to the nearest thousandth.

14.

x	1	2	3	4	5	6	7
y	37	30	33	31	25	29	22

18.

x	4.1	7.5	11.2	17.9	20.4	24.3	27.6
y	18.7	16.9	24.1	30.4	29.1	34.8	35.2

15.

x	1	2	3	4	5	6	7
y	96	81	60	65	47	50	33

19.

x	11.8	27.1	41.9	52.3	68.4	83.7	97.0
y	112.9	124.1	140.6	135.9	171.3	163.4	189.2

16.

x	7	9	11	13	15	17	19
y	13.6	16.9	14.1	21.7	23.4	29.8	27.9

20.

x	120	140	160	180	200	220	240
y	24	22	16	21	17	10	12

17.

x	11	15	19	23	27	31	35
y	30.8	32.1	40.7	37.5	43.2	40.3	46.9

21.

x	50	80	110	140	170	200	230
y	49	35	40	29	20	24	12

22. Let x represent the number of years since 2010.

Year	2011	2012	2013	2014	2015	2016
Number of picnic tables at Rainyday Park	18	23	26	24	32	29

23. Let x represent the number of years since 2010, and let y represent the number of ants, in millions.

Year	2011	2012	2013	2014	2015	2016
Number of ants in the picnic areas at Rainyday Park	2,400,000	3,200,000	4,100,000	3,600,000	4,700,000	3,900,000

24. Let x represent the number of years since 1985.

Year	1985	1986	1987	1988	1989	1990
Percentage of kids who played Super Mario Brothers	21	39	43	40	46	52

25. Let x represent the number of years since 1970.

Year	1975	1977	1979	1981	1983	1985
Percentage of people who thought bell-bottom pants were fashionable	73	79	65	49	23	11

Practice C — Answers

5. a. In 2024, the ladybug population will be 177,568. b. In 2060, the ladybug population will be 125,836.

6. Equation: $y = 0.152x + 2.861$. a. In 2021, the Seahawks will have 4,533,000 followers on Facebook.
    b. In 2044, the Seahawks will have 8,029,000 followers on Facebook.

For the following problems, use the equation that models a given situation to make predictions. If an equation is not provided, then you will need to use a graphing calculator to find the equation.

26. Morgan's Mediterranean Restaurant has found that the number of falafel plates sold can be modeled by the equation $y = 1.3x + 8.4$, where x represents the number of years since 2010 and y represents the number of falafel plates sold, in thousands.

    a. Predict the number of falafel plates the restaurant will sell in 2021.

    b. Predict the number of falafel plates the restaurant will sell in 2028.

27. Phillip believes that his score on an upcoming test can be modeled by the equation $y = 7.5x + 62$, where x represents the number of hours spent studying and y represents Phillip's test score.

    a. Assuming this equation is accurate, predict Phillip's test score if he studies for 2 hours.

    b. Assuming this equation is accurate, predict Phillip's test score if he studies for 5 hours.

28. Patrice's Plant Works estimates that the number of yards it can landscape can be modeled by the equation $y = 1.35x + 6.7$, where x represents the number of workers the company hires, and y represents the number of yards that can be landscaped in one week.

    a. Predict how many yards the company will be able to landscape in a week if 38 workers are hired. Round to the nearest whole number, if necessary.

    b. Predict how many yards the company will be able to landscape in a week if 53 workers are hired. Round to the nearest whole number, if necessary.

29. Genevieve's Gym estimates that the number of gym memberships can be modeled by the equation $y = 0.62x + 2.7$, where x represents the number of years since 2015 and y represents the number of gym memberships, in hundreds.

    a. Predict how many gym memberships there will be in the year 2026.

    b. Predict how many gym memberships there will be in the year 2040.

30. Simone's Sinfully Delicious Smoothies estimates that the number of happy customers can be modeled by $y = 57.6x + 68.4$, where x represents the number of smoothie flavors offered, and y represents the number of happy customers.

    a. Predict the number of happy customers Simone's Sinfully Delicious Smoothies will have if they offer 8 smoothie flavors. Round to the nearest whole number, if necessary.

    b. Predict the number of happy customers Simone's Sinfully Delicious Smoothies will have if they offer 20 smoothie flavors. Round to the nearest whole number, if necessary.

31. Yasmin believes that the balance of her investment portfolio can be modeled by the equation $y = 27.6x + 87.4$, where x represents the number of years since 2015 and y represents the value of the portfolio, in thousands of dollars.

    a. Assuming this equation is accurate, predict the value of Yasmin's investment portfolio in 2024.

    b. Assuming this equation is accurate, predict the value of Yasmin's investment portfolio in 2033.

32. The table gives the number of stray dogs in Poochville. Let *x* represent the number of years since 2010. Use the regression capabilities of a graphing calculator to find the equation of the line of best fit. Round values to the nearest thousandth. Then, make the following predictions.

Year	2011	2012	2013	2014	2015	2016
Number of stray dogs	87	99	95	102	113	111

a. Predict the number of stray dogs in 2027. Round to the nearest whole number, if necessary.

b. Predict the number of stray dogs in 2044. Round to the nearest whole number, if necessary.

33. The table gives the average number of hours of television per day watched by America's impressionable youth. Let *x* represent the number of years since 2000. Use the regression capabilities of a graphing calculator to find the equation of the line of best fit. Round values to the nearest thousandth. Then, make the following predictions.

Year	2002	2004	2006	2008	2010	2012
Average daily hours of television watched by impressionable youth in the U.S.	1.7	2.4	2.3	2.6	3	2.8

a. Predict the number of hours of television per day that will be watched by impressionable youth in the U.S. in 2023. Round to the nearest tenth, if necessary.

b. Predict the number of hours of television per day that will be watched by impressionable youth in the U.S. in 2036. Round to the nearest tenth, if necessary.

34. The table gives the percentage of computer science majors who are women. Letting *x* represent the number of years since 1990, use the regression capabilities of a graphing calculator to find the equation of the line of best fit. Round values off to the nearest thousandth. Then, make the following predictions.

Year	1990	1993	1996	1999	2002	2005
Percent of computer science majors who are women.	31	28	27	27	26	23

a. Predict the percentage of computer science majors who will be women in the year 2022. Round to the nearest tenth, if necessary.

b. Predict the percentage of computer science majors who will be women in the year 2041. Round to the nearest tenth, if necessary.

35. The table gives the amount of money spent on football by the University of Oregon. Let *x* represent the number of years since 2009, and let *y* represent the amount of money spent on football in thousands of dollars. Use the regression capabilities of a graphing calculator to find the equation of the line of best fit. Round values off to the nearest thousandth. Then, make the following predictions.

Year	2009	2010	2011	2012	2013	2014
Amount of money spent on football by the University of Oregon	$170,000	$194,000	$197,000	$226,000	$229,000	$311,000

a. Predict the amount of money that the University of Oregon will spend on football in the year 2026.

b. Predict the amount of money that the University of Oregon will spend on football in the year 2045.

2.8 Parallel and Perpendicular Lines

Overview

In the past, when we were told to find the equation of a line, we were either given the slope of that line or were given two points on the line that allowed us to calculate the slope. However, there are times when we must use information about a *different* line to determine the equation of a line.

Here's a problem:

Find the equation of the line that passes through (6, 1) and is perpendicular to $y = -2x - 5$.

In this problem, we are trying to find the equation for a line and are given only one point. However, we are also given the equation of another line that is *perpendicular* to the line we're looking for. As you will see in this section, that bit of information is enough for us to determine the slope of our line.

In this section, we will look at situations involving parallel lines and perpendicular lines. By studying this, you should learn how to:

- ◆ find the equation of a line that passes through a given point and is parallel to a given line

- ◆ find the equation of a line that passes through a given point and is perpendicular to a given line

A. Parallel Lines

Two lines are parallel if they are both in the same coordinate plane and never intersect. Graphically, we can see that parallel lines do not get closer to or farther away from each other as they move along. As you see in Figure 1, they remain equidistant.

In addition to the property of being equidistant, parallel lines also have the *same slope*.

It's important to remember that if two linear equations indicate the same slope *and* the same y-intercept, then the equations give coincident lines — lines that lie one on top of the other. Coincident lines are *not* parallel.

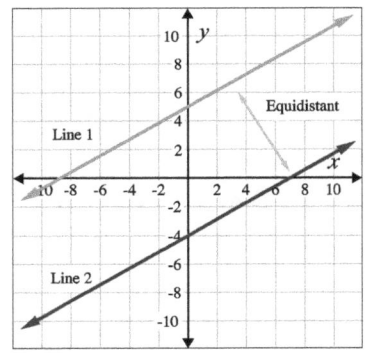

Figure 1.

Parallel Lines and Slope

Parallel lines have the *same slope*.

Example 1

Determine which of the equations, when graphed, will give a line that is parallel to $y = -\frac{1}{2}x + 7$.

1. $y = -\frac{1}{3}x - 5$ 3. $y = -\frac{1}{2}x - 3$ 5. $x + 2y = 8$

2. $y = \frac{1}{2}x + 16$ 4. $x + 2y = 14$

Solutions

Any line that's parallel to the given line will have the same slope, $-\frac{1}{2}$, and a y-intercept other than 7. Let's analyze the equations:

1. $y = -\frac{1}{3}x - 5$

   This line has a slope of $-\frac{1}{3}$, and $-\frac{1}{3}$ is not the same as $-\frac{1}{2}$, so it is not parallel to $y = -\frac{1}{2}x + 7$.

2. $y = \frac{1}{2}x + 16$

   This line has a slope of $\frac{1}{2}$, and $\frac{1}{2}$ is not the same as $-\frac{1}{2}$, so it is not parallel to $y = -\frac{1}{2}x + 7$.

3. $y = -\frac{1}{2}x - 3$

   This line has a slope of $-\frac{1}{2}$, which is the same as the given line. The y-intercept, -3, is not the same as the y-intercept of the given line. This line is parallel to $y = -\frac{1}{2}x + 7$.

4. $x + 2y = 14$

   This equation isn't in slope-intercept form yet, so we begin by solving for y.

$$x + 2y = 14$$
$$\underline{-x \qquad\quad -x}$$
$$2y = -x + 14$$
$$\frac{2y}{2} = \frac{-x + 14}{2}$$
$$y = -\frac{1}{2}x + 7$$

   This line has the same slope *and* y-intercept as $y = -\frac{1}{2}x + 7$. The lines are coincident, not parallel.

5. $x + 2y = 8$

   Once again, we solve for y.

$$x + 2y = 8$$
$$\underline{-x \qquad\quad -x}$$
$$2y = -x + 84$$
$$\frac{2y}{2} = \frac{-x + 8}{2}$$
$$y = -\frac{1}{2}x + 4$$

   We can now see that the lines have the same slope but different y-intercepts, so this line is parallel to $y = -\frac{1}{2}x + 7$.

It's good to remember that horizontal lines are parallel to each other and that vertical lines are also parallel to each other. For example, the lines whose equations are $y = 9$ and $y = 5$ are parallel to each other. The lines whose equations are $x = -7$ and $x = 10$ are also parallel to each other.

Knowing that parallel lines have the same slope can be beneficial to us in certain situations that require us to find the equation of a line. If we are asked to find the equation of a line that is *parallel* to a given line, we will use the slope of the given line in our calculations.

Example 2

Find the equation of the line that passes through $(15, 2)$ and that is parallel to $y = \frac{2}{3}x + 17$.

Solution

Because parallel lines have the same slope, we use the slope of the given line, $\frac{2}{3}$, in our calculations. For this example, we'll use point-slope form to help us generate the equation.

$$y - y_1 = m(x - x_1)$$

$$y - 2 = \frac{2}{3}(x - 15)$$

$$y - 2 = \frac{2}{3}x - 10$$

$$\underline{+2 \qquad\qquad +2}$$

$$y = \frac{2}{3}x - 8$$

Notice that the y-intercept of the other line, 17, was not used at all in our calculations.

Example 3

Find the equation of the line that passes through $(7, -3)$ and is parallel to the line passing through $(12, 9)$ and $(15, 3)$.

Solution

We'll use the slope formula to determine the slope of the other line.

$$m = \frac{y_2 - y_1}{x_2 - x_1} = \frac{3 - 9}{15 - 12} = \frac{-6}{3} = -2$$

Now we'll use slope-intercept form to help us generate the equation of our line.

$$y = mx + b$$

$$-3 = -2(7) + b$$

$$-3 = -14 + b$$

$$\underline{+14 \qquad\qquad +14}$$

$$11 = b$$

$$y = -2x + 11$$

In the previous two examples, the choice of methods was arbitrary. We could have used slope-intercept form to help us generate the equation in Example 2 or point-slope form to help us generate the equation in Example 3. The results would have been the same.

Practice A

Now it's time for you to tackle some problems involving parallel lines. Remember, if you're asked to find the equation of a line and the form is not specified, your final answer should be in slope-intercept form. When you are done, turn the page and check your solutions.

1. Which of the following is an equation of a line that is *parallel* to $y = 4x + 9$?

   a. $y = 2x + 9$ b. $y = 4x - 7$ c. $12x - 3y = 6$ d. $-20x + 5y = 45$

2. Find the equation of the line that passes through $(-12, 8)$ and is parallel to $y = \frac{3}{4}x - 13$.

3. Find the equation of the line that passes through $(4, 1)$ and is parallel to the line passing through $(7, 11)$ and $(10, 20)$.

B. Perpendicular Lines

Two lines are perpendicular if they intersect to form a right angle, which is an angle measuring 90 degrees. See Figure 2.

If perpendicular lines are graphed on a coordinate plane, then the product of their slopes is -1. If the product of two numbers is -1, then the two numbers must be opposite-reciprocals. In other words, the numbers must be reciprocals with opposite signs. For example, $-\frac{3}{5}$ and $\frac{5}{3}$ are opposite-reciprocals, as are 3 and $-\frac{1}{3}$. With this in mind, we have the following property for perpendicular lines.

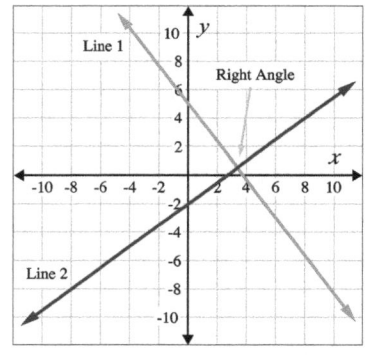

Figure 2.

Perpendicular Lines and Slope

Perpendicular lines have slopes that are *opposite-reciprocals*.

If two lines have different slopes, they cannot possibly be coincident. Therefore, we don't need to worry about analyzing the y-intercepts when determining whether or not two equations will give perpendicular lines. All we have to do is determine whether or not the slopes are opposite-reciprocals.

Example 4

Determine which of the given equations, when graphed, will give a line that is perpendicular to $y = \frac{2}{3}x + 16$.

1. $y = \frac{3}{2}x - 1$ **3.** $y = -\frac{3}{2}x - 9$ **5.** $-9x + 6y = 8$

2. $y = -\frac{2}{3}x + 7$ **4.** $6x + 4y = 12$

Solutions

The given line has a slope of $\frac{2}{3}$. The opposite-reciprocal of $\frac{2}{3}$ is $-\frac{3}{2}$. Therefore, any line that is perpendicular to the given line will have a slope of $-\frac{3}{2}$. Let's analyze the equations.

1. $y = \frac{3}{2}x - 1$

The slope of this line is $\frac{3}{2}$, which is the reciprocal — not the *opposite-reciprocal* — of the slope of the given line, $\frac{2}{3}$. These are not perpendicular lines.

2. $y = -\frac{2}{3}x + 7$

The slope of this line is $-\frac{2}{3}$, which is the opposite — not the *opposite-reciprocal* — of the slope of the given line. These lines are not perpendicular.

3. $y = -\frac{3}{2}x - 9$

This line has a slope of $-\frac{3}{2}$. This is the opposite-reciprocal of the slope of the given line. That means that these lines are perpendicular.

4. $6x + 4y = 12$

This equation is not in slope-intercept form, so we begin by solving for y.

$$6x + 4y = 12$$
$$\underline{-6x -6x }$$
$$4y = -6x + 12$$
$$\frac{4y}{4} = \frac{-6x + 12}{4}$$
$$y = -\frac{3}{2}x + 3$$

The slope of this line is the opposite-reciprocal of the slope of the given line, so these are perpendicular lines.

5. $-9x + 6y = 8$

Once again, we need to solve for y first.

$$-9x + 6y = 8$$
$$\underline{+ 9x \qquad\qquad + 9x}$$
$$6y = 9x + 8$$
$$\frac{6y}{6} = \frac{9x+8}{6}$$
$$y = -\frac{3}{2}x + 3$$

Because both slopes are positive, the slope of this line is not the opposite-reciprocal of the slope of the given line. They are not perpendicular.

It's important to remember that a horizontal line is always perpendicular to a vertical line. This means, for example, that the lines whose equations are $y = 4$ and $x = 3$ are perpendicular to each other.

If we are asked to find the equation of a line that is perpendicular to a given line, then we will use the opposite-reciprocal of the slope of the given line in our calculations.

Example 5

Find the equation of the line that passes through $(6, 1)$ and that is perpendicular to $y = -2x - 5$.

Solution

Because we are finding the equation of a perpendicular line, we will need to use the opposite-reciprocal slope in our calculations. The opposite-reciprocal of -2 is $\frac{1}{2}$. We'll use slope-intercept form to help us generate the equation of our line.

$$y = mx + b$$
$$1 = \frac{1}{2}(6) + b \quad \text{Notice that } m \text{ has been replaced by } \frac{1}{2}.$$
$$1 = 3 + b$$
$$\underline{-3 \qquad\qquad -3}$$
$$-2 = b$$
$$y = \frac{1}{2}x - 2$$

Practice A — Answers

1. b and c

2. $y = \frac{3}{4}x + 17$

3. $y = 3x - 11$

Example 6

Find the equation of the line that passes through $(12, 9)$ and that is perpendicular to the line passing through $(8, -1)$ and $(17, 5)$.

Solution

We use the slope formula to determine the slope of the other line.

$$m = \frac{y_2 - y_1}{x_2 - x_1} = \frac{5 - (-1)}{17 - 8} = \frac{6}{9} = \frac{2}{3}$$

Now we'll use point-slope form, with a slope of -3_2, to help us generate the equation of our line.

$$y - y_1 = m(x - x_1)$$

$$y - 9 = -\frac{3}{2}(x - 12)$$

$$y - 9 = -\frac{3}{2}x + 18$$

$$\underline{+9 \qquad\qquad +9}$$

$$y = -\frac{3}{2}x + 27$$

Once again, the choice of method is arbitrary. We used slope-intercept form in Example 5 and point-slope form was used in Example 6. We could have done the opposite. Either method, when used correctly, will give the correct equations.

Practice B

It's your turn to work a few problems — this time the problems deal with perpendicular lines. Remember, if you are asked to find the equation of a line and the form is not specified, then your final answer should be in slope-intercept form. When you are done, turn the page and check your solutions.

4. Which of the following is an equation of a line that is *perpendicular* to $y = 3x + 18$?

    a. $y = -\frac{1}{3}x + 1$ b. $y = \frac{1}{3}x + 4$ c. $15x + 5y = 60$ d. $7x + 21y = 105$

5. Find the equation of the line that passes through $(10, 6)$ and is perpendicular to $y = \frac{2}{5}x - 9$.

6. Find the equation of the line that passes through $(1, 7)$ and is perpendicular to the line passing through $(9, -6)$ and $(13, -7)$.

Exercises 2.8

For the following problems, determine which of the equations, when graphed, will give a line that is parallel to the given line.

1. $y = -2x + 7$
   a. $y = 2x + 13$
   b. $y = -2x - 1$
   c. $y = \frac{1}{2}x - 9$
   d. $6x + 3y = 21$
   e. $16x + 8y = 48$

2. $y = \frac{4}{5}x - 2$
   a. $y = \frac{4}{5}x - 7$
   b. $y = \frac{5}{4}x - 8$
   c. $4x + 5y = 45$
   d. $12x - 15y = 90$
   e. $-8x + 10y = -20$

3. $y = -\frac{1}{3}x + 6$
   a. $y = -\frac{1}{3}x - \frac{7}{8}$
   b. $y = 3x + \frac{5}{6}$
   c. $2x + 6y = 17$
   d. $-5x - 15y = 30$
   e. $3x + 9y = 54$

4. $y = \frac{9}{10}x + \frac{4}{5}$
   a. $y = -\frac{9}{10}x - \frac{3}{4}$
   b. $y = \frac{10}{9}x + \frac{5}{4}$
   c. $18x - 20y = -16$
   d. $9x - 10y = 40$
   e. $-36x + 40y = 28$

5. $3x + 4y = 56$
   a. $y = -\frac{3}{4}x + 1$
   b. $y = -\frac{3}{4}x + 14$
   c. $9x + 12y = 168$
   d. $6x - 8y = -16$
   e. $-6x - 8y = -56$

6. $2x - y = 3$
   a. $y = 2x - 3$
   b. $y = 2x + 6$
   c. $y = \frac{1}{2}x - 7$
   d. $14x - 7y = 35$
   e. $8x + 4y = 28$

For the following problems, find the equation of the line described.

7. Passes through $(7, 5)$;
   Parallel to $y = 3x + 20$

8. Passes through $(10, -13)$;
   Parallel to $y = -2x + 1$

9. Passes through $(-5, 11)$;
   Parallel to $y = -4x + 13$

10. Passes through $(-7, -2)$;
    Parallel to $y = -x - 3$

11. Passes through $(16, 9)$;
    Parallel to $y = \frac{1}{2}x - 4$

12. Passes through $(28, -3)$;
    Parallel to $y = \frac{1}{4}x + 15$

Practice B — Answers

4. a and d

5. $y = -\frac{5}{2}x + 31$

6. $y = 4x + 3$

13. Passes through $(12, 9)$;
    Parallel to $y = \frac{4}{3}x + 2$

14. Passes through $(-10, 5)$;
    Parallel to $y = -\frac{5}{2}x - 37$

15. Passes through $(-18, 1)$;
    Parallel to $y = -\frac{2}{9}x + 14$

16. Passes through $(-6, 12)$;
    Parallel to $y = -\frac{7}{3}x - 7$

17. Passes through $(4, 7)$;
    Parallel to $y = \frac{2}{3}x + 9$

18. Passes through $(12, -3)$;
    Parallel to $y = \frac{2}{5}x - 18$

19. Passes through $(7, 1)$;
    Parallel to $2x + y = 11$

20. Passes through $(-3, 16)$;
    Parallel to $6x + 2y = 1$

21. Passes through $(-8, 5)$;
    Parallel to $3x + 4y = -12$

22. Passes through $(9, -7)$;
    Parallel to $8x - 12y = -14$

23. Passes through $(4, -11)$; Parallel to the
    line passing through $(17, 20)$ and $(21, 28)$

24. Passes through $(2, 1)$; Parallel to the
    line passing through $(9, 15)$ and $(11, 29)$

25. Passes through $(-5, 17)$; Parallel to the
    line passing through $(-2, 13)$ and $(6, 5)$

26. Passes through $(-6, -3)$; Parallel to the
    line passing through $(11, 3)$ and $(19, 15)$

27. Passes through $(14, 4)$; Parallel to the
    line passing through $(-20, 3)$ and $(8, -5)$

28. Passes through $(-25, 2)$; Parallel to the
    line passing through $(-8, -14)$ and $(-18, -10)$

For the following problems, determine which of the equations, when graphed, will give a line that is perpendicular to the given line.

29. $y = 5x + 9$
    a. $y = -5x + 1$
    b. $y = \frac{1}{5}x - 10$
    c. $y = -\frac{1}{5}x - 3$
    d. $x - 5y = 20$
    e. $3x + 15y = 75$

30. $y = -3x + 29$
    a. $y = -\frac{1}{3}x$
    b. $y = \frac{1}{3}x - 5$
    c. $y = 3x + 17$
    d. $4x - 12y = 60$
    e. $7x + 21y = 84$

31. $y = -\frac{9}{5}x + \frac{1}{2}$
    a. $y = \frac{5}{9}x$
    b. $y = -\frac{5}{9}x - \frac{3}{4}$
    c. $y = \frac{5}{9}x - \frac{13}{21}$
    d. $5x - 9y = 72$
    e. $10x + 18y = 198$

32. $y = \frac{3}{8}x - 14$
    a. $y = -\frac{3}{8}x - 10$
    b. $y = -\frac{8}{3}x + \frac{8}{27}$
    c. $6x - 16y = 32$
    d. $48x + 18y = 108$
    e. $24x - 9y = -45$

33. $4x - 2y = 26$
    a. $y = \frac{1}{2}x - 8$ c. $y = -2x + 3$ e. $11x + 22y = 22$
    b. $y = -\frac{1}{2}x + 19$ d. $7x - 14y = 56$

34. $4x + 20y = 70$
    a. $y = 5x - 3$ c. $y = \frac{1}{5}x + 1$ e. $45x + 9y = -27$
    b. $y = -5x + 12$ d. $30x - 6y = 144$

For the following problems, find the equation of the line described.

35. Passes through $(8, 1)$;
    Perpendicular to $y = 2x + 27$

36. Passes through $(10, -11)$;
    Perpendicular to $y = 5x - 9$

37. Passes through $(12, -15)$;
    Perpendicular to $y = 4x - 3$

38. Passes through $(-9, -2)$;
    Perpendicular to $y = -3x - 10$

39. Passes through $(14, 37)$;
    Perpendicular to $y = -7x + 6$

40. Passes through $(5, 13)$;
    Perpendicular to $y = -\frac{1}{6}x + 64$

41. Passes through $(-11, 4)$;
    Perpendicular to $y = \frac{1}{5}x - 17$

42. Passes through $(6, 2)$;
    Perpendicular to $y = \frac{2}{3}x - \frac{9}{10}$

43. Passes through $(20, -1)$;
    Perpendicular to $y = -\frac{5}{4}x - \frac{19}{20}$

44. Passes through $(-24, -16)$;
    Perpendicular to $y = -\frac{8}{3}x + 19$

45. Passes through $(6, 7)$;
    Perpendicular to $y = \frac{4}{3}x + 83$

46. Passes through $(10, -3)$;
    Perpendicular to $y = -\frac{3}{2}x - 49$

47. Passes through $(-11, 25)$;
    Perpendicular to $2x - 4y = 18$

48. Passes through $(3, 29)$;
    Perpendicular to $7x + 28y = 84$

49. Passes through $(7, 31)$;
    Perpendicular to $5x + 15y = 135$

50. Passes through $(-8, -1)$;
    Perpendicular to $-24x + 6y = -42$

51. Passes through $(36, 25)$; Perpendicular to the line passing through $(7, 19)$ and $(12, 13)$

52. Passes through $(12, 19)$; Perpendicular to the line passing through $(-1, 11)$ and $(2, 7)$

53. Passes through $(-3, 4)$; Perpendicular to the line passing through $(-7, -16)$ and $(3, -11)$

54. Passes through $(-5, 3)$; Perpendicular to the line passing through $(9, -12)$ and $(-6, -7)$

55. Passes through $(8, 15)$; Perpendicular to the line passing through $(67, 5)$ and $(73, -3)$

56. Passes through $(30, -18)$; Perpendicular to the line passing through $(6, 92)$ and $(21, 110)$

CHAPTER 3
Systems of Linear Equations

There are many situations in life that require us to find values for two or more unknown quantities. For example, we might be given some information about a coffee shop: "The shop sold twice as many small coffees ($3 each) as large coffees ($5 each) and collected $539." In this situation, we would be interested in knowing the number of small coffees sold *and* the number of large coffees sold.

In order to find values for two unknown quantities, it's best to work with two different equations. In other words, we should write and solve a *system* of two equations. In general, the number of equations that are required is the same as the number of unknown quantities, so if we needed to discover values for three unknown quantities, we would need to solve a system of three equations.

This chapter will focus on systems of two equations. You will learn three methods that can be used to solve these systems and, in the process, discover situations in which a particular method may be preferable. We will also tackle application problems that require us to set up and solve systems of equations.

This chapter is comprised of the following sections:

3.1 Using Graphing to Solve Systems

Overview

In Chapter 2, we learned that a linear equation in two variables has infinitely many solutions. In other words, there are infinitely many combinations of variable values (ordered pairs) that will make the equation true. When graphed, all of the solutions make a straight line.

If we need to find a *particular* solution, then we must be given more than one equation.

Here is a problem like that:

Solve by graphing: $\begin{cases} -2x + y = 5 \\ x + y = 2 \end{cases}$

This problem may seem confusing and raise a lot of questions: Exactly what does "solve" mean in this case? How can we solve more than one equation at a time? How can we solve anything by graphing? Before tackling this problem, we need to understand systems of equations and revise our definition of the word "solution."

After completing this section, you should be able to:

◆ Recognize a system of equations and a solution for a system of equations

◆ Graphically interpret systems with one solution, no solution, or infinitely many solutions

◆ Solve a system of linear equations graphically — by hand and with a graphing calculator

A. Solution of a System

A set of two linear equations using two variables is called a system of linear equations in two variables. For short, we can call it a **system of equations**. For example, this pair of equations is a system of equations:

$\begin{cases} 5x - 2y = 5 \\ x + y = 8 \end{cases}$

The big, curvy grouping symbol to the left of the equations is called a *brace*, and it's used to show that the two equations occur together and at the same time.

Before we move forward, let's remember a couple of things from Chapter 2:

◆ We know that an ordered pair is a set of two numbers that are written in a particular order. They are written in parentheses, like this: (2, 7). In this case, the first number, 2, represents the x-coordinate and the second number, 7, represents the y-coordinate.

◆ We also know that if an equation has two variables, and if replacing the variables with the values in the ordered pair causes the equation to be true, then the ordered pair is a solution of the equation.

If we are given a system of equations, our definition of "solution" changes slightly: If an ordered pair is a solution of *both* of the equations in a system, then the ordered pair is a solution of the system.

For example, the ordered pair (3, 5) is a solution of the system above because (3, 5) is a solution of both equations. If we replace the variable x with 3 and the variable y with 5, our solution proves true in both equations, as you can see below.

Check the first equation.   Now check the second equation.

$$5x - 2y = 5 \qquad\qquad\qquad x + y = 8$$
$$5(3) - 2(5) = 5 \qquad\qquad\quad 3 + 5 = 8$$
$$15 - 10 = 5 \qquad\qquad\qquad\quad 8 = 8 \checkmark$$
$$5 = 5 \checkmark$$

Example 1

Determine whether or not each point is a solution of the system $\begin{cases} 7x - 4y = -15 \\ 8x - y = 15 \end{cases}$.

1. $(-1, 2)$ 2. $(2, 1)$ 3. $(3, 9)$

Solution

1. $(-1, 2)$

If both of the equations become true equations when replacing x with -1 and y with 2, then this point is a solution of the system.

Check the first equation.   Now check the second equation.

$$7x - 4y = -15 \qquad\qquad\qquad 8x - y = 15$$
$$7(-1) - 4(2) = -15 \qquad\qquad 8(-1) - (2) = 15$$
$$-7 - 8 = -15 \qquad\qquad\qquad -8 - 2 = 15$$
$$-15 = -15 \checkmark \qquad\qquad\qquad -10 = 15$$

The point $(-1, 2)$ is a solution of the first equation, but it is not a solution of the second equation. Therefore, it is *not* a solution of this system.

2. $(2, 1)$

Check the first equation   And now, we can stop.

$$7x - 4y = -15$$
$$7(2) - 4(1) = -15$$
$$14 - 4 = -15$$
$$10 = -15$$

The point $(2, 1)$ is not a solution of the first equation. There's no need to bother checking the second equation. This point is *not* a solution of this system.

3. $(3, 9)$

Check the first equation   Now check the second equation

$$7x - 4y = -15$$ $$8x - y = 15$$
$$7(3) - 4(9) = -15$$ $$8(3) - (9) = 15$$
$$21 - 36 = -15$$ $$24 - 9 = 15$$
$$-15 = -15 \checkmark$$ $$15 = 15 \checkmark$$

The point $(3, 9)$ is a solution of both of the equations in this system. Therefore, it *is* a solution of this system.

Practice A

Now it's your turn. Determine whether or not each point is a solution of the system $\begin{cases} 8x - 2y = -2 \\ 3x = 3y - 12 \end{cases}$
When you're ready to check your answers, turn the page.

1. $(-1, -3)$ **2.** $(1, 5)$ **3.** $(2, 9)$

B. System Types and Graphical Representations

In Chapter 2, we saw that the solutions of a linear equation in two variables can be graphed as a straight line. The graph of a system of linear equations consists of two straight lines because there are two equations. When two straight lines are graphed, one of three possible configurations may occur.

Case 1: Intersecting Lines

Two straight lines can intersect at exactly one point. If two lines intersect at the point (a, b), as in Figure 1, then the point (a, b) is the solution of that system. This point is the only solution of the system because it is the only point that is shared by both lines.

If a system has at least one solution, then the system is consistent. A consistent system that has exactly one solution is called an *independent* system.

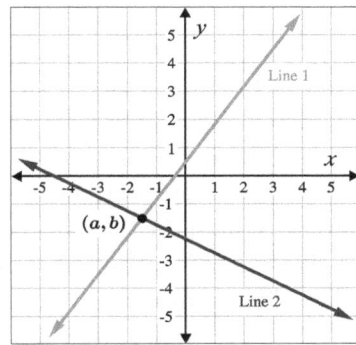

Figure 1.

Case 2: Parallel Lines

The second possibility is that the equations in a system will make parallel lines when graphed. This case is shown in Figure 2. Parallel lines never intersect. In other words, parallel lines don't share any points. In this case, the system has no solution.

A system that has no solution is called an *inconsistent* system. Inconsistent systems occur when equations contradict each other. To put it another way, inconsistent systems occur when the equations in that system cannot both be true at the same time.

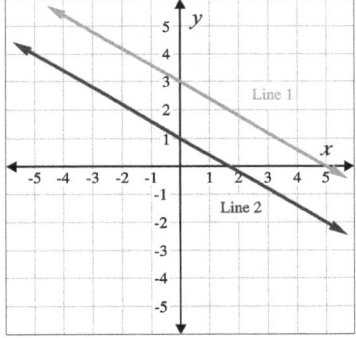

Figure 2.

Case 3: Coincident Lines

The third possibility is that the equations in a system will make coincident lines. Coincident lines are lines that lie exactly on top of each other. In Figure 3, we see that only one line is visible when coincident lines are graphed. When one line lies on top of another line, the lines share an infinite number of points. And so, in this case, the system has infinitely many solutions.

A system that has infinitely many solutions is called a *dependent* system. Dependent systems occur when both equations are equivalent to each other. If both of the equations in a dependent system are rewritten in slope-intercept form, the equations will be identical.

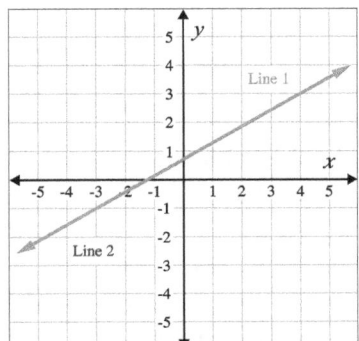

Figure 3.

C. How to Solve a System by Graphing

To solve a system of equations graphically, we need to graph the lines for both equations on the same coordinate plane. We use the same graphing techniques that we used in the previous chapter. Once the equations are graphed, we can interpret the results.

   If the lines intersect, the solution of the system is the ordered pair that corresponds to the point of intersection. If the lines are parallel, then they don't share any points, so there is no solution. In that case, the system is inconsistent. If the lines are coincident, we only end up graphing one line because the lines lie exactly on top of each other. When that happens, there are infinitely many solutions.

   Let's take a look at how this works.

Example 2

Solve each system by graphing.

1. $\begin{cases} -2x + y = 5 \\ x + y = 2 \end{cases}$ 2. $\begin{cases} -x + y = -1 \\ -x + y = 2 \end{cases}$ 3. $\begin{cases} -2x + 3y = -2 \\ -6x + 9y = -6 \end{cases}$

Solutions

1. $\begin{cases} -2x + y = 5 \\ x + y = 2 \end{cases}$

Step 1: It's easiest to graph these equations if they are written in slope-intercept form.

Start with the first equation.   Then solve the second equation for y.

$$-2x + y = 5$$
$$\underline{+2x \qquad\quad + 2x}$$
$$y = 2x + 5$$

$$x + y = 2$$
$$\underline{- x \qquad\quad - x}$$
$$y = -x + 2$$

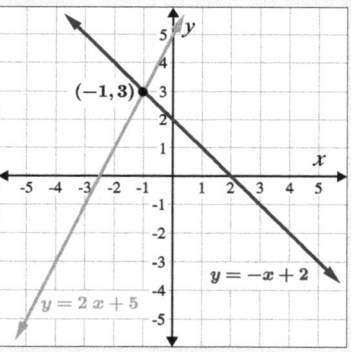

Step 2: Graph both equations on the same coordinate plane.

   The lines in Figure 4 appear to intersect at the point $(-1, 3)$. The solution to this system, then, is $(-1, 3)$. We could also say that $x = -1$ and $y = 3$.

Step 3: Now check your work by substituting -1 for x and 3 for y in each equation.

Figure 4.

Here's the first equation.   The first equation checks out. Here's the second.

$$-2x + y = 5$$
$$-2(-1) + 3 = 5$$
$$2 + 3 = 5$$
$$5 = 5 \checkmark$$

$$x + y = 2$$
$$-1 + 3 = 2$$
$$2 = 2 \checkmark$$

The second equation also checks out. The solution is correct.

2. $\begin{cases} -x + y = -1 \\ -x + y = 2 \end{cases}$

Step 1: Start by writing each equation in slope-intercept form.

Start with the first equation.   Then solve the second equation for y.

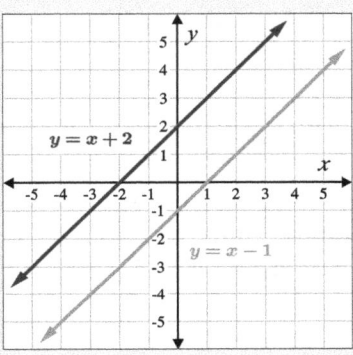

$$-x + y = -1$$
$$\underline{+x \qquad\qquad +x}$$
$$y = x - 1$$

$$-x + y = 2$$
$$\underline{+x \qquad\qquad +x}$$
$$y = x + 2$$

Step 2: Graph both equations.

These lines are parallel, so this system has *no solution.*

It is also acceptable to say that this system is "inconsistent."

We can be sure that these lines are parallel, without intersecting somewhere off in the distance, by noticing that both lines have the same slope, $m = 1$. We know that the lines are not coincident because the y-intercepts are different. See Figure 5.

Figure 5.

3. $\begin{cases} -2x + 3y = -2 \\ -6x + 9y = -6 \end{cases}$

Step 1: Begin by writing each equation in slope-intercept form.

Start with the first equation.   Now write the second equation in slope-intercept form.

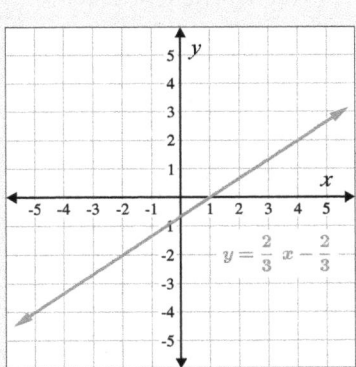

$$-2x + 3y = -2$$
$$\underline{+2x \qquad\qquad +2x}$$
$$3y = 2x - 2$$
$$\frac{3y}{3} = \frac{2x - 2}{3}$$
$$y = \frac{2x}{3} - \frac{2}{3}$$
$$y = \frac{2}{3}x - \frac{2}{3}$$

$$-6x + 9y = -6$$
$$\underline{+6x \qquad\qquad +6x}$$
$$9y = 6x - 6$$
$$\frac{9y}{9} = \frac{6x - 6}{9}$$
$$y = \frac{6x}{9} - \frac{6}{9}$$
$$y = \frac{2}{3}x - \frac{2}{3}$$

Step 2: Now, graph the equations.

In Figure 6, we see that both equations are the same. The two lines are coincident, so this system has infinitely many solutions. We can indicate this fact by writing "dependent."

Figure 6.

Practice A — Answers

1.   No   2.   Yes   3.   No

Practice C

Now that you have a basic understanding of different types of systems of equations and how they can be solved graphically, take a look at the following systems and solve each one by graphing. When you're done, turn the page and check your answers.

4. $\begin{cases} 2x + y = 1 \\ -x + y = -5 \end{cases}$
5. $\begin{cases} -2x + 3y = 6 \\ 6x - 9y = -18 \end{cases}$
6. $\begin{cases} 3x + 5y = 15 \\ 9x + 15y = 15 \end{cases}$
7. $\begin{cases} y = -3 \\ x + 2y = -4 \end{cases}$

D. Solving Graphically with a Graphing Calculator

We can use a graphing calculator to solve a system graphically. This is particularly useful when the solution point does not have integer coordinates. In such cases, finding an exact solution with a hand-drawn graph would be nearly impossible.

Example 3

Use a graphing calculator to solve the system $\begin{cases} 4x - 3y = 2 \\ 6x + 12y = 60 \end{cases}$
Round values off to the nearest thousandth.

Solution

In order to graph an equation on a graphing calculator, the equation must be solved for y, so we must rewrite each equation in slope-intercept form.

Here's the first equation.

$$4x - 3y = 2$$
$$\underline{-4x \qquad\qquad -4x}$$
$$-3y = -4x + 2$$
$$\frac{3y}{-3} = \frac{-4x + 2}{-3}$$
$$y = \frac{-4x}{-3} + \frac{2}{-3}$$
$$y = \frac{4}{3}x - \frac{2}{3}$$

Here's the second.

$$6x + 12y = 60$$
$$\underline{-6x \qquad\qquad -6x}$$
$$12y = -6x + 60$$
$$\frac{12y}{12} = \frac{-6x + 60}{12}$$
$$y = \frac{-6x}{12} + \frac{60}{12}$$
$$y = -\frac{1}{2}x + 5$$

Now we're ready to use our graphing calculator. Below are the steps to follow with a TI-83 or TI-84.

Step 1: On the $\boxed{Y=}$ screen, input the first equation for Y_1 and the second equation for Y_2.

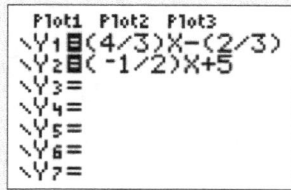

Step 2: View the graph. You may need to adjust your viewing window in order to see the point where the lines intersect. The viewing window for this image is $-10 < x < 10, -10 < y < 10$.

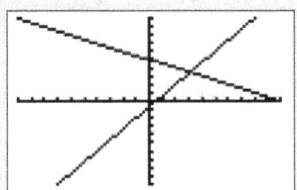

Step 3: Press [CALC]. Do this by pressing [2nd] and then [TRACE]. You will now be viewing a list of calculations that can be done graphically. Select the 5th option, "Intersect."

Step 4: You are now viewing the graph again. At the bottom of the screen, you should see the prompt, "First Curve?"

There should be a little blinking icon on the graph corresponding to Y_1. You may also notice that the equation of this line is in the upper-left corner of the screen. Press the [ENTER] button to select this graph.

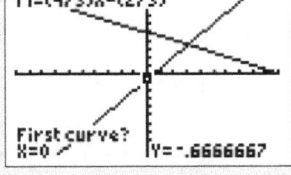

Step 5: Now, the prompt should read "Second Curve?" The blinking icon has moved to the graph corresponding to Y_2, and the equation for this line appears in the upper-left corner of the screen. Press the [ENTER] button to select this graph.

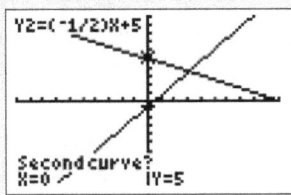

Step 6: At this point, the prompt will read "Guess?" You can use the left/right arrow keys to move the blinking icon. Move this icon until it's close to the point of intersection. Then press the [ENTER] button.

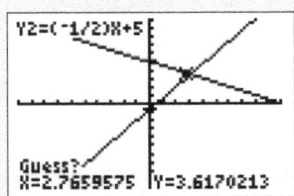

Step 7: The bottom of the screen should now read "Intersection" and give the coordinates of the solution. We round values off to the nearest thousandth when writing the solution:
(3.091, 3.455)

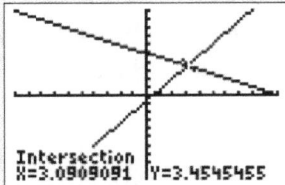

About Steps 4, 5 and 6 in Example 3

Why does the calculator refer to these graphs as "curves"?

> In math, *every* graph that can be drawn without lifting your pencil is called a "curve." We can think of a line, then, as a curve that has a curvature of 0. So your calculator is technically correct in referring to these graphs as "curves."

Why do we need to select these graphs?

> There are only two graphs here, so it seems unnecessary to tell the calculator which graphs to analyze. However, your calculator can graph a lot more than two things at a time. You may have noticed that on the [Y=] screen, there is room to enter several equations. Because of that, the calculator always requires you to verify which two graphs to use when calculating the point of intersection.

Why do we need to "guess"?

> Again, it may seem silly for the calculator to require that a "guess" be made. After all, there is only one point of intersection. Can't the calculator find this point on its own? Keep in mind that while we're focusing on linear graphs right now, not all graphs are linear. With nonlinear graphs, there may be multiple points of intersection. In that case, the calculator needs to know which point of intersection we're trying to find. That is why your calculator will always require a "guess" before calculating the point of intersection.

The intersect feature of a graphing calculator can also be used to help us solve equations in one variable. Let's revisit an equation we solved in Chapter 1.

Example 5

Use a graphing calculator to solve graphically: $3x + 52 = 7x - 12$

Solution

In Chapter 1, we used the table feature of our calculators to solve this equation. This time, we will be using the intersect feature.

Step 1: Begin by storing the left side of the equation as Y_1 and the right side of the equation as Y_2.

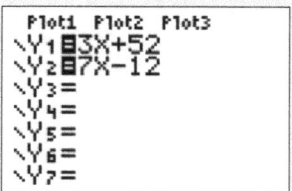

Step 2: View the graph. The viewing window needs to be set so that the point of intersection is visible. In this image, the viewing window is $-10 < x < 30, -10 < y < 150$.

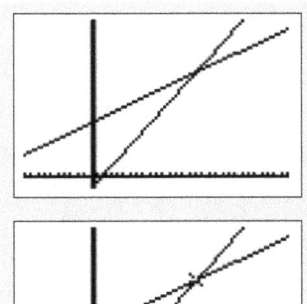

Step 3: Use your calculator to find the point of intersection. Follow the procedure from the previous example.

Step 4: Since the original equation only contained the variable x, we're only interested in the x-coordinate of the point of intersection. This is the solution of the equation.

Solution: $x = 16$

Finally, let's see how the graphing calculator's table feature can help solve a system of equations.

Practice C — Answers

4. $(2, -3)$

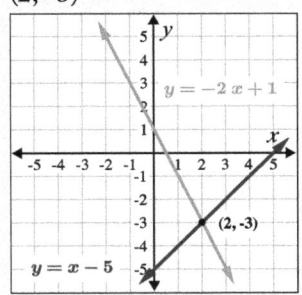

6. no solution (inconsistent)

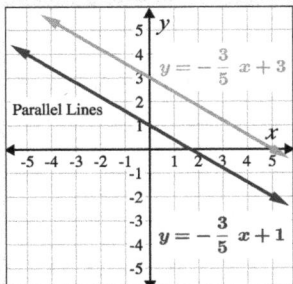

5. infinitely many solutions (dependent)

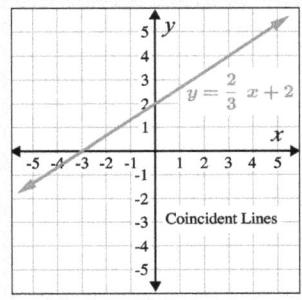

7. $(2, -3)$

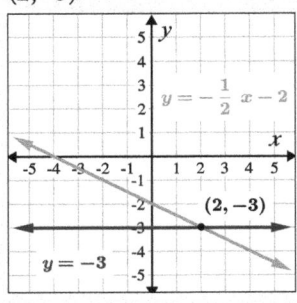

Example 6

Use the table feature of a graphing calculator to solve $\begin{cases} -2x + y = 5 \\ x + y = 2 \end{cases}$

Solution

This is the same system we solved in the first part of Example 2, so the table on our graphing calculators should verify the solution we found. Before we can use our calculators, both equations must be solved for y. We already solved both equations for y in Example 2, so we'll copy the results here.

Step 1: The equations are solved for y.

$$\begin{cases} y = 2x + 5 \\ y = -x + 2 \end{cases}$$

Step 2: Begin by storing the first equation as Y 1 and the second equation as Y 2.

```
Plot1  Plot2  Plot3
\Y1 ■2X+5
\Y2 ■-X+2
\Y3=
\Y4=
\Y5=
\Y6=
\Y7=
```

Step 3: View the table and locate the x-value for which the values of Y_1 and Y_2 are equal.

X	Y1	Y2
-3	-1	5
-2	1	4
-1	3	3
0	5	2
1	7	1
2	9	0
3	11	-1

X=-1

Step 4: When $x = -1$, both of the y-values are 3. This means that the point $(-1, 3)$ solves both equations.

Solution: $(-1, 3)$

Keep in mind that tables can work well for finding solution points with *integer* coordinates. If a solution point does not have integer coordinates, then you will be hard-pressed to find it by using the table feature. In such cases, the intersect feature works much better.

Practice D

It's time for you to try a few problems with your graphing calculator. Turn the page when you're ready to check your answers.

8. Use the intersect feature of a graphing calculator to solve the system. Round values off to the nearest thousandth.
$$\begin{cases} y = 2.718x - 4.965 \\ y = -0.744x + 6.231 \end{cases}$$

9. Use the intersect feature of a graphing calculator to solve the system. Round values off to the nearest thousandth.
$$\begin{cases} 3x + 4y = 28 \\ 7x - 2y = 11 \end{cases}$$

10. Use the intersect feature of a graphing calculator to solve the equation. Round to the nearest thousandth.
$$-4.7x + 9.1 = -0.6x - 1.3$$

11. Use the table feature of a graphing calculator to solve the system.
$$\begin{cases} x + y = 7 \\ 3x - 6y = 3 \end{cases}$$

Exercises 3.1

For the following problems, determine whether or not each point is a solution of the given system.

1. $\begin{cases} 2x + 5y = 32 \\ y = x - 9 \end{cases}$
    a. $(1, 6)$
    b. $(10, 1)$
    c. $(11, 2)$

2. $\begin{cases} 2x - 7y = 13 \\ -4x + 14y = -26 \end{cases}$
    a. $(10, 1)$
    b. $(-11, 5)$
    c. $(-4, -3)$

3. $\begin{cases} 2x + 6y = 8 \\ 3x = -9y - 12 \end{cases}$
    a. $(7, -1)$
    b. $(-7, 1)$
    c. $(4, 0)$

4. $\begin{cases} 2x - 12 = y \\ 5x - 6x = 30 \end{cases}$
    a. $(3, -6)$
    b. $(6, 0)$
    c. $(0, -5)$

For the following problems, use the graphing method to solve these systems of equations by hand, if possible. The solution, when it exists, should be written as an ordered pair.

5. $\begin{cases} y = 3x - 5 \\ y = -\frac{2}{3}x + 6 \end{cases}$

11. $\begin{cases} -3x + y = 5 \\ -x + y = 3 \end{cases}$

17. $\begin{cases} x - 3y = 14 \\ 2x - 6y = 4 \end{cases}$

6. $\begin{cases} y = -x + 1 \\ y = \frac{2}{5}x + 8 \end{cases}$

12. $\begin{cases} x - y = -6 \\ x + 2y = 0 \end{cases}$

18. $\begin{cases} x + 2y = 3 \\ -3x - 6y = -9 \end{cases}$

7. $\begin{cases} y = \frac{3}{4}x - 6 \\ y = -\frac{1}{2}x - 1 \end{cases}$

13. $\begin{cases} 3x + y = 0 \\ 4x - 3y = 12 \end{cases}$

19. $\begin{cases} x - 2y = 6 \\ 3x - 6y = 18 \end{cases}$

8. $\begin{cases} y = 4x + 3 \\ y = x - 3 \end{cases}$

14. $\begin{cases} -4x + y = 7 \\ -3x + y = 2 \end{cases}$

20. $\begin{cases} 2x + 3y = 6 \\ -10x - 15y = 30 \end{cases}$

9. $\begin{cases} x + y = -5 \\ -x + y = 1 \end{cases}$

15. $\begin{cases} 2x + 3y = 6 \\ 3x + 4y = 6 \end{cases}$

10. $\begin{cases} x + y = 4 \\ x + y = 0 \end{cases}$

16. $\begin{cases} x + y = -3 \\ 4x + 4y = -12 \end{cases}$

Practice D — Answers

8.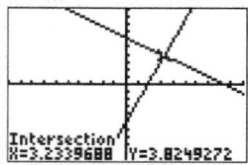

Viewing window: $-10 < x < 10$ and $-10 < y < 10$
Solution: (3.234, 3.825)

10.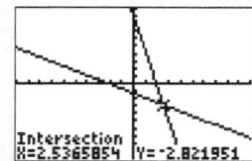

Viewing window: $-10 < x < 10$ and $-10 < y < 10$
Solution: $x = 2.537$

9.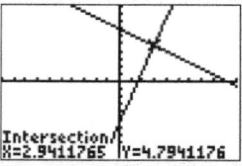

Viewing window: $-10 < x < 10$ and $-10 < y < 10$
Solution: (2.941, 4.794)

11.

Solution: (5, 2)

For the following problems, use the intersect feature of a graphing calculator to solve the system of equations. Round values off to the nearest thousandth.

21. $\begin{cases} y = 1.62x - 2.5 \\ y = -0.31x + 4.9 \end{cases}$

24. $\begin{cases} y = 2.06x + 13.52 \\ y = -1.09x - 4.3 \end{cases}$

27. $\begin{cases} x - 7y = 26 \\ 16x - 3y = 7 \end{cases}$

22. $\begin{cases} y = 3.18x + 2 \\ y = -1.5x - 4.1 \end{cases}$

25. $\begin{cases} 2x - 6y = 9 \\ 3x + 4y = -7 \end{cases}$

28. $\begin{cases} 6x + 11y = -84 \\ 9x + 2y = 13 \end{cases}$

23. $\begin{cases} y = -0.13x - 0.79 \\ y = -2.8x + 9.71 \end{cases}$

26. $\begin{cases} 8x + 5y = 23 \\ 7x - 2y = -5 \end{cases}$

For the following problems, use the intersect feature of a graphing calculator to solve the equation. Round answer off to the nearest thousandth.

29. $-3.2x + 15.9 = 2.15x + 3.7$

30. $-5.39x + 4.17 = 3.82x + 14.48$

31. $9x - (16x - 21) = 5x - 4(2x + 1)$

32. $4.7(2x - 3.1) = 9.4 - 1.7(3 + 2.6x)$

33. $\dfrac{5.2x - 3.1}{1.18} = \dfrac{29.4 - 6.7x}{3.09}$

34. $\dfrac{-15.1x - 8.2}{4.16} = \dfrac{-31.8x - 13.9}{-5.71}$

For the following problems, use the table feature of a graphing calculator to solve the system of equations.

35. $\begin{cases} y = -3x + 33 \\ y = 2x - 2 \end{cases}$

37. $\begin{cases} x + 3y = -10 \\ 2x - 4y = 0 \end{cases}$

39. $\begin{cases} 8x + 2y = 12 \\ 2x - 3y = 24 \end{cases}$

36. $\begin{cases} y = 9x - 31 \\ y = -5x + 11 \end{cases}$

38. $\begin{cases} 3x + 2y = 7 \\ -x + y = 11 \end{cases}$

3.2 Using Substitution to Solve Systems

Overview

We know how to use algebra to solve a linear equation in one variable. Now we'll study an algebraic method for solving a system of two linear equations in two variables.

Here's a problem like that:

> Maurice's Marvelous Coffee Shop sells small coffees for $3 each and large coffees for $5 each. One day, the shop sold twice as many small coffees as large coffees and collected $539. How many small coffees and how many large coffees were sold? Use substitution to help you solve this problem.

You may remember seeing the word "substitute" used to describe replacing a variable with a number. But how can the process of variable replacement be used in a situation like this? We will explore that concept, and then return to this problem and solve it.

This section will help you learn how to:

- Understand substitution and recognize when the substitution method works best

- Use the substitution method to solve a system of linear equations

- Write and solve a system of equations modeling a real-life situation

- Know what to expect when using substitution with a system that gives parallel lines or coincident lines

A. Substitution

When solving a system algebraically, our initial goal is to transform the two equations in two variables into a single equation with one variable. One way to make this transformation is by using the substitution method.

Substitution means replacing a variable with something of equal value. For example, if we know that $x = 3$ and want to find the value of $2x$, we replace the variable x with 3. This gives us $2x = 2(3) = 6$. The value of $2x$ is 6. Informally, many mathematicians refer to this process as "plugging in." In this case, we "plugged 3 in" for x.

In order to replace a variable, we must know a number or expression that is equal to that variable. Therefore, in order to use the substitution method, we must have an equation that is solved for a variable. In other words, we must have an equation in which a variable is isolated.

The Substitution Method

This is the process we follow when using the substitution method to solve a system of equations:

1. Solve one of the equations for one of the variables. In some cases, one of the original equations may already be solved for a variable. If that happens, we can skip to Step 2.

2. In the other equation — the one *not* used to isolate the variable in Step 1 — replace the variable isolated in Step 1 with an equal expression. We use parentheses when replacing a variable with an expression.

3. Simplify and solve this revised equation.

4. Substitute the solution from Step 3 into either of the original equations, or into the equation we formed in Step 1, and then solve. Our work will be the easiest if we use the equation formed in Step 1. At the conclusion of this step, we have number values for both of the variables in the system.

5. Check the solution in both equations.

6. Write the solution as an ordered pair.

This method works best when one of the equations in a system is already solved for a variable.

Example 1

Use substitution to solve the system $\begin{cases} 3x + 9y = 6 \\ x = y - 10 \end{cases}$

Solution

In this system, the second equation is solved for x. The second equation lets us know that x can be replaced with the expression $y - 10$.

Step 1: We skip this step because one of the equations is already solved for a variable.

Step 2: Because the variable is isolated in the second equation, the substitution must take place in the first equation.

$$3x + 9y = 6$$
$$3(y - 10) + 9y = 6 \qquad \text{Replace } x \text{ with } (y - 10).$$

Step 3: Simplify the left side of the equation and solve for y.

$$
\begin{aligned}
3(y - 10) + 9y &= 6 \\
3y - 30 + 9y &= 6 \\
12y - 30 &= 6 \\
\underline{+ 30 \qquad + 30} & \\
12y &= 36 \\
\frac{12y}{12} &= \frac{36}{12} \\
y &= 3
\end{aligned}
$$

Step 4: Now we can replace y with 3 in either of the original equations. The second equation is solved for x, so we'll use that one.

$$x = y - 10$$
$$x = 3 - 10 \qquad \text{Replace } y \text{ with 3 and simplify.}$$
$$x = -7$$

For this step, if we had replaced y with 3 in the first equation and then solved the equation, we would have arrived at the same number value for x. However, this would have required more work.

Step 5: We've found values for x and y. It's time to check our solution by replacing x with -7 and y with 3 in both equations.

$$3x + 9y = 6 \qquad\qquad\qquad x = y - 10$$
$$3(-7) + 9(3) = 6 \qquad\qquad\qquad -7 = 3 - 10$$
$$-21 + 27 = 6 \qquad\qquad\qquad -7 = -7 \checkmark$$
$$6 = 6 \checkmark$$

Step 6: Because the solution checks out, we write it as an ordered pair, $(-7, 3)$.

Based on what we learned in the previous section, we know that if we were to graph both of the equations in this system, the point of intersection would be $(-7, 3)$.

Many systems do not contain an equation that is solved for a variable. In deciding how to get started with these systems, keep in mind that we will do less work if we can locate a variable that already has a coefficient of 1. If none of the variables have a coefficient of 1, then we should look for a variable that can be made to have a coefficient of 1 without introducing fractions into the equation.

Example 2

Use substitution to solve the system $\begin{cases} 2x + 3y = 14 \\ 3x + y = 7 \end{cases}$

Solution

Step 1: Because the coefficient of y in the second equation is 1, we solve the second equation for y.

$$\begin{aligned} 3x + y &= 7 \\ \underline{-3x \qquad\quad -3x} \\ y &= -3x + 7 \end{aligned}$$

Step 2: We used the second equation to isolate a variable. Because of this, the substitution step must happen in the *first* equation.

$$2x + 3y = 14$$
$$2x + 3(-3x + 7) = 14 \qquad\qquad \text{Replace } y \text{ with } (-3x + 7).$$

Step 3: Simplify the left side of the equation and solve for x.

$$2x + 3(-3x + 7) = 14$$
$$2x - 9x + 21 = 14$$
$$-7x + 21 = 14$$
$$\underline{\; -21 \qquad -21}$$
$$-7x = -7$$
$$\frac{-7x}{-7} = \frac{-7}{-7}$$
$$x = 1$$

Step 4: At this point, we can substitute 1 for x in either of the original equations or in the equation we formed in Step 1. We will do the least amount of work if we use the equation formed in Step 1.

$$y = -3x + 7$$
$$y = -3(1) + 7 \quad \text{Replace } x \text{ with } 1 \text{ and simplify.}$$
$$y = 4$$

Step 5: Check the solution in both equations.

Here's the first equation.   Here's the second.

$$2x + 3y = 14 \qquad\qquad 3x + y = 7$$
$$2(1) + 3(4) = 14 \qquad\qquad 3(1) + (4) = 7$$
$$2 + 12 = 14 \qquad\qquad 3 + 4 = 7$$
$$14 = 14 \checkmark \qquad\qquad 7 = 7 \checkmark$$

Step 6: Write the solution as an ordered pair: $(1, 4)$.

The next system doesn't have any variables with a coefficient of 1, so solving it will take a little bit more work.

◤
Example 3

Use substitution to solve the system $\begin{cases} 2x + 4y = 2 \\ 3x + 5y = 1 \end{cases}$

Solution

Step 1: None of the variables in this system have a coefficient of 1. We need to decide which variable to isolate. The variable x in the first equation has a coefficient of 2, and the other terms in the equation are divisible by 2. Therefore, when we solve this equation for x and simplify the results, there will not be any fractions. So let's solve the first equation for x.

$$2x + 4y = 2$$
$$\underline{\quad -4y \qquad\quad -4y \quad}$$
$$2x = -4y + 2$$
$$\frac{2x}{2} = \frac{-4y + 2}{2}$$
$$x = \frac{-4y}{2} + \frac{2}{2}$$
$$x = -2y + 1$$

Step 2: Remember, we used the first equation to isolate a variable. Because of this, the substitution step must take place in the *second* equation.

$$3x + 5y = 1$$
$$3(-2y + 1) + 5y = 1 \qquad\qquad \text{Replace } x \text{ with } (-2y + 1).$$

Step 3: It's time to simplify and solve.

$$3(-2y + 1) + 5y = 1$$
$$-6y + 3 + 5y = 1$$
$$-y + 3 = 1$$
$$\underline{\quad -3 \qquad -3 \quad}$$
$$-y = -2$$
$$\frac{-y}{-1} = \frac{-2}{-1}$$
$$y = 2$$

Step 4: The fastest way to find the number value for x is to substitute 2 for y in the equation formed in Step 1.

$$x = -2y + 1$$
$$x = -2(2) + 1 \qquad \text{Replace } y \text{ with 2 and simplify.}$$
$$x = -4 + 1$$
$$x = -3$$

Step 5: Check the solution in both equations.

Here's the first equation.	Here's the second.
$2x + 4y = 2$	$3x + 5y = 1$
$2(-3) + 4(2) = 2$	$3(-3) + 5(2) = 1$
$-6 + 8 = 2$	$-9 + 10 = 1$
$2 = 2 \checkmark$	$1 = 1 \checkmark$

Step 6: The solution for this system is $(-3, 2)$.

Unfortunately, with some systems, fractions will be unavoidable when solving for a variable. In these cases, the elimination method that we will learn in the next section will typically work much better than the substitution method.

Practice A

On your own paper, use the substitution method to solve the following systems. When you are finished, turn the page and check your answers.

1. $\begin{cases} 2x - 9 = y \\ 4x - 3y = 7 \end{cases}$ 2. $\begin{cases} 5x - 8y = 18 \\ 4x + y = 7 \end{cases}$ 3. $\begin{cases} -2x + 3y = -18 \\ 5x + 10y = 10 \end{cases}$

B. Application Problems Involving Substitution

When we solve practical or applied problems, it is often more convenient to introduce two variables rather than only one variable. However, we should only introduce two variables when two relationships between these variables can be found within the problem. Remember, if we are trying to solve for *two* variables, we must have *two* different equations. Each relationship will produce an equation and when we put these together, we will have a system of two equations in two variables.

For these applied problems, we will use the following five-step process to find our solutions:

1. Introduce two variables, one for each unknown quantity.

2. Find two relationships within the problem and translate each relationship into an equation. If the problem does not contain two relationships, then there's not enough information to solve it by using two variables.

3. Solve the resulting system of equations. In this section, we will use the substitution method to do this.

4. Check the solution in both equations.

5. Translate the mathematical solution into a written statement.

Example 4

The formula for the perimeter (P) of a rectangle is $P = 2L + 2W$, where L represents the length of the rectangle and W represents the width. Use this to help you solve the following problem:

> The length of a rectangle is 13 cm more than the width. The perimeter of the rectangle is 46 cm. Find the dimensions of the rectangle.

Solution

This problem is similar to some problems we looked at in Chapter 1. Back then, we weren't familiar with systems of equations, so we had to try to figure out how to write everything in terms of a single variable. We also drew a diagram to help us. However, it's usually more efficient to tackle problems like this by introducing two variables and writing two equations.

Step 1: We need to figure out the length and width of the rectangle. We'll use the variables L and W, respectively.

Step 2: Next we write *two* equations based on this problem.

◆ We write one equation by replacing P with 46 in the perimeter formula: $46 = 2L + 2W$.

◆ The other equation comes from translating the phrase "the length is 13 cm more than the width": $L = W + 13$.

Now we have a system that can be solved: $\begin{cases} 46 = 2L + 2W \\ L = W + 13 \end{cases}$

Step 3: Now that we have a system of two equations, we use substitution to solve. See if you can follow the process without explanation.

First we find the number value for W.

$$46 = 2(W + 13) + 2W$$
$$46 = 2W + 26 + 2W$$
$$46 = 4W + 26$$
$$\underline{-26 \qquad\qquad -26}$$
$$20 = 4W$$
$$\frac{20}{4} = \frac{4W}{4}$$
$$5 = W$$

Now we find the number value for L.

$$L = W + 13$$
$$L = 5 + 13$$
$$L = 18$$

Step 4: We check the number values of L and W in both equations.

Check the first equation.   Check the second.

$$46 \ = \ 2L + 2W$$
$$46 \ = \ 2(18) + 2(5)$$
$$46 \ = \ 36 + 10$$
$$46 \ = \ 46 \ ✓$$

$$L \ = \ W + 13$$
$$18 \ = \ 5 + 13$$
$$18 \ = \ 18 \ ✓$$

Step 5: We have verified our solution and can translate it into a written statement:

The rectangle is 18 centimeters long and 5 centimeters wide.

Now that we've seen how the five-step process is used to solve an application problem involving a system of equations, let's take a look at the problem from the beginning of the section.

Example 5

Maurice's Marvelous Coffee Shop sells small coffees for $3 each and large coffees for $5 each. One day, the shop sold twice as many small coffees as large coffees and collected $539. How many small coffees and how many large coffees were sold?

Solution

Step 1: Let x represent the number of small coffees sold, and let y represent the number of large coffees sold.

Step 2: We know that the number of small coffees sold, x, is twice as much as the number of large coffees sold, y, so one of our equations is $x = 2y$.

The other equation will involve the value of the coffees sold. The value of the small coffees sold is represented by $3x$, and the value of the large coffees sold is represented by $5y$. The total value of the coffees sold is $539. We use this information to write our second equation:

$$3x + 5y = 539$$

Now we have a system to solve: $\begin{cases} x = 2y \\ 3x + 5y \ = \ 539 \end{cases}$

Step 3: Use substitution to solve the system.

First solve for y.   Now find the number value for x.

$$3(2y) + 5y = 539$$ $$x = 2y$$
$$6y + 5y = 539$$ $$x = 2(49)$$
$$11y = 539$$ $$x = 98$$
$$\frac{11y}{11} = \frac{539}{11}$$
$$y = 49$$

Step 4: Verify that the solution is correct. Check both equations.

$$x = 2y$$ $$3x + 5y = 539$$
$$98 = 2(49)$$ $$3(98) + 5(49) = 539$$
$$98 = 98 \checkmark$$ $$294 + 245 = 539$$
$$539 = 539 \checkmark$$

Step 5: Convert the solution into a written statement:

98 small coffees were sold, and 49 large coffees were sold.

Practice B

Now you try a couple of application problems. Use the five-step process to help you solve the following problems. Turn the page when you are ready to check your answers.

4. Chloe discovers that her rectangular shed is 6 feet longer than three times its width. If the perimeter of the shed is 100 feet, find the dimensions. (Recall: the perimeter of a rectangle is given by the formula $P = 2L + 2W$.)

5. LeRoy scored three times as many points in his first basketball game as he did in the second game. His point total for both games combined was 48 points. How many points did LeRoy score in each game?

C. Solving with Substitution: Parallel Lines and Coincident Lines

We learned in the previous section that not all systems have a solution. When we use the substitution method, we don't always end up with a solution, either. Sometimes in the process of solving, we get rid of both variables and end up with a contradiction. That tells us that the original equations can never be true at the same time.

Example 6

Use substitution to solve the system $\begin{cases} 2x - y = 1 \\ 4x - 2y = 4 \end{cases}$

Solution

Step 1: None of the variables has a coefficient of 1. However, the variable y in the first equation has a coefficient of -1, and that's almost as good. We'll solve the first equation for y.

$$
\begin{array}{rcl}
2x - y &=& 1 \\
\underline{+y} & & \underline{+y} \\
2x &=& y + 1 \\
\underline{-1} & & \underline{-1} \\
2x - 1 &=& y
\end{array}
$$

Step 2: The substitution step must take place in the second equation. Do you remember why?

$$4x - 2y = 4$$
$$4x - 2(2x - 1) = 4$$

Step 3: Simplify the left side of the equation and to solve for x.

$$
\begin{array}{rcl}
4x - 2(2x - 1) &=& 4 \\
4x - 4x + 2 &=& 4 \\
2 &=& 4
\end{array}
$$

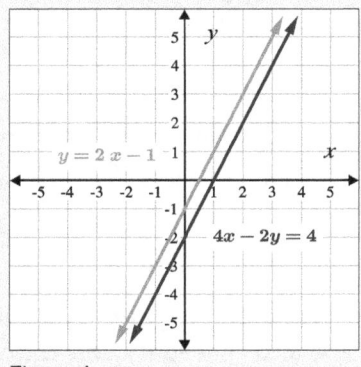

Figure 1.

At this point, we must stop. Both of the variables have disappeared, and we have produced a contradiction. That means that this system has *no solution*.

If we graph the original equations, we see from Figure 1 that the lines are parallel.

Parallel lines aren't the only thing that can come between us and a solution for a system of equations. Sometimes when we are in the process of solving, both variables disappear and the resulting equation is an identity. This means that the original equations are equivalent, and that the two lines of the system are coincident. When this happens, we know that the system has infinitely many solutions.

Example 7

Use substitution to solve the system $\begin{cases} 4x + 8y = 8 \\ 3x + 6y = 6 \end{cases}$

Solution

Step 1: In the first equation, x has a coefficient of 4 and all of the terms in the equation are divisible by 4. We can solve the first equation for x without having fractions in the result.

$$4x + 8y = 8$$
$$\underline{\quad -8y \qquad -8y}$$
$$4x = -8y + 8$$
$$\frac{4x}{4} = \frac{-8y + 8}{4}$$
$$x = \frac{-8y}{4} + \frac{8}{4}$$
$$x = -2y + 2$$

Step 2: The substitution step must take place in the second equation.

$$3x + 6y = 6$$
$$3(-2y + 2) + 6y = 6$$

Step 3: Simplify the left side and attempt to solve for y.

$$3(-2y + 2) + 6y = 6$$
$$-6y + 6 + 6y = 6$$
$$6 = 6$$

We have to stop. Both variables have disappeared and we have produced an identity. This tells us that the lines are coincident, as you can see in Figure 2. This system has infinitely many solutions.

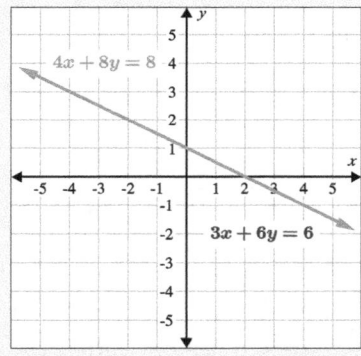

Figure 2.

Practice C

Solve each system of equations using the substitution method. The turn the page to check your solutions.

6. $\begin{cases} 7x - 3y = 2 \\ 14x - 6y = 1 \end{cases}$

7. $\begin{cases} 4x + 3y = 1 \\ -8x - 6y = -2 \end{cases}$

D. Substitutions Involving Fractions

As we mentioned earlier, substitution doesn't work very well when we have a system in which none of the variables have a coefficient of 1 and none of the coefficients can be made to be 1 without introducing fractions into an equation.

This last example will illustrate how the substitution method can get messy when we have to use fractions.

Example 8

Use substitution to solve the system $\begin{cases} 3x + 2y = 1 \\ 4x - 3y = 3 \end{cases}$

Solution

Step 1: None of the variables can be made to have a coefficient of 1 without introducing fractions into an equation. We will solve the first equation for y because the coefficient, 2, will produce the least complicated fractions when it is divided out.

$$3x + 2y = 1$$
$$\underline{\quad -3x \qquad\qquad -3x \quad}$$
$$2y = -3x + 1$$
$$\frac{2y}{2} = \frac{-3x + 1}{2}$$
$$y = \frac{-3x}{2} + \frac{1}{2}$$
$$y = -\frac{3}{2}x + \frac{1}{2}$$

Step 2: The substitution step now takes place in the second equation.

$$4x - 3y = 3$$

$$4x - 3\left(-\frac{3}{2}x + \frac{1}{2}\right) = 3 \qquad \text{Replace } y \text{ with } \left(-\frac{3}{2}x + \frac{1}{2}\right).$$

Practice Set C — Answers

6. There are multiple possible outcomes. The most common results are $4 = 1$ and $\frac{1}{2} = 2$. These results (and all other possible results generated from this system) are contradictions. That means that the system has no solution.

7. There are multiple possible outcomes, the most common of which is $-2 = -2$. This is an identity. All of the other possible results generated from this system are also identities. This system has infinitely many solutions.

Step 3: It's time to simplify and solve. This process is more difficult because of the fractions involved.

$$4x - 3\left(-\frac{3}{2}x + \frac{1}{2}\right) = 3$$

$$4x + \frac{9}{2}x - \frac{3}{2} = 3$$

$$2 \cdot \left(4x + \frac{9}{2}x - \frac{3}{2}\right) = 2 \cdot (3)$$

Now that the parentheses are gone, get rid of these fractions by multiplying both sides by the LCD, which is 2.

$$8x + 9x - 3 = 6$$

$$17x - 3 = 6$$

$$\underline{+3 \qquad +3}$$

$$17x = 9$$

$$\frac{17x}{17} = \frac{9}{17}$$

$$x = \frac{9}{17}$$

Step 4: Substitute $\frac{9}{17}$ for x in the equation formed in Step 1 and simplify.

$$y = -\frac{3}{2}\left(\frac{9}{17}\right) + \frac{1}{2}$$

$$y = -\frac{27}{34} + \frac{1}{2}$$

$$y = -\frac{27}{34} + \frac{17}{34}$$

$$y = -\frac{10}{34}$$

$$y = -\frac{5}{17}$$

Step 5: Substituting these values into each of the original equations will give a rigorous exercise in simplifying expressions involving fractions. If done correctly the process will, in fact, verify that the solution is correct.

Step 6: The solution for this system is $\left(\frac{9}{17}, -\frac{5}{17}\right)$.

Sometimes, the solution to a system of equations involves fractions. Unfortunately, we can't avoid this. However, we often *can* avoid having to use fractions in the process of finding the solution. The elimination method discussed in the next section provides us with an alternative algebraic approach that allows us to avoid fractions. In some cases, we postpone the need to use fractions, and in some cases we avoid the use of fractions altogether.

Practice D

Now it's time for you to work with a messy substitution problem. Solve the system by using substitution. When you're finished, check the answer on the next page.

8. $\begin{cases} 9x - 5y = -4 \\ 2x + 7y = -9 \end{cases}$

Exercises 3.2

For the following problems, solve the system by using substitution. When you have found the solution, write it as an ordered pair. If a single solution cannot be found, indicate whether the system has no solution or infinitely many solutions.

1. $\begin{cases} 3x + 2y = 9 \\ y = -3x + 6 \end{cases}$

2. $\begin{cases} 5x - 3y = -6 \\ y = -4x + 12 \end{cases}$

3. $\begin{cases} 2x + 2y = 0 \\ x = 3y - 4 \end{cases}$

4. $\begin{cases} 3x + 5y = 9 \\ x = 4y - 14 \end{cases}$

5. $\begin{cases} -3x + y = -4 \\ 2x + 3y = 10 \end{cases}$

6. $\begin{cases} -4x + y = -7 \\ 2x + 5y = 9 \end{cases}$

7. $\begin{cases} 6x - 6 = 18 \\ x + 3y = 3 \end{cases}$

8. $\begin{cases} -x - y = 5 \\ 2x + y = 5 \end{cases}$

9. $\begin{cases} -5x + y = 4 \\ 10x - 2y = -8 \end{cases}$

10. $\begin{cases} x + 4y = 1 \\ -3x - 12y = -1 \end{cases}$

11. $\begin{cases} 4x - 2y = 8 \\ 6x + 3y = 0 \end{cases}$

12. $\begin{cases} 2x + 3y = 12 \\ 2x + 4y = 18 \end{cases}$

13. $\begin{cases} 3x - 9y = 6 \\ 6x - 18y = 5 \end{cases}$

14. $\begin{cases} -x + 4y = 8 \\ 3x - 12y = 10 \end{cases}$

15. $\begin{cases} x + y = -6 \\ x - y = 4 \end{cases}$

16. $\begin{cases} 2x + y = 0 \\ x - 3y = 0 \end{cases}$

17. $\begin{cases} 4x - 2y = 7 \\ y = 4 \end{cases}$

18. $\begin{cases} x + 6y = 11 \\ x = -1 \end{cases}$

19. $\begin{cases} 2x - 4y = 10 \\ 3x = 5y + 12 \end{cases}$

20. $\begin{cases} y + 7x + 4 = 0 \\ x = -7y + 28 \end{cases}$

21. $\begin{cases} x + 4y = 0 \\ x + \frac{2}{3}y = \frac{10}{3} \end{cases}$

22. $\begin{cases} x = 24 - 5y \\ x - \frac{5}{4}y = \frac{3}{2} \end{cases}$

23. $\begin{cases} x = 11 - 6y \\ 3x + 18y = -33 \end{cases}$

24. $\begin{cases} 2x + \frac{1}{3}y = 4 \\ 3x + 6y = 39 \end{cases}$

25. $\begin{cases} \frac{4}{5}x + \frac{1}{2}y = \frac{3}{10} \\ \frac{1}{3}x + \frac{1}{2}y = \frac{-1}{6} \end{cases}$

26. $\begin{cases} x - \frac{1}{3}y = \frac{-8}{3} \\ -3x + y = 1 \end{cases}$

Set up and solve a system of equations to help you answer the following questions.

27. The sum of two numbers is 22. One number is 6 more than the other. What are the numbers?

28. The sum of two numbers is 32. One number is 8 more than the other. What are the numbers?

29. The difference of two numbers is 12 and one number is three times as large as the other. What are the numbers?

30. The difference of two numbers is 9 and one number is 10 times larger than the other. What are the numbers?

31. While shopping at Umberto's Ugly Shoes, a mathematically inclined customer can't find any shoes that she likes. However, she discovers that the length of a shoe box is two inches less than twice the width, and that the perimeter of the box is 38 inches. Set up and solve a system of equations to help you find the length and width of the shoe box. (Hint — the perimeter of a rectangle is given by: $P = 2L + 2W$)

32. Hector, an athletic farmer, runs around the perimeter of his rectangular field for exercise. When he runs at a constant rate, it takes him three times as long to run along the length of the field as it takes to run along the width. The perimeter of the field is 688 yards. Set up and solve a system of equations to help you find the dimensions of the field. (Hint — the perimeter of a rectangle is given by: $P = 2L + 2W$)

33. At a recent Bite and Brew festival in Salem, Oregon, four times as many people attended on Sunday as on Thursday. If a total of 14,250 people attended the festival on the two days combined, how many people attended on Sunday and how many attended on Thursday? Set up and solve a system of equations to help you answer this question.

34. Kaylee practiced seven times as many math problems the night before the test as she did two nights before the test. In the two nights combined, she practiced a total of 56 math problems. Set up and solve a system of equations to help you determine how many problems she practiced the night before, and two nights before the test.

35. Gina's Ginormous Sandwiches sells 6.1-inch sandwiches for $4 each and 12.1-inch sandwiches for $7 each. One day, Gina determined that three times as many 6.1-inch sandwiches were sold as 12.1-inch sandwiches. If a total of $551 was collected from sandwich sales, how many of each size of sandwich were sold? Set up and solve a system of equations to help you answer this question.

Practice Set D — Answers

8. $(-1, -1)$.
   The substitution process for this practice problem is quite messy. However, the solution is not — both coordinates of the solution are integers. That means that this system can be solved, without the use of fractions, by using the elimination method discussed in the next section.

3.3 Using Elimination to Solve Systems

Overview

The substitution method gives us an algebraic approach to solving systems of equations that lets us avoid graphing. However, as we saw at the end of the last section, the substitution process can get quite messy in certain situations. In this section, we will learn another algebraic approach to solving systems — the elimination method.

Here's a problem:

> Yvette bought six packages of cheese and eight packages of tortillas for $43. Brandon bought twelve packages of cheese and six packages of tortillas for $51. What is the price for each package of cheese and each package of tortillas? Use elimination to help you solve this problem.

Once we're able to write a system of equations that represents this situation, we will see that substitution won't work very well as a method for solving. This is because none of the variables in the system can be made to have a coefficient of 1 without introducing fractions into an equation. The elimination method will work *much* better in this case. After we've gained a basic understanding of how the elimination method works, we'll return to this problem and solve it.

This section will teach you how to:

- ◆ Understand elimination by addition and recognize when this method works best

- ◆ Use the elimination method to solve a system of linear equations

- ◆ Write and solve a system of equations modeling a real-life situation

- ◆ Know what to expect when using elimination with a system that gives parallel lines or coincident lines

A. Elimination by Addition

Now that we know how to solve a system of two linear equations in two variables by graphing and by substitution, it's time to learn the elimination method. This method is similar to the substitution method because it helps us accomplish our initial goal of transforming two equations in two variables into a single equation with one variable. We can do elimination with either addition or subtraction. However, elimination by subtraction can be more complicated and lead to more errors. Therefore, in this book, we will only use elimination by addition.

With that in mind, we define elimination as the process of adding two equations together in a way that gets rid of a variable. The method of elimination makes use of the following two properties: *addition of equations* and the *additive inverse property*.

Addition of Equations: If A, B, C, and D are algebraic expressions, and if $A = B$ and $C = D$, then $A + C = B + D$. This property is actually a combination of the addition property of equality and substitution. Here's how it looks with variables:

$$A = B$$
$$C = D$$
$$A + C = B + D$$

And here's how it works if we replace a couple of the variables with numbers:

$$A = 8$$
$$C = 3$$
$$A + C = 8 + 3$$

To state this property using words, we say that if we add the left sides of two equations together, and if we add the right sides of the same two equations together, then the resulting sums will be equal.

Additive Inverse Property: This property states that the sum of additive inverses — commonly known as opposites — is zero. Here are two examples:

$$-7 + 7 = 0$$
$$5x + (-5x) = 0$$

Now let's look at how we can use these two properties to solve systems of equations using the elimination method.

The Elimination Method

To solve a system of two linear equations in two variables with this method, we follow seven steps:

1. Make sure that the like terms and the equals signs of the two equations are lined up. It may be necessary to perform some algebraic steps in order to make this happen.

2. If necessary, multiply one or both equations by factors that produce opposite coefficients for one of the variables.

3. Add the equations. The variable with opposite coefficients is eliminated — this is where the name of this method comes from.

4. Now solve the equation obtained in step 3. This yields the number value for one variable.

5. Find the number value for the other variable by doing one of the following:
   a. Substitute the value obtained in Step 4 into either original equation and solve.
   b. Repeat Steps 1 through 4. This time, we eliminate the variable we didn't eliminate the first time through.

6. Check the solution in both equations.

7. Write the solution as an ordered pair.

The elimination method works best when both equations are given to us in standard form, which is $ax + by = c$. This method is particularly beneficial for systems in which the substitution process would involve messy fractions, as we explored in the previous section.

Example 1

Use elimination to solve the system $\begin{cases} x - y = 2 \\ 3x + y = 14 \end{cases}$

Solution

Step 1: Both equations are already in the right form, $ax + by = c$. The like terms and equals signs are also lined up. We don't need to do anything for this step.

Step 2: The coefficients of y are already opposites (1 and −1), so there's no need for us to use multiplication to create opposite coefficients. We don't need to do anything for this step, either.

Step 3: Add the equations.

$$\begin{array}{rcl} x - y & = & 2 \\ + \ 3x + y & = & + \ 14 \\ \hline 4x + 0 & = & 16 \end{array}$$

$$4x \ = \ 16 \qquad \text{We've eliminated the variable } y.$$

Step 4: Solve the equation.

$$4x = 16$$
$$\frac{4x}{4} = \frac{16}{4}$$
$$x = 4$$

We now know the number value for x. We have two choices for finding the number value for y.

Step 5, Option A: Substitute 4 for x in either of the original equations and then solve. We'll use the first equation for this.

$$\begin{array}{rcl} x - y & = & 2 \\ 4 - y & = & 2 \\ -4 & & -4 \\ \hline -y & = & -2 \end{array}$$

$$\frac{-y}{-1} = \frac{-2}{-1}$$
$$y = 2$$

Step 5 — Option B: We start with the original system of equations and eliminate the variable x.

$-3 \cdot (x - y) = -3 \cdot (2)$ Multiply the terms in the first equation by -3.

$\quad -3x + 3y = \qquad -6$

$\underline{+\ 3x + y = \qquad +\ 14}$ Now write the second equation and then add equations.

$\qquad 0 + 4y = \qquad\ 8$ We've eliminated the variable x.

$\qquad\quad 4y = 8$

$\qquad\quad \dfrac{4y}{4} = \dfrac{8}{4}$ Now solve for y.

$\qquad\quad\ y = 2$

Option B gave us the same result as option A: $y = 2$.

Step 6: Check the solution in both equations.

Check the first equation.   Check the second.

$\qquad x - y = 2 \qquad\qquad\qquad 3x + y = 14$

$\qquad 4 - 2 = 2 \qquad\qquad\qquad 3(4) + 2 = 14$

$\qquad\qquad 2 = 2\ \checkmark \qquad\qquad\quad 12 + 2 = 14$

$\qquad\qquad\qquad\qquad\qquad\qquad\quad 14 = 14\ \checkmark$

Step 7: The solution is $(4, 2)$.

In practice, Option A is generally preferred because it's slightly faster. However, if the number value found in Step 4 is a fraction, then working through Option A might be difficult because the computations will involve fractions. If this is the case, Option B is preferable.

Example 2

Use elimination to solve the system $\begin{cases} 6a - 5b = 33 \\ 2a + 3b = 53 \end{cases}$

Solution

Step 1: The equations are already in the proper form, $ax + by = c$, so we move on to the next step.

Step 2: Neither variable has opposite coefficients. We get to decide which variable to eliminate, so let's take the path of least resistance. Notice that the coefficients of the variable a are 6 and 2. Since 6 is a multiple of 2, we only need to modify one of the equations in order to get opposite coefficients. With this in mind, we multiply both sides of the second equation by -3 to create opposite coefficients for the variable a.

$\qquad 2a + 3b = 53$

$-3 \cdot (2a + 3b) = -3 \cdot (53)$

$\qquad -6a - 9b = -159$

The system now looks like this: $\begin{cases} 6a - 5b = 33 \\ -6a - 9b = -159 \end{cases}$

Step 3: Add the equations.

$$6a - 5b = \quad 33$$
$$+ \underline{-6a - 9b = + (-159)}$$
$$0 - 14b = \quad -126$$

With future elimination problems, we'll skip over this step showing that when opposites are added, the result is 0.

$$-14b = -126 \qquad \text{We've eliminated the variable } a.$$

Step 4: Solve the equation.

$$-14b = -126$$
$$\frac{-14b}{-14} = \frac{-126}{-14}$$
$$b = 9$$

Step 5: Substitute 9 for b in either of the original equations and solve. Here's what it looks like with the second equation.

$$2a + 3b = \quad 53$$
$$2a + 3(9) = \quad 53$$
$$2a + 27 = \quad 53$$
$$\underline{\quad -27 \qquad -27}$$
$$2a = \quad 26$$
$$\frac{2a}{2} = \frac{26}{2}$$
$$a = 13$$

Step 6: Verify the solution with both equations.

$$6a - 5b = 33 \qquad\qquad 2a + 3b = 53$$
$$6(13) - 5(9) = 33 \qquad\qquad 2(13) + 3(9) = 53$$
$$78 - 45 = 33 \qquad\qquad 26 + 27 = 53$$
$$33 = 33 \checkmark \qquad\qquad 53 = 53 \checkmark$$

Step 7: Ordered pairs are typically written in alphabetical order. In this case, a represents the first coordinate in our ordered pair, and b represents the second coordinate. The solution is $(13, 9)$.

Order of Coordinates

In story problems dealing with independent and dependent variables, the independent variable always represents the *first* coordinate in the ordered pair. If, for example, the number of calories burned by lifting weights, c, depends on the amount of weight being lifted, w, then the ordered pairs would take the form (w, c) even though c comes before w in the alphabet.

Example 3

Use elimination to solve the system $\begin{cases} 3x + 2y = -4 \\ 4x = 5y + 10 \end{cases}$

Solution

Step 1: The like terms appear to be lined up, but notice that the equals signs are *not* lined up. This means that we have like terms — in this case, the terms containing y — on opposite sides of the equation. If we line up the equals signs, we can see this:

$$\begin{cases} 3x + 2y = -4 \\ 4x \quad\;\; = 5y + 10 \end{cases}$$

We need to write the system so that both equations are structured in the same way. We will do this by subtracting $5y$ from both sides of the second equation.

$$4x = 5y + 10$$
$$\underline{-5y = -5y}$$
$$4x - 5y = \qquad 10$$

The system is now structured correctly:

$$\begin{cases} 3x + 2y = -4 \\ 4x - 5y = 10 \end{cases}$$

Step 2: Looking at the coefficients of x, we note that 4 is not a multiple of 3. Looking at the coefficients of y, we note that 5 is not a multiple of 2. And so, regardless of the variable we choose to eliminate, we will have to modify both equations. Because the coefficients of y already have opposite signs, we will eliminate y. We will multiply both sides of the first equation by 5 and multiply both sides of the second equation by 2 in order to create opposite coefficients.

Here's the first equation:

$$3x + 2y = -4$$
$$5 \cdot (3x + 2y) = 5 \cdot (-4)$$
$$15x + 10y = -20$$

Here's the second:

$$4x - 5y = 10$$
$$2 \cdot (4x - 5y) = 2 \cdot (10)$$
$$8x - 10y = 20$$

The system now looks like this:

$$\begin{cases} 15x + 10y = -20 \\ 8x - 10y = 20 \end{cases}$$

Step 3: Add equations.

$$\begin{array}{r} 15x + 10y = -20 \\ + \ 8x - 10y = +\ 20 \\ \hline 23x = 0 \end{array}$$

Step 4: Solve.

$$23x = 0$$
$$\frac{23x}{23} = \frac{0}{23}$$
$$x = 0$$

Step 5: Substitute 0 for x in either of the original equations, and solve. We'll use the first equation.

$$3x + 2y = -4$$
$$3(0) + 2y = -4$$
$$2y = -4$$
$$\frac{2y}{2} = \frac{-4}{2}$$
$$y = -2$$

Step 6: Verify the solution. Check both equations.

$$3x + 2y = -4 \qquad\qquad\qquad 4x = 5y + 10$$
$$3(0) + 2(-2) = -4 \qquad\qquad\qquad 0 = 5(-2) + 10$$
$$0 - 4 = -4 \qquad\qquad\qquad 0 = -10 + 10$$
$$-4 = -4 \checkmark \qquad\qquad\qquad 0 = 0 \checkmark$$

Step 7: The solution is $(0, -2)$.

Practice A

Now it's your turn to put this into practice. Use the elimination method to solve each system of equations. Then turn the page to check your answers.

1. $\begin{cases} x + y = 6 \\ 2x - y = 0 \end{cases}$
2. $\begin{cases} x + 4y = 1 \\ x - 2y = -5 \end{cases}$
3. $\begin{cases} 2x + 3y = -10 \\ -x + 2y = -2 \end{cases}$
4. $\begin{cases} 5x - 3y = 1 \\ 8x - 6y = 4 \end{cases}$

B. Application Problems Involving Elimination

Sometimes word problems translate into systems of equations that are both in standard form. In these situations, the elimination method works well when solving. The problem we saw at the beginning of this section is a good example of that. Let's revisit that problem now.

Example 4

Yvette bought six packages of cheese and eight packages of tortillas for $43. Brandon bought twelve packages of cheese and six packages of tortillas for $51. What is the price for each package of cheese and each package of tortillas?

Solution

This problem involves value, but this time the value per item is not given.

Step 1: We are asked to determine the price per package for cheese and tortillas. Let c represent the price per package of cheese, and let t represent the price per package of tortillas.

Step 2: We need *two* equations based on this problem:

$$\begin{cases} 6c + 8t = 43 \\ 12c + 6t = 51 \end{cases}$$

The first equation represents the value of the items purchased by Yvette. The second equation represents the value of the items purchased by Brandon

Step 3: We have a system of two equations, both of which are in standard form. We will use elimination to solve the system. See if you can follow along without the help of explanations.

$$
\begin{aligned}
6c + 8t &= 43 \\
-2 \cdot (6c + 8t) &= -2 \cdot (43) \\
-12c - 16t &= -86 \\
\underline{+\ 12c + 6t} &= \underline{+\ 51} \\
-10t &= -35 \\
\frac{-10t}{-10} &= \frac{-35}{-10} \\
t &= 3.5
\end{aligned}
$$

Now we find the number value for c.

$$
\begin{aligned}
6c + 8(3.5) &= 43 \\
6c + 28 &= 43 \\
\underline{-\ 28} \quad &\ \underline{-\ 28} \\
6c &= 15 \\
\frac{6c}{6} \quad &\ \frac{15}{6} \\
c &= 2.5
\end{aligned}
$$

Step 4: We check the number values of c and t in the first and second equations.

$$6c + 8t \;=\; 43 \qquad\qquad\qquad 12c + 6t \;=\; 51$$
$$6(2.5) + 8(3.5) \;=\; 43 \qquad\qquad 12(2.5) + 6(3.5) \;=\; 51$$
$$15 + 28 \;=\; 43 \qquad\qquad\qquad 30 + 21 \;=\; 51$$
$$43 \;=\; 43 \;\checkmark \qquad\qquad\qquad\quad 51 \;=\; 51 \;\checkmark$$

Step 5: Now that we have verified our solution, we translate it into a written statement:

Cheese costs $2.50 per package, and tortillas cost $3.50 per package.

Practice B

Use the five-step process — this time with the elimination method — to help you solve the following problem. Turn the page when you're ready to check your answer.

5. Henrietta buys twelve pounds of bananas and ten pounds of apples for $12. Gustavo buys eight pounds of bananas and five pounds of apples for $7. What is the price per pound of bananas and apples?

C. Solving with Elimination: Parallel and Coincident Lines

When the lines of a system are parallel or coincident, the method of elimination produces results that are similar to the results produced by the substitution method. If both of the variables are eliminated and we're left with a contradiction, then the two lines of the system are parallel, and the system has no solution. If both of the variables are eliminated and we are left with an identity, then the two lines of the system are coincident, and the system has infinitely many solutions.

Let's look at a couple of systems and see how this works.

Example 5

Use elimination to solve each system.

1. $\begin{cases} 2x - y = 1 \\ 4x - 2y = 4 \end{cases}$

2. $\begin{cases} 4x + 8y = 8 \\ 3x + 6y = 6 \end{cases}$

Solutions

1. $\begin{cases} 2x - y = 1 \\ 4x - 2y = 4 \end{cases}$

Step 1: The equations are in the proper form, so we can skip this step.

Step 2: We can create opposite coefficients for the variable x by multiplying both sides of the first equation by –2.

$$2x - y \;=\; 1$$
$$-2 \cdot (2x - y) \;=\; -2 \cdot (1)$$
$$-4x + 2y \;=\; -2$$

Step 3: Now we can add the two equations together.

$$
\begin{array}{rcr}
-4x + 2y &=& -2 \\
+\,4x - 2y &=& +\,4 \\
\hline
0 &=& 2
\end{array}
$$

This equation is false — a contradiction. Therefore, the lines of this system are parallel and the system has no solution.

2. $\begin{cases} 4x + 8y \;=\; 8 \\ 3x + 6y \;=\; 6 \end{cases}$

Step 1: The equations are in the proper form, so we move to Step 2.

Step 2

We create opposite coefficients for the variable x by multiplying both sides of the first equation by –3 and multiplying both sides of the second equation by 4.

Here's the first equation: Here's the second:

$$4x + 8y \;=\; 8 \qquad\qquad\qquad 3x + 6y \;=\; 6$$
$$-3 \cdot (4x + 8y) \;=\; -3 \cdot (8) \qquad\qquad 4 \cdot (3x + 6y) \;=\; 4 \cdot (6)$$
$$-12x - 24y \;=\; -24 \qquad\qquad 12x + 24y \;=\; 24$$

Step 3: Add equations.

$$
\begin{array}{rcr}
-12x - 24y &=& -24 \\
+\,12x + 24y &=& +\,24 \\
\hline
0 &=& 0
\end{array}
$$

This equation is always true — an identity. Therefore, the lines of this system are coincident, and the system has infinitely many solutions.

Practice Set C

Solve each of the following systems using the elimination method. Then check your solutions on the next page.

6. $\begin{cases} -x + 2y = 6 \\ -6x + 12y = 1 \end{cases}$

7. $\begin{cases} 4x - 28y = -4 \\ x - 7y = -1 \end{cases}$

Exercises 3.3

For each of the following problems, solve the system using the elimination method. If a system cannot be solved, indicate whether it is inconsistent or dependent.

1. $\begin{cases} x + y = 11 \\ x - y = -1 \end{cases}$

2. $\begin{cases} x + 3y = 13 \\ x - 3y = -11 \end{cases}$

3. $\begin{cases} 3x - 5y = -4 \\ -4x + 5y = 2 \end{cases}$

4. $\begin{cases} 2x - 7y = 1 \\ 5x + 7y = -22 \end{cases}$

5. $\begin{cases} -3x + 4y = -24 \\ 3x - 7y = 42 \end{cases}$

6. $\begin{cases} 8x + 5y = 3 \\ 9x - 5y = -71 \end{cases}$

7. $\begin{cases} -x + 2y = -6 \\ x + 3y = -4 \end{cases}$

8. $\begin{cases} 4x + y = 0 \\ 3x + y = 0 \end{cases}$

9. $\begin{cases} x + y = -4 \\ -x - y = 4 \end{cases}$

10. $\begin{cases} -2x - 3y = -6 \\ 2x + 3y = 6 \end{cases}$

11. $\begin{cases} 3x + 4y = 7 \\ x + 5y = 6 \end{cases}$

12. $\begin{cases} 4x - 2y = 27 \\ x + 4y = 26 \end{cases}$

13. $\begin{cases} 3x + y = -4 \\ 5x - 2y = -14 \end{cases}$

14. $\begin{cases} 5x - 3y = 20 \\ -x + 6y = -4 \end{cases}$

15. $\begin{cases} 6x + 2y = -18 \\ -x + 5y = 19 \end{cases}$

16. $\begin{cases} x - 11y = 17 \\ 2x - 22y = 4 \end{cases}$

17. $\begin{cases} -2x + 3y = 20 \\ -3x + 2y = 15 \end{cases}$

18. $\begin{cases} -5x + 2y = -4 \\ -3x - 5y = 10 \end{cases}$

19. $\begin{cases} -3x - 4y = 2 \\ -9x - 12y = 6 \end{cases}$

20. $\begin{cases} 3x - 5y = 28 \\ -4x - 2y = -20 \end{cases}$

21. $\begin{cases} 6x - 3y = 3 \\ 10x - 7y = 3 \end{cases}$

22. $\begin{cases} -4x + 12y = 0 \\ -8x + 16y = 0 \end{cases}$

23. $\begin{cases} 3x + y = -1 \\ 12x + 4y = 6 \end{cases}$

24. $\begin{cases} 8x + 5y = -23 \\ -3x - 3y = 12 \end{cases}$

25. $\begin{cases} 2x + 8y = 10 \\ 3x + 12y = 15 \end{cases}$

26. $\begin{cases} 4x + 6y = 8 \\ 6x + 8y = 12 \end{cases}$

27. $\begin{cases} 10x + 2y = 2 \\ -15x - 3y = 3 \end{cases}$

28. $\begin{cases} x + \frac{3}{4}y = -\frac{1}{2} \\ \frac{3}{5}x + y = -\frac{7}{5} \end{cases}$

30. $\begin{cases} 8x - 3y = 25 \\ 4x - 5y = -5 \end{cases}$

32. $\begin{cases} 12x + 16y = -36 \\ -10x + 12y = 30 \end{cases}$

29. $\begin{cases} x + \frac{1}{3}y = \frac{4}{3} \\ -x + \frac{1}{6}y = \frac{2}{3} \end{cases}$

31. $\begin{cases} -10x - 4y = 72 \\ 9x + 5y = 39 \end{cases}$

33. $\begin{cases} 25x - 32y = 14 \\ -50x + 64y = -28 \end{cases}$

For each of the following problems, set up and solve a system of equations to help you answer the question. You should use the elimination method to solve your system.

34. Stuart B. Snobb, an A-list movie star, purchased 3 Ferraris and 6 Lamborghinis and paid a total of $2,010,000. Morris T. Meathead, a famous football player, purchased 2 Ferraris and 3 Lamborghinis and paid a total of $1,110,000. If all of the Ferraris cost the same amount of money, and all of the Lamborghinis cost the same amount of money (but not necessarily the same amount as the Ferraris), then determine the cost of a Ferrari and the cost of a Lamborghini.

35. At a farmers' market, Frederick buys 5 pounds of cherries and 7 pounds of grapes for $41. At the same farmers' market, Wilhelmina buys 10 pounds of cherries and 2 pounds of grapes for $46. Determine the price per pound of cherries and grapes at the farmers' market.

36. At a livestock sale, Dolores bought four sheep and two cows for $1,660. At the same sale, Ramsay bought three sheep and five cows for $2,925. What was the price of each sheep and cow at the livestock sale?

37. During the last election year, J. D. Moneybags paid a total of $7,350 to attend 3 fundraising dinners with the governor and 5 fundraising dinners with the local congressperson. H. M. Rich, who lives in the same town as J. D. Moneybags, paid a total of $9,150 to attend 2 fundraising dinners with the governor and 9 fundraising dinners with the local congressperson. Determine the price of attending a fundraising dinner with the governor and with the local congressperson.

Practice B — Answer

5. System: $\begin{cases} 12b + 10a = 12 \\ 8b + 5a = 7 \end{cases}$

Answer: Bananas cost $0.50 per pound, and apples cost $0.60 per pound.

Practice C — Answers

6. no solution (inconsistent)

7. infinitely many solutions (dependent)

3.4 Using Systems to Solve Mixture Problems

Overview

In past sections, we have already looked at several problems involving value. Sometimes value is defined in terms of money. In other situations, we use different criteria to measure value — points scored in a basketball game, for example.

Here's a new kind of problem:

> Lupita needs 40 milliliters (ml) of a 14% acid solution for a chemistry experiment. She has two acid solutions that can be used to create this mixture. Solution A is 10% acid, and Solution B is 20% acid. How much of each solution should she use in her mixture?

In this problem, there are two unknown amounts, so we will need to write a system of two equations. We're given enough information to write *one* equation based on value. In this situation, "value" refers to the amount of acid present. That means that the other equation in our system will be based on quantity —the quantity of liquid in the mixture.

It's common in real life to encounter situations in which we know how the quantities of two things are related as well as how the values of those two things are related. We will refer to problems like these as *mixture problems*. This section will help you learn how to set up and solve a system of equations for a mixture problem.

A. Solving Mixture Problems

When we put two things together, we refer to this as a mixture. When a mathematical problem involves the quantity and value of two things that are put together, we refer to the problem as a *mixture problem*.

In the process of solving mixture problems, we write one equation that communicates the quantity of things that make up the mixture. This equation is typically fairly easy to write. The other equation, though, communicates value, and that equation can be more difficult to write. In order to construct the value equation properly, we need to consider the values of the things being mixed as well as the total value of the mixture.

For example, if we're mixing peanuts that are worth $2 per pound with cashews that are worth $10 per pound in order to create a blend that's worth $7 per pound, then one of the relationships we consider involves the value of the peanuts, the value of the cashews, and the total value of the mixture. When a rate is given in a mixture problem — in the mixed nuts example, the rate is *dollars per pound* — this is typically an indication of value. Problems dealing with percent concentration, such as the problem at the beginning of this section, also involve value.

We also need to consider units when writing the value equation. Every term in the value equation must use the *same units* to convey value. For example, when solving problems involving money, we can't use values conveyed in *cents* and *dollars* within the same equation. Every term in the equation must use *cents* to convey value, or, every term in the equation must use *dollars* to convey value.

We'll see how that works in the first example. In this, and all of the other examples, we'll use the five-step process outlined in section 3.2 to solve the problems.

Example 1

A parking meter contains 27 coins consisting of dimes and quarters only. If the meter contains $4.35, how many of each type of coin are there?

Solution

Step 1: Let D = number of dimes. Let Q = number of quarters.

Step 2: We are dealing with a mixture of dimes and quarters, and there are two relationships we need to be aware of.

First, we have the *quantity* relationship. We know there are a total of 27 coins. In other words, the sum of D and Q is 27. This gives us the equation

$D + Q = 27$

The second relationship is the *value* relationship. We know that the value of a dime is 10 cents and that the value of a quarter is 25 cents. We also know that the value of all the coins put together is $4.35. Unfortunately, this value is given to us in dollars, not cents. We either have to convert the value of each coin into dollars or convert the total value of the mixture into cents.

If we convert the value of each coin into dollars, we get

$0.10D + 0.25Q = 4.35$

If we convert the total value of the mixture into cents, we get

$10D + 25Q = 435$

We'll use the second equation in our system because it contains whole numbers.

Step 3: Now that we have a system of two equations, it's time to solve. We can do so with any of the methods we've learned in this chapter. Because both of the equations are in standard form, the elimination method works well.

$$\begin{cases} D + Q = 27 \\ 10D + 25Q = 435 \end{cases}$$

$$-10 \cdot (D + Q) = -10 \cdot (27) \quad \rightarrow \quad -10D - 10Q = -270 \quad \rightarrow \quad D + 11 = 27$$

$$-10D - 10Q = -270 \qquad\qquad \underline{+\ 10D + 25Q = +435} \qquad\qquad \underline{\quad -11 \quad\ \ -11\quad}$$

$$15Q = 165 \qquad\qquad\qquad D = 16$$

$$\frac{15Q}{15} = \frac{165}{15}$$

$$Q = 11$$

Step 4: We check the number values of D and Q in both equations.

$$D + Q = 27 \qquad\qquad 10D + 25Q = 435$$
$$16 + 11 = 27 \qquad\qquad 10(16) + 25(11) = 435$$
$$27 = 27 \checkmark \qquad\qquad 160 + 275 = 435$$
$$435 = 435 \checkmark$$

Step 5: Now that we have verified our solution, we translate it into a written statement:

There are 16 dimes and 11 quarters in the parking meter.

Now let's get back to our chemist friend as she gets ready for her experiment.

Example 2

Lupita needs 40 milliliters (ml) of a 14% acid solution for a chemistry experiment. She has two acid solutions that can be used to create this mixture. Solution A is 10% acid, and Solution B is 20% acid. How much of each solution should she use in her mixture?

Solution

Step 1: Let x = number of ml of Solution A, and let y = number of ml of Solution B.

Step 2: We need an equation for the *quantity* relationship, and another equation for the *value* relationship.

Quantity: There is a total of 40 ml of liquid in the mixture, so $x + y = 40$.

Value: In this problem, we're given acidity rates. That means that *value* refers to the amount of acid in each solution. Solution A is 10% acid, so the amount of acid present in Solution A can be written as $0.10x$. Solution B is 20% acid, so the amount of acid present in Solution B can be written as $0.20y$. The mixture is 14% acid, and we know that there will be 40 ml of liquid in this mixture. Therefore, the amount of acid present in the mixture can be written as $0.14(40)$. We can use these expressions to write our value equation:

$$0.10x + 0.20y = 0.14(40)$$

We simplify the values in this equation:

$$0.1x + 0.2y = 5.6$$

We can use this equation, or we can multiply both sides of the equation by 10 so that all of the coefficients are whole numbers:

$$10 \cdot (0.1x + 0.2y) = 10 \cdot (5.6)$$
$$x + 2y = 56$$

The equation with whole number coefficients makes a nicer looking and less complicated system to solve.

Step 3: Now we solve the system. Once again, the elimination method works well.

$$\begin{cases} x + y = 40 \\ x + 2y = 56 \end{cases}$$

$$-1 \cdot (x + y) = -1 \cdot (40) \qquad \rightarrow \qquad -x - y = -40 \qquad \rightarrow \qquad x + 16 = 40$$

$$-x - y = -40 \qquad\qquad\qquad \underline{+\quad x + 2y = +56} \qquad\qquad \underline{-16 \quad -16}$$

$$y = 16 \qquad\qquad\qquad x = 24$$

Step 4: Check the solution in both equations.

$$x + y = 40 \qquad\qquad\qquad x + 2y = 56$$

$$24 + 16 = 40 \qquad\qquad 24 + 2(16) = 56$$

$$40 = 40 \checkmark \qquad\qquad 24 + 32 = 56$$

$$56 = 56 \checkmark$$

Step 5: The solution is correct, so we translate it into a written statement:

Lupita should use 24 milliliters of Solution A and 16 milliliters of Solution B.

In the previous example, the total value of the mixture — that is, the total amount of acid present in the mixture — was not given in the problem. We had to calculate it by multiplying the acidity rate (0.14) times the amount of liquid (40). Here's another problem that will require us to calculate total value before writing our system.

Example 3

Patrick's Premium Coffee Shop sells Kona coffee for $22 per pound and Colombian coffee for $8 per pound. The shop offers a Kona-Colombian blend that sells for $12.20 per pound. How much Kona coffee and how much Colombian coffee should be used to make 20 pounds of the coffee blend?

Solution

Step 1: Let x = pounds of Kona coffee used and let y = pounds of Colombian coffee used.

Step 2: There are 20 pounds of coffee in the blend, so this is the *quantity* equation:

$$x + y = 20$$

The blend sells for $12.20 per pound, and there are 20 pounds, so the total value of the blend is $12.20(20) = $244. The value of the Kona coffee in the blend is represented by $22x$, and the value of the Colombian coffee in the blend is represented by $8y$. We can now write the *value* equation:

$$22x + 8y = 244$$

Step 3: Just for variety, we'll use substitution to solve this system, but the elimination method would also work well.

$$\begin{cases} x + y = 20 \\ 22x + 8y = 244 \end{cases}$$

$$x + y = 20 \quad \rightarrow \quad 22x + 8(20 - x) = 244 \quad \rightarrow \quad y = 20 - 6$$

$$\underline{-x \qquad\qquad -x} \qquad\qquad 22x + 160 - 8x = 244 \qquad\qquad y = 14$$

$$y = 20 - x \qquad\qquad 14x + 160 = 244$$

$$\underline{\qquad\qquad -160 \quad -160}$$

$$14x = 84$$

$$\frac{14x}{14} = \frac{84}{14}$$

$$x = 6$$

Step 4: Check the solution in both equations.

$$x + y = 20 \qquad\qquad\qquad 22x + 8y = 244$$

$$6 + 14 = 20 \qquad\qquad\qquad 22(6) + 8(14) = 244$$

$$20 = 20 \checkmark \qquad\qquad\qquad 132 + 112 = 244$$

$$244 = 244 \checkmark$$

▶ *Step 5:* The blend should contain 6 pounds of Kona coffee and 14 pounds of Colombian coffee.

Practice A

Now you get to solve a few mixture problems. Use the five-step process, as we did in the examples, and be sure to write one equation for quantity and another for value. Turn the page when you are ready to check your answers.

1. A bag contains only nickels and dimes. The value of the collection is $2. If there are 26 coins in all, how many of each coin are there?

2. A chemistry student needs 60 ml of a 26% salt solution. He has two salt solutions that can be used to create this mixture — Solution A is 30% salt and Solution B is 20% salt. How much of each solution should be used in the mixture?

3. Nancy's Nut Store sells peanuts for $2 per pound and cashews for $10 per pound. The store sells a peanut-cashew mixture for $7 per pound. How many pounds of peanuts and how many pounds of cashews should be used to create 40 pounds of the mixture?

Exercises 3.4

Set up and solve a system of equations to help you answer each question.

1. Galilahi buys a 14-pound mixture of grapes for $3.10. Green grapes sell for 25¢ a pound, and red grapes sell for 20¢ a pound. How many pounds of each type of grape are in the mixture?

2. Cleaning Chemicals, Inc., sells 80 liters of a blended cleaning solution for $28. To create the blended solution, the company uses Formula 207, which sells for 20¢ per liter and Formula 417, which sells for 40¢ per liter. How many liters of each solution are used to form the 80-liter mixture?

3. Claudia's Chemical Lab has found that the cost of 42 grams of a certain chemical compound is $14.40. The compound was created with Chemical A, which costs 45¢ per gram and Chemical B, which costs 30¢ per gram. How many grams of each chemical were used to form the compound?

4. A local production of *The Importance of Being Earnest* drew 342 people, some adults and some children, who paid a total of $1,033 for admission. If adult tickets were sold for $4 each and child tickets were sold for $1.50 each, how many adults and how many children attended the play?

5. Mid-Willamette Musical College sold 200 tickets for their spring program. Tickets for students cost $2.50 each, and tickets for nonstudents cost $3.50 each. If MWMC collected a total of $537 in ticket sales, how many students and how many nonstudents purchased tickets for the performance?

6. A parking meter contains 42 coins. The total value of the coins is $8.40. If the meter contains only dimes and quarters, how many of each type of coin are there?

7. When cleaning the coins out of Cheapskate Wishing Well, a worker discovers that there are only pennies and nickels. If the worker collects 153 coins, and the total value of the coins is $4.25, how many of the coins are pennies and how many are nickels?

8. Hoshi's piggy bank contains 78 coins. The coins are only nickels and dimes. If the value of the coins is $5.60, how many of each type of coin are there?

9. A magician tells members of the audience to put nickels and quarters into his top hat. After the coins are put in, the magician shakes his hat and announces that there are 44 coins worth $5.40 in the hat. Assuming that the magician is correct, how many of the coins in the hat are nickels and how many are quarters?

10. A coin machine at the U.S. Mint malfunctions, spewing dimes and quarters everywhere. The cleanup crew gathers 1,000 coins worth $185.20. How many of the coins are dimes and how many are quarters?

11. Paulina needs 22 ml of a 38% acid solution for a chemistry experiment. She has two acid solutions, A and B, that can be mixed together to form the solution. Solution A is 40% acid, and Solution B is 30% acid. How much of each solution should she use to create the mixture?

12. Tegrin needs 50 ml of a 72% salt solution. He has two salt solutions, A and B, that he can mix together to form the solution. Solution A is 60% salt, and Solution B is 80% salt. How much of each solution should he use to create the mixture?

13. Aaron needs 2 liters of an 18% acid solution for a science fair demonstration. He has two solutions, A and B, that can be mixed together to form the solution. Solution A is 10% acid and Solution B is 15% acid.
    a. Can Aaron form the needed 18% acid solution? (Verify this algebraically.)
    b. Now, suppose that Aaron is able to locate Solution C, a 20% acid solution. How much of Solutions A and C would he have to use in order to create 2 liters of the needed 18% solution? How much of Solutions B and C would have to be used to create 2 liters of the 18% solution?

14. Kaia needs 3 liters of a 12% acid solution for an experiment. She has two acid solutions, A and B, that she can mix together to form the solution. Solution A is 14% acid, and Solution B is 20% acid.
    a. Can Kaia form the needed 12% solution? (Verify this algebraically.)
    b. Now, suppose that Kaia locates Solution C, a 4% acid solution. How much of Solutions A and C should be used to create 3 liters of the needed mixture? How much of Solutions B and C should be used to create 3 liters of the needed mixture?

15. Victor needs 100 ml of a 16% acid solution. He has a bottle of 20% acid solution. How much pure water and how much of the 20% acid solution should be mixed to create 100 ml of a 16% acid solution?

16. Chanel needs 1 liter of a 78% salt solution. She has a bottle of 80% salt solution. How much pure water and how much of the 80% salt solution should be mixed to create 1 L of a 78% salt solution?

17. Brazil nuts are sold for $8 per pound, and walnuts are sold for $5 per pound. How much of each type of nut should be used to create 12 pounds of a mixture worth $7.25 per pound?

18. Jenalyn's Java House sells Kona coffee for $20 per pound, and a hazelnut house blend for $9 per pound. How much Kona coffee and how much hazelnut coffee should be used to create 75 pounds of a mixture that will sell for $12.08 per pound?

19. Frank's Farm Stand sells homemade fruit juice. Apple juice sells for $1.50 per quart, and grape juice sells for $3.50 per quart. The stand sells an apple-grape mixture for $2.20 per quart. How much apple juice and how much grape juice should be used to create 10 gallons of apple-grape mixture? (Hint: there are 4 quarts in 1 gallon.)

20. Deion, a video game programmer, is creating a role-playing game in which players can find treasure chests filled with gold or silver. Players receive 75 points for finding a treasure chest filled with gold and 40 points for finding a treasure chest filled with silver. Deion needs to put 150 treasure chests into the game, and the average treasure chest must be worth 54 points. How many treasure chests should be filled with gold, and how many should be filled with silver?

Practice A — Answers

1. System: $\begin{cases} N+D = 26 \\ 5N + 10D = 200 \end{cases}$ or $\begin{cases} N + D = 26 \\ 0.05N + 0.10D = 2 \end{cases}$

   Answer: The bag contains 12 nickels and 14 dimes.

2. System: $\begin{cases} x + y = 60 \\ 0.3x + 0.2y = 0.26(60) \end{cases}$

   Answer: The student should use 36 ml of Solution A and 24 ml of Solution B.

3. System: $\begin{cases} x + y = 40 \\ 2x + 10y = 7(40) \end{cases}$

   Answer: The mixture must contain 15 pounds of peanuts and 25 pounds of cashews.

CHAPTER 4

Polynomials

By now, you should feel comfortable adding and subtracting linear expressions, graphing linear equations, solving linear equations, and solving linear systems of equations. It's time to expand our skills so that we are able to deal with higher-degree expressions and equations.

Before we do that, however, we'll need to briefly introduce the concepts of polynomial and degree. We will also review concepts such as terms and coefficients. We'll examine how to add, subtract, and multiply polynomial expressions. We'll also learn rules for working with exponents. Finally, we will conclude the chapter by applying what we learn about exponents to situations involving conversion of units and scientific notation.

This chapter is divided into the following sections:

4.1 Addition and Subtraction of Polynomials

Overview

You may not realize it, but you've seen many polynomials already in this course. Now it's time to formally introduce you to the concepts of *polynomials*, *terms*, and *degree*. We'll then look at how the operations of addition and subtraction work with polynomials.

Here's a polynomial problem for you:

Subtract $8x^2 - 5x + 2$ from $3x^2 + x - 12$.

The variable x is raised to different powers in different terms. Two terms contain x^2, two terms contain x, and two terms don't contain any variable at all. The different powers of the variable must be taken into consideration when adding or subtracting. We'll discuss why this is true and then return to this problem. This section will show you how to:

◆ identify polynomials, terms, and degree

◆ identify and combine like terms

◆ simplify expressions containing parentheses

◆ add and subtract polynomials

A. Introduction to Polynomials

What makes an expression a *polynomial?* **Polynomials** are algebraic expressions that *do not* contain variables in the denominators of fractions and that *do* have whole numbers as the exponents on the variable quantities. The table below shows some examples of expressions that *are* polynomials.

Expression	Why It's a Polynomial
$3x^4$	The exponent on the variable is a whole number.
$\frac{2}{5}x^2y^5$	There is a fraction, but no variable appears in the denominator. The exponents on the variables are whole numbers.
$5x^3 + 3x^2 - 2x + 1$	The exponents on the variables are whole numbers.

The table below offers some expressions that are *not* polynomials:

Expression	Why It's Not a Polynomial
$\frac{3}{x} - 16$	A variable appears in the denominator of the fraction
$4x^2 - 5x + x^{-3}$	A negative exponent appears on the variable in the third term.
$7x^{2/3}$	The exponent on the variable is not a whole number.

The **terms** of a polynomial are the expressions being added together. When listing the terms of a polynomial, keep in mind that subtracting is the same as adding the opposite. For example, the expression $4x - 5$ is mathematically equivalent to the expression $4x + (-5)$. A polynomial consists of only one term if it does not involve addition or subtraction.

The **coefficient** refers to the constant factor of a term. For example, in the term $7x$, the coefficient is 7. If a variable doesn't appear to have a coefficient, then the coefficient is 1.

The table below presents some examples of terms and coefficients.

Polynomial	Terms	Coefficients
$7x - 16$	$7x$ and -16	7 and -16
$3x^2 + 5x + 11$	$3x^2, 5x$ and 11	$3, 5$ and 11
$9x^3 - x^2 + x - 8$	$9x^3, -x^2, x,$ and -8	$9, -1, 1$ and -8

Classifying a Polynomial by the Number of Terms

There are special names we use when classifying a polynomial by the number of terms. Let's take a look at those names.

Number of Terms	Name	Example	Comment
One	Monomial	$4x^2$	"Mono" means "one" in Greek.
Two	Binomial	$4x^2 - 7x$	"Bi" means "two" in Latin.
Three	Trinomial	$4x^2 - 7x + 3$	"Tri" means "three" in Greek.
Four or more	Polynomial	$4x^3 - 7x^2 + 3x - 1$	We don't have specific names for polynomials with more than three terms.

The Degree of a Polynomial

The **degree of a term** is the total number of variables being multiplied together in that term. When determining the total number of variables being multiplied together, it is helpful to consider what the term looks like when written in expanded form.

Let's look at some monomials — polynomials with just one term.

Monomial	Degree	Comment
$5x^3$	3	This monomial can be written as "$5 \cdot x \cdot x \cdot x$." Three x's are multiplied together, so this is a monomial of degree 3.
$60a^5$	5	By rewriting the expression as "$60 \cdot a \cdot a \cdot a \cdot a \cdot a$," we see that it is a monomial of degree 5.
13	0	This monomial doesn't contain any variables, so it is a monomial of degree 0.

In the table above, you may notice that the degree of the monomial for the first two monomials is the same as the exponent on the variable. This is always the case with terms that contain a single letter.

If a polynomial contains more than one term, we look at all of the terms to determine the degree of the polynomial. The degree of the term of highest degree is equal to the degree of the polynomial.

Polynomial	Degree	Comment
$2x^3 + 6x - 5$	3	The first term, $2x^3$, is a third-degree term. $6x$ is a first-degree term. -5 is a term of degree zero. The term of highest degree determines the degree of the whole polynomial, so this is a polynomial of degree 3.
$7y - 10y^4$	4	This is a fourth-degree polynomial because the term of highest degree, $-10y^4$, is a fourth-degree term.
$2x^5 - x^7 - 8x^3 + x$	7	The term of highest degree is $-x^7$, so this is a seventh-degree polynomial.

B. Combining Like Terms

Consider the polynomial $4x^2 + 3x^2 + x^2$. The variable parts of the terms, including the exponents on the variables, are identical. When this happens, we call the terms **like terms**. This is a good name for them, too, because terms like these with identical variable parts and different numerical coefficients represent different amounts of the same type of quantity.

When we have quantities of the same type, we can combine them using addition and subtraction. This should be done when simplifying algebraic expressions. We've already had some practice combining like terms with linear expressions — for example, $5x + 3x$ is rewritten as $8x$. Now we'll expand our skills to include terms within higher-degree expressions.

Example 1

Combine the like terms: 6 houses + 4 houses.

Solution

6 and 4 of the same type of thing give 10 of that type, so 6 houses + 4 houses = 10 houses.

Suppose we let the letter x represent "house." The expression 6 houses + 4 houses becomes $6x + 4x$. We can simplify this by combining the terms: $6x + 4x = 10x$.

☠ Warning! Incorrect Approach! ☠

It is important to note that 6 houses + 4 houses does *not* give us 10 motels. When we add or subtract, the type of quantity in our result must match the type of quantity in the terms we are combining! When simplifying the above expression, some people make the mistake of writing a result that has a different type of quantity:

$$6x + 4x = 10x^2$$

This is incorrect because x^2 is a *different type of quantity* than x.

Let's look at an example where we begin with two different types of quantity.

Example 2

Combine the like terms: 6 houses + 4 houses + 2 motels.

Solution

Houses and motels are not the same thing. We can combine the "house" terms, but we can't combine them with the "motel" term. 6 houses + 4 houses + 2 motels = 10 houses + 2 motels.

Suppose that we let x represent "house" and x^2 represent "motel." We know that x is not the same thing as x^2, just as a house is not the same thing as a motel. In this case, our equation is

$$6x + 4x + 2x^2 = 10x + 2x^2.$$

In looking through the previous examples, we've seen the following:

1. Like terms can and should be combined with addition and subtraction because they represent amounts of the same type of quantity.

2. When combining like terms, the resulting term will be the same type of quantity as the terms we are combining.

3. Variables with different exponents are *not* like terms.

Whenever we're asked to *simplify* an expression, we must make sure that we combine like terms. Let's take a look at a few more expressions.

Example 3

Simplify each expression.

1. $7x + 8y - 3x$
2. $4a^3 - 2a^2 + 8a^3 + a^2 - 2a^3$
3. $-13c^2d - 5cd^2 + 7c^2d - 18c^2d^2 + 11cd^2 + 9c^2d - c^2d^2$

Solutions

1. $7x + 8y - 3x$

   $7x + 8y - 3x$ The variable x represents a different type of quantity than the variable y, so we combine the terms containing x and leave the term containing y alone.

   $4x + 8y$ This is the simplified expression.

2. $4a^3 - 2a^2 + 8a^3 + a^2 - 2a^3$

$$4a^3 + 8a^3 - 2a^3 = 10a^3 \qquad \text{We start by combining the } a^3 \text{ terms.}$$

$$-2a^2 + a^2 = -a^2 \qquad \text{Then we combine the } a^2 \text{ terms.}$$

Using these results, we write the simplified expression, $10a^3 - a^2$. We can't combine any more because a^3 and a^2 represent different types of quantities.

3. $-13c^2d - 5cd^2 + 7c^2d - 18c^2d^2 + 11cd^2 + 9c^2d - c^2d^2$

This expression has three different types of quantities: c^2d, cd^2, and c^2d^2. This means that there are three different types of terms to combine.

$$-13c^2d + 7c^2d + 9c^2d = 3c^2d \qquad \text{We combine the } c^2d \text{ terms first.}$$

$$-5cd^2 + 11cd^2 = 6cd^2 \qquad \text{Next we combine the } cd^2 \text{ terms.}$$

$$-18c^2d^2 - c^2d^2 = -19c^2d^2 \qquad \text{Finally, we combine the } c^2d^2 \text{ terms.}$$

Using these results, we write the simplified expression:

$$-13c^2d - 5cd^2 + 7c^2d - 18c^2d^2 + 11cd^2 + 9c^2d - c^2d^2 = 3c^2d + 6cd^2 - 19c^2d^2$$

Practice B

Now it's your turn to practice combining like terms. Simplify each of the following expressions. When you're finished, turn the page to check your work.

1. $3x + 6x + 11x$

2. $5a + 2b + 4a - b - 7b$

3. $10x^3 - 4x^3 + 3x^2 - 12x^3 + 5x^2 + 2x + x^3 + 8x$

4. $2a^5 - a^5 + 1 - 4ab - 9 + 9ab - 2 - 3 - a^5$

5. $11mn^2 - 3m^2n^2 - 9m^2n + 5m^2n^2 - 7m^2n - mn^2$

C. Simplifying Expressions Containing Parentheses

When parentheses occur in an expression, they should be removed before the like terms are combined. To remove the parentheses, we use the distributive property:

$$a(b + c) = ab + bc \qquad \text{and} \qquad a(b-c) = ab-ac$$

In the example below, we'll use the distributive property to remove the parentheses and then simplify the expression by combining like terms.

Example 4

Simplify the following expressions.

1. $9a - 5(a + 3)$ **2.** $4x + 9(x^2 - 6x - 2) + 5$

Solutions

1. $9a - 5(a + 3)$

$9a - 5(a + 3)$ First we apply the distributive property. Because there's a minus sign before the 5, we multiply −5 times a and −5 times 3.

$9a - 5a - 15$ Now we combine the like terms, $9a$ and $-5a$.

$4a - 15$

2. $4x + 9(x^2 - 6x - 2) + 5$

$4x + 9(x^2 - 6x - 2) + 5$ Remove the parentheses with the distributive property.

$4x + 9x^2 - 54x - 18 + 5$ Now combine the like terms with addition and subtraction.

$-50x + 9x^2 - 13$

It's common practice to write the terms in a polynomial so that variable powers are descending from left to right. The term of highest degree comes first, the term of second-highest degree comes second, and so on. So it's best to write our final answer like this:

$9x^2 - 50x - 13$

The second solution in Example 4 illustrates the standard form of a polynomial. A polynomial in one variable is written in **standard form** if the degrees of the terms are descending from left to right.

Example 5

Simplify $2 + 2[5 + 4(1 + a)]$.

Solution

$2 + 2[5 + 4(1 + a)]$	Eliminate the innermost set of parentheses first by using the distributive property.
$2 + 2[5 + 4 + 4a]$	Then, simplify the expression inside the brackets.
$2 + 2[9 + 4a]$	Next, remove this set of brackets with the distributive property.
$2 + 18 + 8a$	Combine like terms.
$20 + 8a$	Finally, rewrite the polynomial in standard form.
$8a + 20$	

Practice C

Using the examples as a guide, simplify each expression completely. Write your answers in standard form. When you're finished, you can turn the page and check your answers.

6. $4(x + 6) + 3(2 + x + 3x^2) - 2x^2$

7. $7(x + x^3) - 4x^3 - x + 1 + 4(x^2 - 2x^3 + 7)$

8. $5(a + 2) + 6a - 7 + (8 + 4)(a + 3a + 2)$

9. $2[8 - 3(x - 3)]$

10. $x^2 + 3x + 7[x + 4x^2 + 3(x + x^2)]$

D. Adding and Subtracting Polynomials

When adding and subtracting polynomials, we regularly encounter multiplications in this form:

$+1(a + b)$ or $-1(a + b)$

However, these expressions don't appear quite like that. Instead, they appear as:

$+ (a + b)$ or $- (a + b)$

Using the distributive property, we can remove the parentheses in this way:

$$+(a + b) = +1(a + b)$$
$$= (+1)(a) + (+1)(b)$$
$$= a + b$$

The parentheses have been removed, and the sign of each term remains the same. This means that any time we see a plus sign (+) immediately in front of parentheses, we may rewrite the expression with identical terms and without using parentheses.

In Chapter 1, we saw how the distributive property allows us to see the effect of putting a minus sign in front of an expression in parentheses. Let's look at that again.

$$-(a + b) = -1(a + b)$$
$$= (-1)(a) + (-1)(b)$$
$$= -a - b$$

The parentheses have been removed, and now the sign of each term within the parentheses changes to its opposite. This means that any time we see a minus sign (–) immediately in front of parentheses, we may rewrite the expression using terms of the *opposite* sign and without using parentheses.

Let's look at some more examples of applying a plus sign or a minus sign to parentheses.

Example 6

Simplify each expression

1. $+(14 a^2 b^3 - 6 a^3 b^2 + a b^4)$ 			2. $-(21 a^2 + 7a - 18)$

Solution

1. $+(14 a^2 b^3 - 6 a^3 b^2 + a b^4)$

This set of parentheses is preceded by a plus sign. All we do is remove the parentheses. The terms inside are unchanged. Our simplified expression is:

$$14 a^2 b^3 - 6 a^3 b^2 + a b^4$$

2. $-(21 a^2 + 7a - 18)$

This set of parentheses is preceded by a minus sign. We drop the parentheses, but we have to change the sign of *each* term inside the parentheses. The simplified expression is:

$$-21 a^2 - 7a + 18$$

If a set of parentheses does not have any sign in front of it, we treat it the same way we would treat a set of parentheses with a plus sign in front. We remove the parentheses and keep the sign of each term inside the same. For example, $(5x + 7) = 5x + 7$.

Now let's use these new skills to add and subtract polynomials.

Practice B — Answers

1. $20x$ 			3. $-5 x^3 + 8 x^2 + 10x$ 			5. $10 m n^2 + 2 m^2 n^2 - 16 m^2 n$

2. $9a - 6b$ 			4. $5ab - 13$

Example 7

Simplify each expression.

1. $(3x + 7) + (x - 3)$ **2.** $(5y^3 + 11) - (12y^3 - 2)$

Solution

1. $(3x + 7) + (x - 3)$

$(3x + 7) + (x - 3)$ Begin by removing the parentheses.

$3x + 7 + x - 3$ The signs of the terms should not change.

$4x + 4$ Now combine like terms to arrive at the simplified expression.

2. $(5y^3 + 11) - (12y^3 - 2)$

$(5y^3 + 11) - (12y^3 - 2)$ Start by removing the parentheses.

$5y^3 + 11 - 12y^3 + 2$ Change the sign of each term in the second set of parentheses.

$-7y^3 + 13$ Finish by combining like terms.

When an addition or subtraction problem involving polynomials is stated in words, we typically enclose the polynomials in parentheses when writing down the problem. The next two examples illustrate this.

Example 8

Add $4x^2 + 2x - 8$ to $3x^2 - 7x - 10$.

Solution

The phrase "add A to B" implies that we begin with B and then add A. In this case, we begin our expression with $3x^2 - 7x - 10$.

$(3x^2 - 7x - 10) + (4x^2 + 2x - 8)$ Both of the polynomials are written with parentheses.

$3x^2 + 7x - 10 + 4x^2 + 5x - 2$ Remove the parentheses. Leave the sign of each term the same.

$7x^2 - 5x - 18$ Combine the like terms.

Practice C — Answers

6. $7x^2 + 7x + 30$ **8.** $59a + 27$ **10.** $50x^2 + 31x$

7. $-5x^3 + 4x^2 + 6x + 29$ **9.** $-6x + 34$

In Example 15, it really doesn't matter which polynomial we begin the expression with because we are adding. As you know, the commutative property of addition tells us that the order of things being added doesn't matter. $(3x^2 - 7x - 10) + (4x^2 + 2x - 8)$ gives the same result as $(4x^2 + 2x - 8) + (3x^2 - 7x - 10)$.

Now it's time to revisit the problem from the beginning of this section. Because the problem is instructing us to subtract polynomials, the order in which we write those polynomials definitely matters.

Example 9

Subtract $8x^2 - 5x + 2$ from $3x^2 + x - 12$.
The phrase "Subtract A from B" implies that we begin with B and then subtract A. We must begin our expression with $3x^2 + x - 12$.

$(3x^2 + x - 12) - (8x^2 - 5x + 2)$ With the first set of parentheses, don't change the signs of the terms. With the second set of parentheses, write each term

$3x^2 + x - 12 - 8x^2 + 5x - 2$ with opposite signs.

$-5x^2 + 6x - 14$ Finish by combining like terms.

☠ Warning! Incorrect Approach! ☠

There are a couple of common mistakes associated with the problem in Example 9.

One common mistake occurs when people attempt to write the initial algebraic expression:

Subtract $8x^2 - 5x + 2$ from $3x^2 + x - 12$ ➡ $(8x^2 - 5x + 2) - (3x^2 + x - 12)$

This is not correct! As we saw in Chapter 1, whenever a sentence states "Subtract A from B," we must translate it as $B - A$.

The second common mistake occurs when people remove the parentheses:

$(3x^2 + x - 12) - (8x^2 - 5x + 2)$ ➡ $3x^2 + x - 12 - 8x^2 - 5x + 2$

This is incorrect because the initial algebraic expression indicates that the minus sign in front of the second set of parentheses must be applied to *every* term inside of those parentheses, not just the first term. The last three terms above *should* be "$-8x^2 + 5x - 2$"

Practice D

It's now your turn to add and subtract polynomials. Simplify each expression. When you're finished, you can check your answers on the next page.

11. $(6y^2 + 2y - 1) + (5y^2 - 18)$

12. $(9m - n) - (10m + 12n)$

13. Add $2r^2 + 4r - 1$ to $3r^2 - r - 7$.

14. Subtract $4s - 3$ from w $7s + 8$.

Exercises 4.1

For the following problems, simplify each of the algebraic expressions. If only one letter is used in the expression, then your final answer should be written in standard form.

1. $x + 3x$

2. $9a + 12a$

3. $5m - 3m$

4. $3a + 5a + 2a$

5. $7y - 10y + 2y$

6. $5mt + 7mt - 8mt$

7. $hp - 3hp - 5hp$

8. $3a^2 + 6a^2 + 2a^2$

9. $14a^2b + 4a^2b + 19a^2b$

10. $-7n + 41 + 16n$

11. $16x + 47x - 33$

12. $47 - 19y - y - 21$

13. $13x - 14 - 12x + 3$

14. $7x^3 - 2x^2 - 10x + 1 - 5x^2 - 3x^3 - 12 + x$

15. $21y - 15x + 40xy - 6 - 11y + 7 - 12x - xy$

16. $x + y - x - y + x - y$

17. $5x^2 - 3x - 7 + 2x^2 - x$

18. $-2z^3 + 15z + 4z^3 + z^2 - 6z^2 + z$

19. $18x^2y - 14x^2y - 20x^2y$

20. $-9w^5 - 9w^4 - 9w^5 + 10w^4$

21. $2x^4 + 4x^3 - 8x^2 + 12x - 1 - 7x^3 - x^4 - 6x + 2$

22. $17d^3r + 3d^3r - 5d^3r + 6d^2r + d^3r - 30d^2r$

23. $2a^3b^2c + 3a^2b^2 + 4a^2b^2 - a^3b^2c$

24. $210ab^4 + 412ab^4 + 100a^4b$

25. $5x^2y^3 + 3x^3y^2 + 2x^2y^3 + 1$

26. $3(x + 5) + 2x$

27. $7(a + 2) + 4$

28. $y + 5(y + 6)$

29. $2b + 6(3 - 5b)$

30. $5a - 7c + 3(a - c)$

31. $8x - 3x + 4(2x + 5) + 3(6x - 4)$

32. $2z + 4ab + 5z - ab + 12(1 - ab - z)$

33. $(a + 5)4 + 6a - 20$

34. $(4a + 5b - 2)3 + 3(4a + 5b - 2)$

35. $(8x + 3y^2)4 + 4(10x + 5y^2)$

36. $3(2a + 2a^2) + 8(3a + 3a^2)$

37. $5[4(r - 2s) - 3r - 5s] + 12s$

38. $5[4(6x - 3) + x] - 2x - 25x + 4$

39. $-8(3a + 2)$

40. $-4(2x - 3y)$

41. $7n - 3(9n - 11)$

42. $12 - 8c - 6(7 + 5c)$

43. $(3x^2 + 5x - 2) + (4x^2 - 10x - 5)$

44. $(-2x^3 + 4x^2 + 5x - 8) - (x^3 - 3x^2 - 11x + 1)$

45. $(-5x - 12xy + 4y^2) - (-7x + 7xy - 2y^2)$

46. $(6a^2 - 3a + 7) - (4a^2 + 2a - 8)$

47. $(5x^2 - 24x - 15) + (x^2 - 9x + 14)$

48. $(3x^3 - 7x^2 + 2) + (x^3 + 6)$

49. $(9a^2b - 3ab + 12ab^2) + (ab^2 + 2ab)$

50. $(6x^2 - 12x) - (4x^2 - 3x - 1) + (4x^2 - 10x - 4)$

51. $(5a^3 - 2a - 26) + (4a^3 - 11a^2 + 2a) - (7a + 8a^3 + 20)$

52. $(2xy - 15) - (5xy + 4)$

53. Add $4x + 6$ to $8x - 15$.

54. Add $5y^2 - 5y + 1$ to $-9y^2 + 4y - 2$.

55. Add $3(x + 6)$ to $4(x - 7)$.

56. Add $-2(x^2 - 4)$ to $5(x^2 + 3x - 1)$.

57. Add four times $5x + 2$ to three times $2x - 1$.

58. Add five times $-3x + 2$ to seven times $4x + 3$.

59. Add -4 times $9x + 6$ to -2 times $-8x - 3$.

60. Subtract $6x^2 - 10x + 4$ from $3x^2 - 2x + 5$.

61. Subtract $11x^2 + 14x - 9$ from $5x^2 - 20x - 13$

62. Subtract $a^2 - 16$ from $a^2 - 16$

4.2 Multiplication of Polynomials

Overview

When adding and subtracting, we may *only* combine like terms. When multiplying, on the other hand, we may combine any two terms — whether they're like terms or not. Here's a problem:

Multiply: $(x^2 + 4)(x^2 + 7x + 2)$.

Both polynomials in this problem contain multiple terms. Before we tackle this problem, we'll need to cover some of the basics of polynomial multiplication, starting with multiplication by a monomial.

By studying this section, you will learn how to:

- ◆ Multiply a polynomial by a monomial
- ◆ Multiply a polynomial by a polynomial

A. Multiplying a Polynomial by a Monomial

Multiplying a polynomial with multiple terms by a monomial is a direct application of the distributive property. We know how this works when the monomial is a constant; we multiply every term of the polynomial by the monomial.

Original Expression	Applying the Distributive Property	Simplified Expression
$3(x + 9)$	$(3)(x) + (3)(9)$	$3x + 27$
$-2(x^3 - 3x)$	$(-2)(x^3) - (-2)(3x)$	$-2x^3 + 6x$

Let's start by analyzing a multiplication problem involving powers of x.

Example 1

Simplify: $x^2 \cdot x^4$

Solution

$$
\begin{aligned}
x^2 \cdot x^4 &= xx \cdot xxxx && \text{2 factors of } x \text{ and 4 factors of } x \\
&= xxxxxx && \text{6 factors of } x \text{ (expanded form)} \\
&= x^6 && \text{6 factors of } x \text{ (exponential form)}
\end{aligned}
$$

Factors

When multiplying, the result is known as the product. The things being multiplied are known as factors. For example, in the equation $5 \cdot 2 = 10$, 10 is the product. 5 and 2 are the factors.

Example 1 suggests the following rule:

If x is a real number and n and m are natural numbers, then $x^n x^m = x^{n+m}$.

This is called the **product rule for exponents**. To put it in simpler words, this rule says that when we multiply two exponential quantities that have the same base — in this case, the exponential quantities both have a base of x — we add the exponents in order to find the exponent on the answer. Note that the answer has the same base as the exponential quantities being multiplied.

Let's look at another example.

Example 2

Simplify: $x^3 \cdot x^5$

Solution

$$x^3 \cdot x^5 = x^{3+5}$$
$$= x^8$$

The exponential quantities being multiplied both have a base of x, so the product has a base of x. The exponents on the quantities being multiplied, 3 and 5, are added together to find the exponent for the product.

Now we'll look at some examples of multiplying a polynomial with multiple terms by a monomial that contains a variable.

Example 3

Multiply the following:

1. $x(x - 7)$

2. $2x^3(x^4 - x^3 - x)$

Solutions

1. $x(x - 7)$

$$x(x - 7) = x \cdot x + x(-7)$$
$$= x^{1+1} - 7x$$
$$= x^2 - 7x$$

2. $2x^3(x^4 - x^3 - x)$

$$2x^3(x^4 - x^3 - x) = 2x^3 \cdot x^4 - 2x^3 \cdot (-x^3) - 2x^3 \cdot (-x)$$
$$= 2x^{3+4} - 2x^{3+3} - 2x^{3+1}$$
$$= 2x^7 - 2x^6 - 2x^4$$

When we multiply terms containing variables, we can apply the commutative and associative properties of multiplication to re-order or regroup the numerical coefficients and the variables when we simplify the product. In the next example, you'll see how this works.

▶ **Example 4**

Multiply the following:

   1. $2x^3 \cdot 7x^5$ **2.** $3x^3y^4z \cdot 5x^2yz^8$

Solutions

1. $2x^3 \cdot 7x^5$

$$2x^3 \cdot 7x^5 = 2 \cdot x^3 \cdot 7 \cdot x^5$$ The commutative property allows us to switch the order of x^3 and 7.

$$= 2 \cdot 7 \cdot x^3 \cdot x^5$$ The associative property allows us to group the 2 and the 7 together and group the x^3 and x^5 together.

$$= (2 \cdot 7) \cdot (x^3 \cdot x^5)$$ Now we finish simplifying the expression.

$$= 14x^8$$ In this example, we skip the step of writing the variable expression x^{3+5}. In practice, this step is typically done with mental math.

2. $3x^3y^4z \cdot 5x^2yz^8$

Once again, we begin by reordering and regrouping the numbers and letters in the expression.

$$3x^3y^4z \cdot 5x^2yz^8 = 3 \cdot 5 \cdot x^3 \cdot x^2 \cdot y^4 \cdot y \cdot z \cdot z^8$$ The coefficient of the product is found by multiplying 3 and 5.

$$= 15 \cdot x^3 \cdot x^2 \cdot y^4 \cdot y \cdot z \cdot z^8$$ The variable parts of the product are found by applying the product rule for exponents

$$= 15x^5y^5z^9$$

In future problems, we'll shorten this process of multiplying monomials to two steps:

1. Multiply the numerical coefficients of the terms to determine the numerical coefficient of the answer.

2. Use the product rule for exponents to combine any variables present in the terms.

This shortened process can often be done mentally, and it will help us move more quickly as we multiply polynomials by monomials.

Example 5

Multiply: $8\,a^2(3\,a^4 - 5\,a^3 + a)$

Solution

The thought process for completing the two-step process should be as follows:

Step 1: Coefficients: $8 \cdot 3 = 24$, $8 \cdot (-5) = -40$, and $8 \cdot 1 = 8$

Step 2: Variables: $a^2 \cdot a^4 = a^6$, $a^2 \cdot a^3 = a^5$, and $a^2 \cdot a = a^3$

Working through the two steps mentally allows us arrive at the final answer in just one line of writing:

$$8\,a^2(3\,a^4 - 5\,a^3 + a) = 24\,a^6 - 40\,a^5 + 8\,a^3$$

When multiple letters such as x, y and z are involved, the second step is a little more complicated. However, if you are comfortable with what we did in the previous example, it shouldn't be difficult.

Example 6

Multiply: $-4x^2y^7z(x^6yz + 8x^3y^2z^2)$

Solution

The thought process for completing the two-step process should be as follows:

Step 1: Coefficients: $-4 \cdot 1 = -4$ and $-4 \cdot 8 = -32$

Step 2: Variables: $x^2y^7z \cdot x^6yz = x^8y^8z^2$ and $x^2y^7z \cdot x^3y^2z^2 = x^5y^9z^3$

$$-4x^2y^7z(x^6yz + 8x^3y^2z^2) = -4x^8y^8z^2 - 32x^5y^9z^3$$

Once again, by working through the two-step thought process, we arrive at the final answer in just one line of writing

If Step 2 is too complicated to do mentally, some people find it helpful to work through the problem in two lines of writing 1) multiplying the numerical coefficients and grouping the variables in each term and then 2) simplifying the variables.

$$\begin{aligned} -4x^2y^7z(x^6yz + 8x^3y^2z^2) &= -4(x^2x^6)(y^7y)(zz) - 32(x^2x^3)(y^7y^2)(zz^2) \\ &= -4x^8y^8z^2 - 32x^5y^9z^3 \end{aligned}$$

We've seen that if a variable is present in only one of two terms being multiplied, the variable doesn't change when writing the answer. For example, in the equation $3 \cdot 2x^4 = 6x^4$, the first factor, 3, doesn't contain a variable while the second factor, $2x^4$, does. The variable part of the product is thus identical to the variable part of the second factor. Both contain x^4. That same principle applies in the next example.

Example 7

Multiply: $(9x^2z + 4w)(5zw^3)$

Solution

If it helps to rewrite the problem so that the monomial comes first, then feel free to do so:

$$(5zw^3)(9x^2z + 4w)$$

Notice that in the first multiplication, x and w only appear in one of the terms. Therefore, their appearances don't change when we write the product: $9x^2z \cdot 5zw^3 = 45x^2z^2w^3$.

Similarly, in the second multiplication, the variable z only appears in one of the terms. This means that its appearance does not change when writing the product: $4w \cdot 5zw^3 = 20zw^4$.

With this in mind, we complete the problem.

$$\begin{aligned}(9x^2z + 4w)(5zw^3) &= 45x^2z^2w^3 + 20zw^4 \quad \text{This is mathematically correct.} \\ &= 45w^3x^2z^2 + 20w^4z \quad \text{However, it's customary to write the variables} \\ &\qquad\qquad\qquad\qquad\qquad\quad \text{of a term in alphabetical order.}\end{aligned}$$

Practice A

Now it's your turn. Use the distributive property to rewrite these expressions without parentheses. Make sure your answers are simplified as much as possible. When you are finished, check your work on the next page.

1. $3(x + 8)$

2. $(2 + a)4$

3. $(a^2 - 2b + 6)2a$

4. $8a^2b^3(2a + 7b + 3)$

5. $4x(2x^5 + 6x^4 - 8x^3 - x^2 + 9x - 11)$

6. $(3a^2b)(2ab^2 + 4b^3)$

7. $5mn(m^2n^2 + m + n^3)$

8. $6.03(2.11a^3 + 8.00a^2b)$

B. Multiplying a Polynomial by a Polynomial

When multiplying two polynomials that each have multiple terms, we again use the distributive property. We do this by considering one of the expressions enclosed in parentheses as a single entity. Let's consider a somewhat unusual example of applying the distributive property:

$$☺(c + d) = ☺c + ☺d$$

In this case, we multiply the smiley face by c, and also multiply the smiley face by d. Now let's replace the smiley face with a binomial in parentheses:

$$(a + b)(c + d)$$

We're going to treat $(a + b)$ as a single entity when applying the distributive property.

$$(a + b)(c + d) = (a + b)c + (a + b)d$$ Multiply $(a + b)$ by c and multiply $(a + b)$ by d.

$$= ac + bc + ad + bd$$ Use the distributive property to remove parentheses from the expression.

$$= ac + ad + bc + bd$$ The previous expression was mathematically correct, but it's customary to use alphabetical order when writing terms. In this expression, the terms beginning with a are written before the terms beginning with b.

In this example, it doesn't matter which of the original binomials we consider as the single entity. We could have just as easily treated $(c + d)$ as the single entity. If we had, the first step would have yielded the expression $a(c + d) + b(c + d)$, and the final result would have turned out the same.

The FOIL Method

When multiplying two binomials, we can shorten the process somewhat by using the **FOIL method**, which takes its name from this acronym:

First terms   **O**uter terms   **I**nner terms   **L**ast terms

Let's consider the expression $(a + b)(c + d)$ again. Here's how the FOIL method works:

F	The **First** terms are a and c because they come first within each set of parentheses, so we begin by multiplying $a \cdot c$.
O	The **Outer** terms are a and d because when disregarding parentheses and looking at the order in which the terms appear from left to right, a is the first term and d is the last term. In other words, a and d occupy the outer positions. We next multiply $a \cdot d$.
I	The **Inner** terms are b and c because in the ordering of terms, b and c occupy the inner positions. We next multiply $b \cdot c$.
L	The **Last** terms are b and d because they come last within each set of parentheses, so we end by multiplying $b \cdot d$.

Using the FOIL method, we have:

$$(a + b)(c + d) = ac \qquad + ad \qquad + bc \qquad + bd$$
$$\text{First} \quad + \quad \text{Outer} \quad + \quad \text{Inner} \quad + \quad \text{Last}$$

This is the same result we got when we multiplied these binomials the first time. The FOIL method shows us that in general, when we multiply two polynomials together, we must multiply every term of one polynomial by every term of the other polynomial. After we finish multiplying, we then combine any like terms.

Let's take a look at how this works with a few examples. In each case, we'll use the FOIL method to multiply the polynomials and then combine like terms to simplify the expression.

Example 8

Multiply the following expressions:

$(a + 6)(a + 3)$

$(a - 4)(a - 3)$

$(x + y)(2x + 4y)$

$(n - 7)(n + 7)$

Solutions

1. $(a + 6)(a + 3)$

$$
\begin{aligned}
(a + 6)(a + 3) &= a \cdot a \quad + a \cdot 3 \quad + 6 \cdot a \quad + 6 \cdot 3 \\
&\quad\ \ \text{F} \quad\quad + \text{O} \quad\quad + \text{I} \quad\quad + \text{L} \\
&= a^2 \quad\ + 3a \quad\ + 6a \quad\ + 18 \\
&= a^2 + 9a + 18
\end{aligned}
$$

2. $(a - 4)(a - 3)$

$$
\begin{aligned}
(a - 4)(a - 3) &= a(a) \quad + a(-3) \quad + (-4)(a) \quad + (-4)(-3) \\
&= a^2 \quad\ - 3a \quad\quad - 4a \quad\quad + 12 \\
&= a^2 + 7a + 12
\end{aligned}
$$

3. $(x + y)(2x + 4y)$

$$
\begin{aligned}
(x + y)(2x + 4y) &= x \cdot 2x \quad + x \cdot 4y \quad + y \cdot 2x \quad + y \cdot 4y \\
&= 2x^2 \quad\ + 4xy \quad\ + 2xy \quad\ + 4y^2 \\
&= 2x^2 + 6xy + 4y^2
\end{aligned}
$$

4. $(n - 7)(n + 7)$

$$
\begin{aligned}
(n - 7)(n + 7) &= n(n) \quad + n(7) \quad + (-7)(n) \quad + (-7)(-7) \\
&= n^2 \quad + 7n \quad\ - 7n \quad\quad - 49 \\
&= n^2 - 42
\end{aligned}
$$

Practice A — Answers

1. $3x + 24$
2. $4a + 8$
3. $2a^3 - 4ab + 12a$
4. $16a^3b^3 + 56a^2b^4 + 24a^2b^3$
5. $8x^6 + 24x^5 - 32x^4 - 4x^3 + 36x^2 - 44x$
6. $6a^3b^3 + 12a^2b^4$
7. $5m^3n^3 + 5m^2n + 5mn^4$
8. $12.7233a^3 + 48.24a^2b$

Number 4 of the previous example features binomials that are called conjugates. Conjugates are binomials that have identical first terms and opposite last terms, such as 7 and –7. When simplifying the result of conjugates, the "outer" and "inner" products cancel each other out, so our final expression only has two terms.

Some people prefer to use a vertical layout when multiplying polynomials. This alternative approach allows us to use a process that is similar to what we do when multiplying large numbers by hand. Let's see how it works with the following example.

Example 9

Multiply: $(3x + 7)(4x - 5)$

Solution

Step 1: Write the problem using a vertical layout.

Step 2: Multiply the expression $3x + 7$ by the term -5.

Step 3: Place a zero as a place holder. Then multiply $3x + 7$ by $4x$.

Step 4: Finish by combining like terms to form the final answer.

$$
\begin{array}{r}
3x + 7 \\
\times \quad 4x - 5 \\
\hline
-15x - 35 \\
+ \quad 12x^2 + 28x + 0 \\
\hline
12x^2 + 13x - 35
\end{array}
$$

We can use the FOIL method to verify our answer:

$$
\begin{aligned}
(3x + 7)(4x - 5) &= 3x(4x) \quad + \quad 3x(-5) \quad + \quad 7(4x) \quad + \quad 7(-5) \\
&= 12x^2 \quad\quad - 15x \quad\quad + 28x \quad\quad - 35 \\
&= 12x^2 + 13x - 35
\end{aligned}
$$

Next, we'll look at an expression that is often simplified incorrectly.

Example 10

Multiply: $(m - 3)^2$

Solution

In order to correctly square an expression, we must multiply the expression by itself:

$$(m - 3)^2 = (m - 3)(m - 3)$$

$$
\begin{aligned}
(m - 3)(m - 3) &= m(m) \quad\quad + \quad m(-3) \quad + \quad (-3)(m) \quad + \quad (-3)(-3) \\
&= m^2 \quad\quad\quad - 3m \quad\quad - 3m \quad\quad\quad + 9 \\
&= m^2 - 6m + 9
\end{aligned}
$$

> ### ☠ Warning! Incorrect Approach! ☠
>
> With the expression from the previous example, $(m - 3)^2$, many people incorrectly assume that they can just square the first and last terms and then write the simplified expression:
>
> $$(m - 3)^2 = (m)^2 + (-3)^2$$
> $$= m^2 + 9$$
>
> This is not correct! As we saw in Example 10, there is more to the problem than this. Never attempt to square a binomial by simply squaring the first and last terms.

When presented with a problem like Example 10, you can avoid the common error described above by learning this simple truth:

When you square a binomial, the answer is a *trinomial*.

Now it's time to multiply a binomial times a trinomial. We revisit the problem from the beginning of this section.

Example 11

Multiply: $(x^2 + 4)(x^2 + 7x + 2)$

Solution

When multiplying a binomial and a trinomial, we're required to find six products before we combine like terms, so the FOIL acronym doesn't really apply here. We will treat one expression, $(x^2 + 7x + 2)$, as a single entity and then apply the distributive property.

$$(x^2 + 4)(x^2 + 7x + 2) = x^2(x^2 + 7x + 2) + 4(x^2 + 7x + 2)$$
$$= x^2 \cdot x^2 + x^2 \cdot 7x + x^2 \cdot 2 + 4 \cdot x^2 + 4 \cdot 7x + 4 \cdot 2$$
$$= x^4 + 7x^3 + 2x^2 + 4x^2 + 28x + 8$$
$$= x^4 + 7x^3 + 6x^2 + 28x + 8$$

The choice of which expression to treat as the single entity is arbitrary. We could start by treating $(x^2 + 4)$ as the single entity. The problem would then looks like this:

$$(x^2 + 4)(x^2 + 7x + 2) = (x^2 + 4)x^2 + (x^2 + 4)7x + (x^2 + 4)2$$
$$= x^2 \cdot x^2 + 4 \cdot x^2 + x^2 \cdot 7x + 4 \cdot 7x + x^2 \cdot 2 + 4 \cdot 2$$
$$= x^4 + 4x^2 + 7x^3 + 28x + 2x^2 + 8$$
$$= x^4 + 7x^3 + 6x^2 + 28x + 8$$

The simplified expression is the same.

The vertical layout can also be used to work through this problem:

$$
\begin{array}{r}
x^2 + 7x + 2 \\
\times \qquad x^2 + 4 \\
\hline
4x^2 + 28x + 8 \\
+ \quad x^4 + 7x^3 + 2x^2 + 0x + 0 \\
\hline
x^4 + 7x^3 + 6x^2 + 28x + 8
\end{array}
$$

Example 12

Multiply: $(x + 5)^3$

Solution

When cubing a binomial, we must avoid the temptation to cube the first and last terms and write a simplified expression consisting of just two terms. We need to take the same approach we took in Example 10. In this case, to cube an expression, we write the expression as a factor three times, and then proceed to multiply.

$(x + 5)^3 \; = \; (x + 5)(x + 5)(x + 5)$ Begin by writing $(x + 5)$ as a factor three times.

$= \; [(x + 5)(x + 5)](x + 5)$ Pick two of the factors to multiply. We'll multiply the first two binomials and leave the third binomial for later.

$= \; [x^2 + 5x + 5x + 25](x + 5)$ Use the FOIL method to multiply the first two binomials.

$= \; [x^2 + 10x + 25](x + 5)$ Simplify the expression in brackets by combining like terms.

$= \; x^2(x + 5) + 10x(x + 5) + 25(x + 5)$ Treating $(x + 5)$ as a single entity, apply the distributive property.

$= \; x^2 \cdot x + x^2 \cdot 5 + 10x \cdot x + 10x \cdot 5 + 25 \cdot x + 25 \cdot 5$ Remove parentheses by applying the distributive property yet again.

$= \; x^3 + 5x^2 + 10x^2 + 50x + 25x + 125$ Now simplify each term.

$= \; x^3 + 15x^2 + 75x + 125$ Finish the problem by combining like terms.

Practice B

Now that you've seen how this works in our examples, it's your turn. Find the following products. Make sure your answers are simplified as much as possible. When you're finished, check your work on the next page.

9. $(a + 1)(a + 4)$

10. $(m - 9)(m - 2)$

11. $(2x + 4)(x + 5)$

12. $(x + y)(2x - 3y)$

13. $(3a^2 - 1)(5a^2 + a)$

14. $(2x^2y^3 + xy^2)(5x^3y^2 + x^2y)$

15. $(a + 4)(a - 4)$

16. $(r - 7)(r - 7)$

17. $(x + 6)^2$

18. $(y - 8)^2$

19. $(a + 3)(a^2 + 3a + 6)$

Exercises 4.2

For the following problems, perform the multiplications and combine any like terms.

1. $x(x + 6)$

2. $y(y + 7)$

3. $m(m - 4)$

4. $k(k - 11)$

5. $3x(x + 2)$

6. $4y(y + 7)$

7. $6a(a - 5)$

8. $9x(x - 3)$

9. $3x(5x + 4)$

10. $4m(2m + 7)$

11. $2b(b - 1)$

12. $7a(a - 4)$

13. $3x^2(5x^2 + 4)$

14. $9y^3(3y^2 + 2)$

15. $4a^4(5a^3 + 3a^2 + 2a)$

16. $2x^4(6x^3 - 5x^2 - 2x + 3)$

17. $-5x^2(x + 2)$

18. $-6y^3(y + 5)$

19. $2x^2y(3x^2y^2 - 6x)$

20. $8a^3b^2c(2ab^3 + 3b)$

21. $b^5x^2(2bx - 11)$

22. $4x(3x^2 - 6x + 10)$

23. $9y^3(2y^4 - 3y^3 + 8y^2 + y - 6)$

24. $-a^2b^3(6ab^4 + 5ab^3 - 8b^2 + 7b - 2)$

25. $(a + 4)(a + 2)$

26. $(x + 1)(x + 7)$

27. $(y + 3)(y - 3)$

28. $(t + 8)(t - 2)$

29. $(i - 3)(i + 5)$

30. $(x - y)(2x + y)$

31. $(3a - 1)(2a - 6)$

32. $(5a - 2)(6a - 8)$

33. $(6y + 11)(3y + 10)$

34. $(2t + 6)(3t + 4)$

35. $(4 + x)(3 - x)$

36. $(6 - a)(6 + a)$

37. $(x^2 + 2)(x + 1)$

38. $(x^2 + 5)(x + 4)$

39. $(3x^2 - 5)(2x^2 + 1)$

40. $(4a^2b^3 - 2a)(5a^2b - 3b)$

41. $(6x^3y^4 + 6x)(2x^2y^3 + 5y)$

42. $5(x - 7)(x - 3)$

43. $4(a + 1)(a - 8)$

44. $a(a - 3)(a + 5)$

45. $x(x + 1)(x + 4)$

46. $x^2(x + 5)(x + 7)$

47. $y^3(y - 3)(y - 2)$

48. $2a^2(a + 4)(a + 3)$

49. $5y^6(y + 7)(y + 1)$

50. $ab^2(a^2 - 2b)(a + b^4)$

51. $x^3y^2(5x^2y^2 - 3)(2xy - 1)$

52. $6(a^2 + 5a + 3)$

53. $8(c^3 + 5c + 11)$

54. $3a^2(2a^3 - 10a^2 - 4a + 9)$

55. $6a^3b^3(4a^2b^6 + 7ab^8 + 2b^{10} + 14)$

56. $(a - 4)(a^2 + a - 5)$

57. $(x - 7)(x^2 + x - 3)$

58. $(2x + 1)(5x^3 + 6x^2 + 8)$

59. $(7a^2 + 2)(3a^5 - 4a^3 - a - 1)$

60. $(x + y)(2x^2 + 3xy + 5y^2)$

61. $(2a + b)(5a^2 + 4a^2b - b - 4)$

62. $(x + 3)^2$

63. $(x + 1)^2$

64. $(x - 5)^2$

65. $(a + 2)^2$

66. $(a - 9)^2$

67. $-(3x - 5)^2$

68. $-(8t + 7)^2$

69. $(n + 4)^3$

70. $(y - 2)^3$

Practice B — Answers

9. $a^2 + 5a + 4$

10. $m^2 - 11m + 18$

11. $2x^2 + 14x + 20$

12. $2x^2 - xy - 3y^2$

13. $15a^4 + 3a^3 - 5a^2 - a$

14. $10x^5y^5 + 7x^4y^4 + x^3y^3$

15. $a^2 - 16$

16. $r^2 - 14r + 49$

17. $x^2 + 12x + 36$

18. $y^2 - 16y + 64$

19. $a^3 + 6a^2 + 15a + 18$

4.3 Rules of Exponents

Overview

In the last section, you were introduced to the product rule for exponents. Now it's time to explore exponents more thoroughly. We'll revisit the product rule, and learn a lot of other useful rules as well.

Before we do that, here's a problem to consider:

$$\text{Simplify: } \frac{3x^3y^4}{4xy^5} \cdot \frac{8x^6y^9}{9x^8y^2}$$

There's a lot happening in this problem. It is stated as a multiplication problem, but the fraction bars indicate that division will also be involved. And then, of course, there are all those exponents to consider. We need to learn some of the rules of exponents before we can simplify this expression.

When you're finished with this section, you will be able to:

◆ Use the product rule for exponents

◆ Use the quotient rule for exponents

◆ Work with zero as an exponent

A. The Product Rule for Exponents

An **exponent** records the number of identical factors in a multiplication. So, if x is any real number and n is a natural number, then

$$x^n = \underbrace{x \cdot x \cdot x \cdot \ldots \cdot x}_{n \text{ factors of } x}$$

For example, $x^3 = x \cdot x \cdot x$.

In the expression x^n, x is called the **base**, and n is the **exponent**. The number represented by x^n is called a **power**. The term x^n is read as "x to the nth power." For example, x^3 can be read "x to the third power" and x^5 can be read as "x to the fifth power." Some powers have alternative language associated with them. For example, x^2 can be read as "x squared" instead of "x to the second power," and x^3 can be read as "x cubed" instead of "x to the third power." This brings us to a rule that we used in the previous section when we were multiplying polynomials.

The Product Rule for Exponents

If A is a real number, and if m and n are natural numbers, then $A^m \cdot A^n = A^{m+n}$

The product rule tells us that when we multiply two powers that have the same base, we add the exponents in order to find the exponent on the answer. Note that the answer has the same base as the powers being multiplied.

Let's look at a couple of expressions involving the product rule for exponents.

Example 1

Use the product rule for exponents to simplify the following expressions.

1. $n^6 \cdot n^{14}$
2. $5^2 \cdot 5^4$

Solution

1. $n^6 \cdot n^{14}$

$$n^6 \cdot n^{14} = n^{6+14}$$
$$= n^{20}$$

2. $5^2 \cdot 5^4$

$$5^2 \cdot 5^4 = 5^{2+4}$$
$$= 5^6$$

☠ Warning! Incorrect Approach! ☠

With the second expression in Example 1, some people make the mistake of multiplying the bases and writing the answer with a base of 25:

$$5^2 \cdot 5^4 = 25^6$$

This is incorrect! Using a calculator, we can verify the following:

$$5^2 \cdot 5^4 = 25 \cdot 625 = 15{,}625$$

and

$$25^6 = 244{,}140{,}625$$

The differing final results illustrate that $5^2 \cdot 5^4$ is *not* the same thing as 25^6.

When applying the product rule for exponents, we must be careful to write our answer with the same base as the expressions being multiplied.

Sometimes, an expression in parentheses can be the base, as in the next example.

Example 2

Use the product rule for exponents to simplify: $(x - 2y)^8 (x - 2y)^5$

Solution

$$(x - 2y)^8 (x - 2y)^5 = (x - 2y)^{8+5}$$
$$= (x - 2y)^{13}$$

In the previous section, we saw how the commutative and associative properties of multiplication allow us to reorder and regroup numbers and variables when simplifying expressions. We also saw how the process can be shortened into two basic steps:

1. Multiply the numerical coefficients of the terms to determine the numerical coefficient of the answer.

2. Use the product rule for exponents to combine exponential expressions that have the same base.

Here are a few more examples:

Example 3

Simplify the following expressions.

1. $4y^3 \cdot 6y^2$

2. $5(a+6)^2 \cdot 3(a+6)^8$

3. $4a^{\triangle} \cdot 5a^{\triangledown}$

4. $9a^2 b^6 (8ab^4)(2b^3)$

Solutions

1. $4y^3 \cdot 6y^2$

$$4y^3 \cdot 6y^2 = (4 \cdot 6) \cdot (y^3 \cdot y^2)$$
$$= 24y^5$$

2. $5(a+6)^2 \cdot 3(a+6)^8$

$$5(a+6)^2 \cdot 3(a+6)^8 = (5 \cdot 3) \cdot [(a+6)^2 \cdot (a+6)^8]$$
$$= 15(a+6)^{10}$$

3. $4a^{\triangle} \cdot 5a^{\triangledown}$

$$4a^{\triangle} \cdot 5a^{\triangledown} = (4 \cdot 5) \cdot (a^{\triangle} \cdot a^{\triangledown})$$
$$= 20a^{\triangle + \triangledown}$$

4. $9a^2 b^6 (8ab^4)(2b^3)$

$$9a^2 b^6 (8ab^4)(2b^3) = (9 \cdot 8 \cdot 2) \cdot (a^2 \cdot a) \cdot (b^6 \cdot b^4 \cdot b^3)$$
$$= 144a^3 b^{13}$$

Practice A

Use the product rule for exponents to find each product. When applicable, use the commutative and associative properties of multiplication, too. Then turn the page and check your answers.

1. $x^2 \cdot x^5$

2. $x^9 \cdot x$

3. $7^6 \cdot 7^2$

4. $(-3)^{12} \cdot (-3)^8$

5. $(x+2)^3 \cdot (x+2)^5$

6. $4a^3 b^2 \cdot 9a^2 b$

7. $x^4 \cdot 4y^2 \cdot 2x^2 \cdot 7y^6$

8. $(x-y)^3 \cdot 4(x-y)^2$

9. $4x^3 y^2 (3xy^4 \cdot 7x^6 y^3)$

10. $8x^4 y^2 x x^3 y^5$

11. $2aaa^3 (ab^2 a^3) b6ab^2$

12. $a^n \cdot a^m \cdot a^r$

13. $x^n \cdot x^3$

B. Using the Quotient Rule for Exponents

The second rule we want to examine is the rule for dividing two exponential quantities that have the same base. Take a look at how the following examples suggest a rule for quotients.

Example 4

Simplify the following expressions.

1. $\dfrac{x^5}{x^2}$

2. $\dfrac{a^8}{a^3}$

Solution

1. $\dfrac{x^5}{x^2}$

$\dfrac{x^5}{x^2} = \dfrac{xxxxx}{xx}$ We begin by writing the numerator and denominator in expanded notation.

$\quad = \dfrac{(\cancel{xx})xxx}{(\cancel{xx})}$ The numerator and denominator both have factors of (xx). These factors can be canceled.

$\quad = xxx$ We can now write our result in exponential form.

$\quad = x^3$

Look at the original expression. Notice that $5 - 2 = 3$.

2. $\dfrac{a^8}{a^3}$

$\dfrac{a^8}{a^3} = \dfrac{aaaaaaaa}{aaa}$

$\quad = \dfrac{(\cancel{aaa})aaaaa}{(\cancel{aaa})}$

$\quad = aaaaa$

$\quad = a^5$

Take a look at the original expression and notice that $8 - 3 = 5$.

This example suggests the following rule:

The Quotient Rule for Exponents

If A is any real number except 0, and if m and n are natural numbers, then $\dfrac{A^m}{A^n} = A^{m-n}$.

The quotient rule tells us that when we divide two powers having the same nonzero base, we subtract the exponent of the denominator from the exponent of the numerator. Just as we saw with the product rule for exponents, we must remember that in order for this rule to apply, the following conditions must be met:

1. The powers we are dividing must have the same base.

2. The simplified expression will have the same base as the original powers.

We'll use the quotient rule to simplify a couple of expressions.

Example 5

Use the quotient rule for exponents to simplify each expression.

   1. $\dfrac{x^6}{x^2}$ **2.** $\dfrac{8^5}{8^2}$

Solutions

1. $\dfrac{x^6}{x^2}$ **2.** $\dfrac{8^5}{8^2}$

    $\dfrac{x^6}{x^2} = x^{6-2}$ $\dfrac{8^5}{8^2} = 8^{5-2}$

        $= x^4$ $= 8^3$

☠ Warning! Incorrect Approach! ☠

There are two common mistakes that some people make when attempting to apply the quotient rule for exponents. In the first expression of Example 5, some mistakenly *divide exponents*:

$$\frac{x^6}{x^2} = x^{\frac{6}{2}} = x^3$$

By substituting almost any number for x and simplifying, we can see that this result is incorrect. When we replace x with 2, for example, the original expression gives us $\frac{2^6}{2^2} = \frac{64}{4} = 16$, but the incorrectly simplified expression gives $2^3 = 8$.

In the second expression of Example 5, some mistakenly *divide bases*:

$$\frac{8^5}{8^2} = \left(\frac{8}{8}\right)^{5-2} = 1^3$$

Using a calculator to help us, we can evaluate the original expression: $\frac{8^5}{8^2} = \frac{32{,}768}{64} = 512$. This result is certainly not equal to 1^3, but it *is* equal to 8^3.

In the next example, we divide monomials with numerical coefficients that are not written in exponential form. When this happens, we use the following approach:

1. Divide the numerical coefficients of the terms to determine the numerical coefficient of the answer.

2. Use the quotient rule for exponents to combine the variables written in exponential form.

This is essentially the same approach we took when multiplying monomials.

Example 6

Simplify the following expressions.

1. $\dfrac{27\,a^3\,b^6\,c^2}{3\,a^2\,bc}$

2. $\dfrac{15\,x^{\square}}{3\,x^{\triangle}}$

Solutions

1. $\dfrac{27\,a^3\,b^6\,c^2}{3\,a^2\,bc}$

$\dfrac{27\,a^3\,b^6\,c^2}{3\,a^2\,bc} = \left(\dfrac{27}{3}\right)\cdot(a^{3-2}\,b^{6-1}\,c^{2-1})$

$= 9a\,b^5\,c$

2. $\dfrac{15\,x^{\square}}{3\,x^{\triangle}}$

$\dfrac{15\,x^{\square}}{3\,x^{\triangle}} = \left(\dfrac{15}{3}\right)\cdot(x^{\square-\triangle})$

$= 5\,x^{\square-\triangle}$

In the second expression, the bases are the same, so we subtract the exponents. We don't know exactly what $\square-\triangle$ is, but the notation $\square-\triangle$ indicates the subtraction.

Practice B

Using the quotient rule for exponents, find each quotient. Then turn the page to check your work.

14. $\dfrac{y^9}{y^5}$

16. $\dfrac{(x+6)^5}{(x+6)^3}$

18. $\dfrac{x^9}{x^n}$

15. $\dfrac{a^7}{a}$

17. $\dfrac{26\,x^4\,y^6\,z^2}{13\,x^2\,y^2\,z}$

Practice A — Answers

1. x^7
2. x^{10}
3. 7^8
4. $(-3)^{20}$ By the way, $(-3)^{20}$ is equal to 3^{20}.

5. $(x+2)^8$
6. $36\,a^5\,b^3$
7. $56\,x^6\,y^8$
8. $4\,(x-y)^5$
9. $84\,x^{10}\,y^9$

10. $8\,x^8\,y^7$
11. $12\,a^{10}\,b^5$
12. a^{n+m+r}
13. x^{n+3}

C. Zero as an Exponent

When we use the quotient rule for exponents on the expression $\frac{A^m}{A^n}$, there are three possibilities for the value of the exponent on our answer:

1. If $m > n$, then $m - n$ gives us a positive exponent on our answer.

2. If $m < n$, then $m - n$ gives us a negative exponent on our answer.

3. If $m = n$, then $m - n$ gives us an exponent of zero on our answer.

In the problems we've looked at so far, we've seen Case 1. The exponents of the numerators have been greater than the exponents of the denominators, so when applying the quotient rule for exponents, we've ended up with positive exponents on our answers.

We'll skip Case 2 for now and take a closer look at that later in this chapter.

That leaves us with Case 3. Let's consider what happens when the exponents in the numerator and denominator are the same. Since they are the same, we can use the letter n to represent both exponents. The quotient rule for exponents gives us:

$$\frac{A^n}{A^n} = A^{n-n} = A^0$$

But what real number, if any, does A^0 represent?

Let's think about our experience with division in arithmetic. We know that *any* nonzero number divided by itself gives us an answer of 1. For example:

$$\frac{8}{8} = 1, \frac{43}{43} = 1 \text{ and } \frac{-258}{-258} = 1$$

If $A \neq 0$, then $A^n \neq 0$, and that means that $\frac{A^n}{A^n} = 1$. However, by using the quotient rule for exponents, we saw that $\frac{A^n}{A^n} = A^0$. These results give us our next rule of exponents.

The Zero-Exponent Rule

If A is any real number except 0, then $A^0 = 1$.

Example 7

Use the zero-exponent rule to simplify each of the following expressions.

1. 6^0

2. $(2a + 5)^0$

3. $4y^0$

4. $(4y)^0$

Solutions

1. $6^0 = 1$

    This is a simple illustration of the zero-exponent rule: Any number raised to the power of 0 is 1.

2. $(2a + 5)^0 = 1$

    In this case, the value represented by $2a + 5$ is raised to the power of 0, so the result is 1.

3. $4y^0$

    $$4y^0 = 4 \cdot 1$$
    $$= 4$$

    In this example, the zero-exponent only applies to the y. The coefficient, 4, is not affected.

4. $(4y)^0 = 1$

    Because of the parentheses, the zero-exponent applies to the entire expression.

    $$(4y)^0 = 1$$

The last two expressions from the previous example illustrate the importance of parentheses when applying exponents. Based on the order of operations, an exponent must be applied before multiplication takes place — unless the multiplication is inside of parentheses. In other words, we should only apply an exponent to the thing that appears immediately to the left of it.

You may recall seeing this in previous math classes with expressions involving negative numbers. Remember that a negative sign signifies that something is being *multiplied* by -1. In the absence of parentheses, we must apply an exponent before a negative sign. Let's see how this looks with the number -3:

$-3^2 = -9$ The number 3 is immediately to the left of the exponent. Therefore, we *only* square the number 3: $-3^2 = -(3)(3) = -9$

$(-3)^2 = 9$ Parentheses are immediately to the left of the exponent. Therefore, we apply the exponent to the parentheses. That means *everything* inside of the parentheses gets squared: $(-3)^2 = (-3)(-3) = 9$

Now, let's look at a couple of expressions involving the quotient rule *and* the zero-exponent rule.

Example 8

Simplify the following expressions.

1. $\dfrac{8x^2}{4x^2}$

2. $\dfrac{5(x+4)^8(x-1)^5}{5(x+4)^3(x-1)^5}$

Solutions

1. $\dfrac{8x^2}{4x^2}$

   In this problem, we use the two-step process described before example 6.

   $\dfrac{8x^2}{4x^2} = \dfrac{8}{4} \cdot x^{2-2}$ Divide the numerical coefficients, and use the quotient rule to combine the variables.

   $= 2x^0$ Now apply the zero-exponent rule.

   $= 2 \cdot 1$

   $= 2$

2. $\dfrac{5(x+4)^8(x-1)^5}{5(x+4)^3(x-1)^5}$

   In this problem, we treat $(x+4)$ and $(x-1)$ as single entities and use the same approach we used in the first expression. We use the quotient rule to combine the $(x+4)$ and $(x-1)$ expressions.

   $\dfrac{5(x+4)^8(x-1)^5}{5(x+4)^3(x-1)^5} = \dfrac{5}{5} \cdot (x+4)^{8-3}(x-1)^{5-5}$

   $= 1 \cdot (x+4)^5(x-1)^0$

   $= 1 \cdot (x+4)^5 \cdot 1$

   $= (x+4)^5$

Now we're ready to tackle the problem from the beginning of this section. This problem requires us to use all of the rules of exponents we've learned so far.

Example 9

Simplify: $\dfrac{3x^3y^4}{4xy^5} \cdot \dfrac{8x^6y^9}{9x^8y^2}$

Solution

We will start by multiplying these fractions together. This will require us to use the product rule for exponents. After that, as we work through the rest of the problem, the quotient rule and the zero-exponent rule will be used as well.

$\dfrac{3x^3y^4}{4xy^5} \cdot \dfrac{8x^6y^9}{9x^8y^2} = \dfrac{(3x^3y^4)(8x^6y^9)}{(4xy^5)(9x^8y^2)}$ After multiplying the numerators and denominators to combine fractions, use the product rule to help simplify.

$= \dfrac{24x^9y^{13}}{36x^9y^7}$ Use the quotient rule to help simplify.

$= \dfrac{24}{36}x^0y^6$ Now simplify the fraction and use the zero-exponent rule.

$= \dfrac{2}{3} \cdot 1 \cdot y^6$

$= \dfrac{2}{3}y^6$

In the previous example, the numerical coefficients gave us the fraction $\frac{24}{36}$ when multiplied. This fraction then had to be simplified. However, when multiplying fractions, we can use cross-canceling to reduce the numerical coefficients before combining the fractions. In the previous example, that step would look like this:

$\dfrac{3x^3y^4}{4xy^5} \cdot \dfrac{8x^6y^9}{9x^8y^2} = \dfrac{1(3)x^3y^4}{1(4)xy^5} \cdot \dfrac{2(8)x^6y^9}{3(9)x^8y^2}$ 3 and 9 are replaced with 1 and 3. 4 and 8 are replaced with 1 and 2.

$= \dfrac{2x^9y^{13}}{3x^9y^7}$

By cross-canceling, we immediately arrive at the simplified fraction $\frac{2}{3}$ when the reduced numerical coefficients are multiplied. Whenever possible, use cross-canceling to reduce the numerical coefficients when multiplying fractions.

Practice C

Use the rules of exponents that we've studied to simplify each expression. Assume that none of the bases are zero. After you have simplified each expression, you can find the answers below.

19. 7^0

20. $9x^2y^0$

21. $(9x^2y)^0$

22. -5^0

23. $(-5)^0$

24. $\dfrac{36a^4b^3c^8}{8ab^3c^6}$

25. $\dfrac{51(a-4)^3(a+11)^{12}}{17(a-4)(a+11)^{12}}$

26. $\dfrac{52a^7b^3(a+b)^8}{26a^2b(a+b)^8}$

27. $\dfrac{9a^3b^4}{4a^2b} \cdot \dfrac{2ab^5}{27a^2b^3}$

Exercises 4.3

Use the product rule to help you simplify each expression.

1. $d^4 \cdot d^8$
2. $x^4 \cdot x^{13}$
3. $3^3 \cdot 3^5 \cdot y^3$
4. $d^2 \cdot 2^2 \cdot d \cdot 2^8$
5. $db^2 \cdot 9db$
6. $8\,m^2 v^3 \cdot 3\,m^7 v$

7. $p^2 (5\,q^6)(5\,p^6)$
8. $3p\,q^2 (2\,p^6 q^3)(4\,p^4)$
9. $8\,r^2 w^6 \cdot 8\,w^4 r^3$
10. $3\,b^3 c^2\, bb\, c^7$
11. $7zzz^3 z^2 z$
12. $(a-3)^2 \cdot (a-3)^4$

13. $(v-u) \cdot (v-u)^2$
14. $7\,(m+n)^5 \cdot 6\,(m+n)^{11}$
15. $2\,(3+c^2)^2 \cdot 7\,(3+c^2)^3$
16. $p^a \cdot q^{a+b} \cdot p^b$
17. $c^w \cdot c^7 \cdot c^{2w}$

Use the quotient rule to help you simplify each expression. Assume that none of the bases are zero.

18. $\dfrac{a^{12}}{a^2}$
19. $\dfrac{p^7}{p^6}$
20. $\dfrac{b^4}{b}$
21. $\dfrac{36\,x^9}{4\,x^3}$

22. $\dfrac{96\,n^{18}}{8\,n^9}$
23. $\dfrac{8\,x^6 y^{72}}{2\,x^2 y^{33}}$
24. $\dfrac{121\,b^7 c^5 d^2}{11\,b^2 c^3}$
25. $\dfrac{400\,r^5 s^2 t^4}{20\,r^3 s t^3}$

26. $\dfrac{(8+k)^{15}}{(8+k)^6}$
27. $\dfrac{(7+y)^5}{(7+y)}$
28. $\dfrac{x^3 (y+1)^{10}}{x\,(y+1)^2}$
29. $\dfrac{y^p}{y^q}$

30. $\dfrac{p^n}{p^m}$
31. $\dfrac{d^7}{d^n}$
32. $\dfrac{(v+u)^n}{(v+u)^{p+1}}$

Use the rules of exponents to help you simplify each expression. Assume that none of the bases are zero.

33. a. $(7v)^0$ b. $7v^0$
34. a. $(13c)^0$ b. $13c^0$
35. a. $(-12)^0$ b. -12^0
36. $2^4 \cdot 2^5$
37. $4y^5 \cdot 5y^6$

38. $2a^3b^2 \cdot 3ab$
39. $12xy^3 z^2 \cdot 4x^2 y^2 z \cdot 3x$
40. $(3ab)(2\,a^2 b)$
41. $(4x^2)(8xy^3)$
42. $\left(\frac{1}{4}a^2 b^4\right)\left(\frac{1}{2}b^4\right)$

43. $\left(\frac{3}{4}xy\right)\left(\frac{16}{9}y\right)\left(\frac{3}{2}x^2 y^5\right)$
44. $\dfrac{12x^5 y^6}{7x^3 y^5} \cdot \dfrac{21x^{10}y}{16xy^2}$
45. $\dfrac{25\,a^{11} z^{16}}{18\,a^3 z} \cdot \dfrac{27\,a z^8}{5\,a^9 z^3}$
46. $3\,a^2 b^3 \left(\dfrac{14\,a^2 b^5}{2b}\right)$
47. $\dfrac{40x^5 z^{10}(z-x^4)^{12}(x+z)^2}{10z^7(z-x^4)^5}$

Practice C — Answers

19. 1
20. $9x^2$
21. 1

22. -1
23. 1
24. $\frac{9}{2}a^3 c^2$

25. $3\,(a-4)^2$
26. $2\,a^5 b^2$
27. $\frac{1}{6}b^5$ or $\frac{b^5}{6}$

4.4 Power Rules of Exponents

Overview

We know how to simplify a product of powers, a quotient of powers, and even something to the power of zero. Now, it's time to explore what happens when we take the power of a power.

Here's a related problem:

Simplify: $\left(7\,a^4\,b^2\,c^8\right)^2$

This problem contains an expression involving powers contained within parentheses, and then an exponent outside of the parentheses. We haven't seen anything like this yet. We need to learn how to simplify the power of a power before we are ready to tackle this problem.

As you read through this section, you will learn how to:

◆ Use the power rule for powers

◆ Use the power rule for products

◆ Use the power rule for quotients

A. The Power Rule for Powers

Now that we can use the product and quotient rules for exponents to multiply and divide exponential expressions with the same base, it's time to explore situations in which we raise an exponential expression to a power. Some people think of this as putting two exponents on the same base. When we do this, parentheses keep the exponents separated.

Example 1

Simplify the following expressions.

1. $\left(a^2\right)^3$ **2.** $\left(x^9\right)^4$

Solutions

1. $\left(a^2\right)^3$

$\left(a^2\right)^3 = a^2 \cdot a^2 \cdot a^2$ Raising a quantity to the third power means the same thing as writing the quantity as a factor three times.

$= a^{2+2+2}$ Apply the product rule for exponents.

$= a^6$ Finish by simplifying the expression.

Look at the exponents on the beginning and the simplified expressions. Notice that $2 \cdot 3 = 6$.

2. $(x^9)^4$

$\quad(x^9)^4 \;=\; x^9 \cdot x^9 \cdot x^9 \cdot x^9 \qquad$ Write the quantity x^9 as a factor four times.

$\qquad\qquad = \; x^{9+9+9+9} \qquad\qquad$ Apply the product rule for exponents.

$\qquad\qquad = \; x^{36}$

Again, look at the exponents on the beginning and the simplified expressions: $9 \cdot 4 = 36$.

Example 1 illustrates the following rule.

The Power Rule for Powers

If A is a real number, and if m and n are natural numbers, then $\left(A^m\right)^n = A^{m+n}$.

In other words, if we raise a power to a power, then we multiply the exponents. Let's look at more expressions involving the power rule for powers.

Example 2

Simplify the following expressions.

   1. $(x^3)^4$ **2.** $(d^{20})^6$ **3.** $(x^{\square})^{\triangle}$

Solutions

1. $(x^3)^4$

$\quad(x^3)^4 \;=\; x^{3 \cdot 4}$

$\qquad\qquad = \; x^{12}$

Multiplying exponents can often be done mentally, leaving just one step of writing.

2. $(d^{20})^6$

$\quad(d^{20})^6 \;=\; d^{120} \qquad$ Mentally multiply the exponents on the original expression: $20 \cdot 6 = 120$

3. $(x^{\square})^{\triangle}$ $\qquad\qquad$ We have no idea what numbers $\square$ and $\triangle$ represent, but we know the power rule for powers.

$\quad(x^{\square})^{\triangle} \;=\; x^{\square\triangle} \qquad$ The notation $\square\triangle$ indicates that we are multiplying $\square$ and $\triangle$.

Practice A

Simplify these two expressions using the power rule for powers. Then check your work.

1. $(x^5)^4$ **2.** $(y^7)^7$

B. The Power Rule for Products

Now it's time to see what happens when a product is raised to a power.

Example 3

Simplify the following expressions.

1. $(ab)^3$ **2.** $(4xyz)^2$

Solutions

1. $(ab)^3$

$(ab)^3 = ab \cdot ab \cdot ab$	Raising a quantity to the third power means the same thing as writing the quantity as a factor three times.
$= aaabbb$	Reorder the variables so that the a's are grouped together and the b's are grouped together.
$= a^3 b^3$	Apply the product rule for exponents and simplify.

2. $(4xyz)^2$

$(4xyz)^2 = 4xyz \cdot 4xyz$	Write the quantity $4xyz$ as a factor two times.
$= 4 \cdot 4 \cdot xx \cdot yy \cdot zz$	Reorder the numbers and variables.
$= 16\,x^2 y^2 z^2$	Multiply the numbers and use the product rule for exponents on the variables.

Example 2 suggests the following rule.

The Power Rule for Products

If A and B are real numbers, and if n is a natural number, then $(AB)^n = A^n B^n$.

The power rule for products tells us that in order to raise a product to a power, we apply the exponent outside the parentheses to each and every factor inside the parentheses. This rule is sometimes referred to as the "distributive property for exponents."

Notice that in order for this rule to apply, the product AB must be written inside of parentheses. When written without parentheses, the expression AB^n represents $A^1 B^n$.

In the following example, we'll use the power rule for products to help us simplify the expressions.

Example 4

Simplify the following expressions.

   1. $(ab)^7$ **2.** $(axy)^4$ **3.** $(5ab)^2$ **4.** $(2st)^5$

Solutions

1. $(ab)^7$

   $(ab)^7 = a^7 b^7$

2. $(axy)^4$

   $(axy)^4 = a^4 x^4 y^4$

3. $(5ab)^2$

   $(5ab)^2 = 5^2 a^2 b^2$

         $= 25\, a^2 b^2$

4. $(2st)^5$

   $(2st)^5 = 2^5 s^5 t^5$

         $= 32\, s^5 t^5$

Notice that there are three factors inside of the parentheses: 5, a and b.

☠ Warning! Incorrect Approach! ☠

In parts 3 and 4 of the previous example, some people mistakenly write the following:

$$(5ab)^2 = 5\, a^2 b^2 \quad \text{and} \quad (2st)^5 = 2 s^5 t^5$$

This is not correct! In these expressions, the exponent outside of the parentheses must be applied to all *three* of the factors inside of the parentheses — not just the variable factors.

   In the last few examples, all of the variables inside of the parentheses have had no exponent showing. This means that they have all had an exponent of 1. Let's see what happens when a variable inside of the parentheses has an exponent greater than one.

Example 5

Simplify: $\left(a^4 b^3\right)^2$

Solution

$\left(a^4 b^3\right)^2 = \left(a^4\right)^2 \left(b^3\right)^2$ We begin by using the power rule for products. Apply the outside

        $= a^8 b^6$ exponent of 2 to each of the factors inside of the parentheses.
                            Finish by using the power rule for powers on each set of parentheses
                            separately.

   In this problem, people often use mental math and write the final result in just one step.

$\left(a^4 b^3\right)^2 = a^8 b^6$ Use mental math to multiply each of the inside exponents by 2:

                       $4 \cdot 2 = 8$ and $3 \cdot 2 = 6$. Use the results to write the final answer.

Now that we've seen how the power rule for products works when the factors inside the parentheses have exponents, let's simplify a couple more expressions.

Example 6

Simplify each expression:

1. $(7a^4 b^2 c^8)^2$

2. $(-3x^7 y^2)^4$

Solutions

1. $(7a^4 b^2 c^8)^2$

This is the problem we saw at the beginning of the section. Notice that there are *four* factors inside of the parentheses, so the outside exponent of 2 needs to be applied to each factor.

$(7a^4 b^2 c^8)^2 = 7^2 a^8 b^4 c^{16}$ Use mental math to apply the outside exponent of 2.

$\qquad\qquad = 49 a^8 b^4 c^{16}$ Finish by simplifying the numerical coefficient.

2. $(-3x^7 y^2)^4$

The coefficient of the expression inside of parentheses is negative. Be careful in situations like this. To correctly apply the outside exponent of 4, use parentheses with the negative coefficient.

$(-3x^7 y^2)^4 = (-3)^4 x^{28} y^8$ Use mental math to apply the outside exponent of 4. Keep the numerical coefficient in parentheses.

$\qquad\qquad = 81 x^{28} y^8$ Finish by simplifying the numerical coefficient: $(-3)^4 = 81$

If we fail to keep the numerical coefficient in parentheses during the first step of this problem, we end up with -3 4 $= -81$. However, -81 is *not* the correct coefficient for our final answer.

In some cases, applying the power rule for products is optional. Consider the following example.

Example 7

Simplify$(6a^3 c^7)^0$. Assume that $a \neq 0$ and $c \neq 0$.

Solution

It's tempting to apply the power rule for products in this example, but it's not necessary. Remember that $A^0 = 1$ for $A \neq 0$. *Any* nonzero expression raised to the power of zero — even if it's in parentheses — will give a result of 1.

$(6a^3 c^7)^0 = 1$

Practice A — Answers

1. x^{20}

2. y^{49}

If we apply the power rule for products in this example, the work looks like this:

$$(6a^3c^7)^0 = 6^0a^0c^0$$
$$1 \cdot 1 \cdot 1$$
$$= 1$$

The result is still 1, but it takes us three steps to get there.

In Example 7, we saw an expression for which the power rule for products was optional. In some cases, this rule simply can't be applied. Keep in mind that the power rule for products applies only to *products*. It can't be applied in cases involving sums or differences. Consider the following example.

Example 8

Simplify the following expressions.

1. $(3x)^2$ **2.** $(3 + x)^2$

Solutions

1. $(3x)^2$

$(3x)^2 = 9x^2$ Apply the outside exponent mentally: $(3)^2 = 9$ and $(x)^2 = x^2$

2. $(3 + x)^2$

$(3 + x)^2 = (3 + x)(3 + x)$ There's a sum inside the parentheses, so we can't apply the power rule for products. We write the expression $3 + x$ as a factor twice.

$= 9 + 3x + 3x + x^2$ Use the FOIL method to write the products.

$= x^2 + 6x + 9$ Combine like terms and write the answer in standard form.

☠ Warning! Incorrect Approach! ☠

Do not, under any circumstance, apply the power rule for products to set of parentheses containing a sum or a difference. Many people incorrectly attempt to do this with the second expression in Example 8:

$$(3 + x)^2 = 3^2 + x^2 = 9 + x^2$$

This is not incorrect! When you square a binomial, the answer is a *trinomial*.

We saw this warning in Section 4.2, and we see it again here. Why? Because this is one of the most common errors made by algebra students. Don't let it happen to you.

As we saw in the previous section, a set of parentheses containing a sum or difference can be treated as a single entity when applying exponent rules. For example, we can apply the product rule as follows: $(n - 4)^3 \cdot (n - 4)^7 = (n - 4)^{10}$

We'll see something similar in the next example.

Example 9

Simplify: $[2(x+1)^4]^6$

Solution

$$[2(x+1)^4]^6 = 2^6(x+1)^{24} \quad \text{The brackets contain two factors: 2 and } (x+1)^4. \text{ The exponent}$$
outside of the parentheses, 6, is applied to both of these factors.

$$= 64(x+1)^{24} \quad \text{At this point, we stop. We cannot apply the power rule for}$$
products to the expression $(x+1)^{24}$ because there is a *sum* inside of the parentheses.

In problems like Example 9, it's customary to stop after applying the power rule for products. We could write the expression $(x+1)$ as a factor twenty-four times and then attempt to use polynomial multiplication to get an expression that doesn't contain parentheses, but doing this by hand is impractical.

Practice B

For these problems, make use of the appropriate power rules to simplify each expression. When you're finished, turn the page to check your work.

3. $(ax)^4$

4. $(3bxy)^2$

5. $(9x^3y^5)^2$

6. $(1a^5b^8c^3d)^6$

7. $[4t(s-5)]^0$

8. $[(a+8)(a+5)]^4$

9. $[(12c^4u^3(w-3)^2]^5$

10. $[10t^4y^7j^3d^2v^6n^4g^8(2-k)^{17}]^4$

11. $(x^3x^5y^2y^6)^9$

12. $(10^6 \cdot 10^{12} \cdot 10^5)^{10}$

C. The Power Rule for Quotients

Let's explore what happens when we raise a quotient to a power.

Example 10

Simplify: $\left(\frac{a}{b}\right)^3$

Solution

$$\left(\frac{a}{b}\right)^3 = \frac{a}{b} \cdot \frac{a}{b} \cdot \frac{a}{b} \quad \text{Begin by writing the expression } \frac{a}{b} \text{ as a factor three times.}$$

$$= \frac{a \cdot a \cdot a}{b \cdot b \cdot b} \quad \text{Use fraction multiplication to write a single fraction.}$$

$$= \frac{a^3}{b^3} \quad \text{Finish by simplifying the result.}$$

The result we obtained in the previous example suggests the following rule.

The Power Rule for Quotients

If A and B are real numbers and n is a natural number, and if $B \neq 0$, then $\left(\frac{A}{B}\right)^n = \frac{A^n}{B^n}$.

To state this rule using words, we can say that if a quotient is raised to a power, then the exponent on the outside of parentheses must be applied to both the numerator and the denominator of the expression inside parentheses. In order for this rule to be applicable, the quotient must be in parentheses. The power rule for quotients does not apply to expressions like $\frac{A^n}{B}$.

Example 11

Simplify the following expressions.

1. $\left(\frac{a^3}{b^5}\right)^7$
2. $\left(\frac{2x^3}{b^2}\right)^4$

Solutions

1. $\left(\frac{a^3}{b^5}\right)^7$

$\left(\frac{a^3}{b^5}\right)^7 = \frac{(a^3)^7}{(b^5)^7}$ Begin by applying the power rule for quotients.

$= \frac{a^{21}}{b^{35}}$ Use the power rule for powers to simplify the variable expressions.

We know that $3 \cdot 7 = 21$ and that $5 \cdot 7 = 35$, so we can use mental math to write the simplified expression immediately.

$\left(\frac{a^3}{b^5}\right)^7 = \frac{a^{21}}{b^{35}}$

2. $\left(\frac{2x^3}{b^2}\right)^4$

$\left(\frac{2x^3}{b^2}\right)^4 = \frac{(2x^3)^4}{b^8}$ Apply the power rule for quotients. In the denominator, use mental math to write the simplified variable expression.

$= \frac{2^4 x^{12}}{b^8}$ Next, apply the power rule for products in the numerator. Again, use mental math to write the simplified variable expression.

$= \frac{16 x^{12}}{b^8}$ Finish by simplifying the numerical coefficient in the numerator.

In the second expression from Example 11, we took two steps to write the numerator without parentheses. We started by applying the power rule for quotients, and then we applied the power rule for products. In the next example, we'll combine these steps. When we apply the power rule for quotients, we'll distribute the exponent to each factor in the numerator and in the denominator.

Example 12

Simplify: $\left(\dfrac{3\,c^4\,r^2}{2^3\,g^5}\right)^3$

Solution

$\left(\dfrac{3\,c^4\,r^2}{2^3\,g^5}\right)^3 = \dfrac{3^3\,c^{12}\,r^6}{2^9\,g^{15}}$ Apply the power rule for quotients. Distribute the exponent outside the parentheses to each factor in the numerator and the denominator.

$\qquad = \dfrac{27\,c^{12}\,r^6}{512\,g^{15}}$ Finish by simplifying the numerical expressions.

Now let's look at a couple of expressions involving parentheses that contain sums or differences. As we've done in the past, we treat parenthetical expressions as single entities when applying exponent rules.

Example 13

Simplify the following expressions.

3. $\left[\dfrac{(a-2)}{(a+7)}\right]^4$ 4. $\left[\dfrac{6x(4-x)^4}{2a(y-4)^6}\right]^2$

Solutions

1. $\left[\dfrac{(a-2)}{(a+7)}\right]^4$

$\left[\dfrac{(a-2)}{(a+7)}\right]^4 = \dfrac{(a-2)^4}{(a+7)^4}$ Apply the power rule for quotients.

At this point, we have to stop because the remaining parentheses contain sums and differences.

2. $\left[\dfrac{6x(4-x)^4}{2a(y-4)^6}\right]^2$

$\left[\dfrac{6x(4-x)^4}{2a(y-4)^6}\right]^2 = \dfrac{6^2 x^2(4-x)^8}{2^2 a^2(y-4)^{12}}$ Apply the power rule for quotients.

$\qquad = \dfrac{36\,x^2(4-x)^8}{4\,a^2(y-4)^{12}}$ Simplify the numerical expressions.

$\qquad = \dfrac{9\,x^2(4-x)^8}{a^2(y-4)^{12}}$ Finish by reducing the fraction.

Practice B — Answers

3. $a^4 x^4$

4. $9\,b^2 x^2 y^2$

5. $81\,x^6 y^{10}$

6. $a^{30}\,b^{48}\,c^{18}\,d^6$

7. 1

8. $(a+8)^4(a+5)^4$

9. $12^5\,c^{20}\,u^{15}\,(w-3)^{10}$

10. $10^4\,t^{16}\,y^{28}\,j^{12}\,d^8\,v^{24}\,n^{16}\,g^{32}\,(2-k)^{68}$

11. $x^{72} y^{72}$

12. 10^{230}

In the second expression from Example 13, we could have started the process by reducing the original fraction inside of the brackets. The next example will illustrate that approach.

Example 14

Simplify the following expression.

1. $\left(\dfrac{a^3 b^5}{a^2 b}\right)^3$

   2. $\left(\dfrac{36 x^6}{54 y^7}\right)^4$

Solutions

1. $\left(\dfrac{a^3 b^5}{a^2 b}\right)^3$

There are two ways to simplify this expression. One way is to begin by simplifying within the parentheses.

$$\left(\dfrac{a^3 b^5}{a^2 b}\right)^3 = (a b^4)^3 \qquad \text{Reduce the fraction with the quotient rule for exponents.}$$

$$= a^3 b^{12} \qquad \text{Finish by applying the power rule for products.}$$

The other approach is to use the power rule for quotients first and finish by using the quotient rule to reduce the fraction.

$$\left(\dfrac{a^3 b^5}{a^2 b}\right)^3 = \dfrac{a^9 b^{15}}{a^6 b^3} \qquad \text{Begin by applying the power rule for quotients.}$$

$$= a^3 b^{12} \qquad \text{Then reduce the fraction with the quotient rule for exponents.}$$

2. $\left(\dfrac{36 x^6}{54 y^7}\right)^4$

For this problem, we'll use the first approach from the previous expression.

$$\left(\dfrac{36 x^6}{54 y^7}\right)^4 = \left(\dfrac{2 x^6}{3 y^7}\right)^4 \qquad \text{First simplify the fraction inside of the parentheses.}$$

$$= \dfrac{2^4 x^{24}}{3^4 y^{28}} \qquad \text{Then apply the power rule for quotients.}$$

$$= \dfrac{16 x^{24}}{81 y^{28}} \qquad \text{Finish by simplifying the numerical coefficients.}$$

If we attempt to use the second approach, the process begins like this:

$$\left(\dfrac{36 x^6}{54 y^7}\right)^4 = \dfrac{36^4 x^{24}}{54^4 y^{28}} \qquad \begin{array}{l}\text{Begin by applying the power rule for quotients. Then, simplify the} \\ \text{numerical coefficients.}\end{array}$$

$$= \dfrac{1,679,616 \; x^{24}}{8,503,056 \; y^{28}} \qquad \text{Finish by reducing the fraction (good luck).}$$

Feel free to reduce this fraction on your own. In the meantime, it's clear that the first approach allows us to arrive at our final answer much more easily.

Our final example illustrates how the power rule for quotients is applied when all of the exponents are variables.

Example 15

Simplify: $\left(\dfrac{a^r b^s}{c^t}\right)^w$

Solution

$$\left(\dfrac{a^r b^s}{c^t}\right)^w = \dfrac{a^{rw} b^{sw}}{c^{tw}}$$

Practice C

Use the power rule for quotients — and other rules for exponents you've learned — to simplify each expression. When you're finished, you can view the next page to check your answers.

13. $\left(\dfrac{a}{c}\right)^5$

14. $\left(\dfrac{2x}{3y}\right)^3$

15. $\left(\dfrac{x^2 y^4 z^7}{a^5 b}\right)^9$

16. $\left[\dfrac{2a^4(b-1)}{3b^3(c+6)}\right]^4$

17. $\left(\dfrac{8a^3 b^2 c^6}{4a^2 b}\right)^3$

18. $\left[\dfrac{(9+w)^2}{(3+w)^5}\right]^{10}$

19. $\left[\dfrac{5x^4(y+1)}{5x^4(y+1)}\right]^6$

20. $\left(\dfrac{16x^3 v^4 c^7}{12x^2 v c^6}\right)^0$

Exercises 4.4

Use the power rules for exponents to simplify the following problems. Assume that all bases are nonzero and that all variable exponents are natural numbers.

1. $(w^3)^4$

2. $(x^7)^{12}$

3. $(n^\blacktriangle)^\Phi$

4. $(\Delta^\pi)^\square$

5. $(ac)^5$

6. $(nm)^7$

7. $(2a)^3$

8. $(-7m)^2$

9. $(-2v)^5$

10. $(-5q)^3$

11. $(-3j)^4$

12. $(6mn)^2$

13. $(7y^3)^2$

14. $(3m^3)^4$

15. $(5x^6)^3$

16. $(2p^3)^6$

17. $(-10a^3 b^9)^4$

18. $(-8x^2 y^{11})^2$

19. $(x^2 y^3 z^5)^4$

20. $(2a^5 b^{11})^0$

21. $(x^3 y^2 z^4)^5$

22. $(m^6 n^2 p^5)^5$

23. $(a^4 b^7 c^6 d^8)^8$

24. $(x^2 y^3 z^9 w^7)^3$

25. $(9xy^3)^0$

26. $\left(\dfrac{1}{2}f^2 r^6 s^5\right)^4$

27. $\left(\dfrac{1}{8}c^{10} d^8 e^4 f^9\right)^2$

28. $\left(\dfrac{3}{5}a^3 b^5 c^{10}\right)^3$

29. $(xy)^4 (x^2 y^4)$

30. $(2a^2)^4 (3a^5)^2$

31. $(3m^2 t^4)^3 (2mt^6)^2$

32. $(h^3 k^5)^2 (h^2 k^4)^3$

33. $(x^4 y^3 z)^4 (x^5 y z^2)^2$

34. $(ab^3 c^2)^5 (a^2 b^2 c)^2$

35. $\dfrac{(x^6 y^5)^3}{(x^2 y^3)^5}$

36. $\dfrac{(a^8 b^{10})^3}{(a^7 b^5)^3}$

37. $\dfrac{(m^5 n^6 p^4)^4}{(m^4 n^5 p)^4}$

38. $\dfrac{(x^8 y^3 z^2)^5}{(x^6 yz)^6}$

39. $\dfrac{(10 x^4 y^5 z^{11})^3}{(xy^2)^4}$

40. $\dfrac{(9a^4 b^5)(2b^2 c)}{(3a^3 b)(6bc)}$

41. $\dfrac{(2x^3 y^3)^4 (5x^6 y^8)^2}{(4x^5 y^3)^2}$

42. $\left(\dfrac{3x}{5y}\right)^2$

43. $\left(\dfrac{3ab}{4xy}\right)^3$

44. $\left(\dfrac{x^2 y^2}{2z^3}\right)^5$

45. $\left(\dfrac{3a^2 b^3}{c^4}\right)^3$

46. $\left(\dfrac{4^2 a^3 b^7}{b^5 c^4}\right)^2$

47. $\left(\dfrac{18\,x^{13} y^8}{12\,x^{10} z^4}\right)^3$

48. $\left(\dfrac{35\,a^5 c^7}{40\,b^3 c}\right)^2$

49. $\left(\dfrac{-320\,n^{11} p}{160\,n^2 x^7}\right)^4$

50. $\left[\dfrac{(n-9)^2}{(y+2)^{12}}\right]^5$

51. $\left[\dfrac{x^2 (y-1)^3}{(x+6)}\right]^4$

52. $(x^n t^{2m})^4$

53. $\dfrac{(x^{n+2})^3}{x^{2n}}$

54. $(xy)^\triangle$

55. $(2ab)^\triangledown$

56. $\dfrac{(3a^\triangle b^\triangledown)^\square}{(5xy^\lozenge)^\blacktriangledown}$

57. $\dfrac{10\,m^\triangle}{5\,m^\triangledown}$

58. $\dfrac{4^3 a^\triangle a^\square}{4\,a^\triangledown}$

59. $\left(\dfrac{4x^\triangle}{2y^\triangledown}\right)^\square$

60. $\left(\dfrac{16\,a^\blacktriangledown b^\triangle}{5\,a^\lozenge b^\triangledown}\right)^0$

Practice C — Answers

13. $\dfrac{a^5}{c^5}$

14. $\dfrac{8x^3}{27y^3}$

15. $\dfrac{x^{18} y^{36} z^{63}}{a^{45} b^9}$

16. $\dfrac{16\,a^{16}(b-1)^4}{81\,b^{12}(c+6)^4}$

17. $8a^3 b^3 c^{18}$

18. $\dfrac{(9+w)^{20}}{(3+w)^{50}}$

19. 1

20. 1

4.5 Negative Exponents

Overview

In the previous few sections, we've used a number of different exponent rules to help us simplify expressions. However, there is one phenomenon we have yet to encounter: negative exponents.

Here's a problem

Evaluate: $\dfrac{1}{10^{-2}} + \dfrac{3}{4^{-3}}$

Before we can evaluate this expression, we must understand what negative exponents mean and learn how to deal with them.

This section will help you be able to:

♦ Understand negative exponents

♦ Work with negative exponents to simplify expressions

A. Understanding Negative Exponents

The concept of negative exponents is more easily understood if we use reciprocals to help us. You may recall working with reciprocals in previous math courses. In this book, at the end of Chapter 2, we used opposite-reciprocals when considering slopes of perpendicular lines.

Two real numbers are *reciprocals* of each other if their product is 1. The number 0 has no reciprocal because it's impossible to multiply something by 0 and get a result of 1. Every non-zero real number has exactly one reciprocal.

The table shows a few examples of reciprocals.

Reciprocals	Explanation
4 and $\frac{1}{4}$	These are reciprocals because $4 \cdot \frac{1}{4} = 1$.
$-\frac{1}{2}$ and -2	These are reciprocals because $-\frac{1}{2} \cdot (-2) = 1$.
$\frac{1}{a}$ and a	As long as $a \neq 0$, these are reciprocals because $\frac{1}{a} \cdot a = \frac{a}{a} = 1$.
x^3 and $\frac{1}{x^3}$	As long as $x \neq 0$, these are reciprocals because $x^3 \cdot \frac{1}{x^3} = \frac{x^3}{x^3} = 1$.

Now that we are comfortable with the concept of *reciprocal*, let's use some rules of exponents to simplify an expression involving a negative exponent.

Example 1

Simplify: $x^3 \cdot x^{-3}$ Assume that $x \neq 0$.

Solution

$$
\begin{aligned}
x^3 \; x^{-3} &= x^{3+(-3)} && \text{Start by applying the product rule for exponents.} \\
&= x^0 && \text{Now apply the zero-exponent rule.} \\
&= 1
\end{aligned}
$$

The product of x^3 and x^{-3} is 1. That means that x^3 and x^{-3} must be reciprocals. We now know the following things:

- Every non-zero real number has exactly one reciprocal.
- As long as $x \neq 0$, x^3 and $\dfrac{1}{x^3}$ are reciprocals.
- As long as $x \neq 0$, x^3 and x^{-3} are reciprocals.

Therefore, we also know that $x^{-3} = \dfrac{1}{x^3}$. This illustrates the following rule.

Rule for Negative Exponents

If A is a nonzero real number, and if n is a natural number, then $A^{-n} = \dfrac{1}{A^n}$.
By extension, the following relationship is also true: $\dfrac{1}{A^{-n}} = A^n$.

It can be helpful to think of the word "reciprocal" when confronted with negative exponents. You can think of the expression A^{-n} as "the reciprocal of A^n."

In the following examples, we'll use the rule for negative exponents to simplify expressions. In general, an expression isn't considered to be simplified if it contains negative exponents, so our job will be to rewrite each of the expressions using only positive exponents.

Example 2

Simplify the following expressions.

1. x^{-6} 2. a^{-1} 3. 7^{-2}

Solutions

1. x^{-6}

$$x^{-6} = \frac{1}{x^6} \qquad \text{To simplify, write the reciprocal of } x^6.$$

2. a^{-1}

$$a^{-1} = \frac{1}{a} \qquad \text{Write the reciprocal of } a^1, \text{ which is the same as the reciprocal of } a.$$

3. 7^{-2}

$7^{-2} = \dfrac{1}{7^2}$ Write the reciprocal of 7^2.

$\phantom{7^{-2}} = \dfrac{1}{49}$ Rewrite the expression as the reciprocal of 49.

The last expression in Example 2 shows us that a negative exponent does *not* denote a negative value. In other words, the expression A^{-n} doesn't necessarily represent a negative number.

Let's move on to a problem that will require us to apply multiple rules of exponents.

Example 3

Simplify: $(3a)^{-6}$

Solution

$(3a)^{-6} = \dfrac{1}{(3a)^6}$ Write the reciprocal of $(3a)^6$. Then apply the power rule for products.

$\phantom{(3a)^{-6}} = \dfrac{1}{3^6 a^6}$

$\phantom{(3a)^{-6}} = \dfrac{1}{729\, a^6}$

If a fraction is raised to a negative power, the fraction will be in parentheses. Simplifying it can be a little complicated. However, some approaches are more complicated than others.

Example 4

Simplify: $\left(\dfrac{4}{p^3}\right)^{-2}$

Solution

Approach 1: In this approach, the reciprocal of $\left(\dfrac{4}{p^3}\right)^2$ is interpreted as $\dfrac{1}{\left(\frac{4}{p^3}\right)^2}$.

$\left(\dfrac{4}{p^3}\right)^{-2} = \dfrac{1}{\left(\frac{4}{p^3}\right)^2}$ Write the reciprocal of $\left(\dfrac{4}{p^3}\right)^2$. Rewrite the expression as a division problem.

$\phantom{\left(\dfrac{4}{p^3}\right)^{-2}} = 1 \div \left(\dfrac{4}{p^3}\right)^2$ Following the correct order of operations, first apply the exponent 2.

$\phantom{\left(\dfrac{4}{p^3}\right)^{-2}} = 1 \div \dfrac{16}{p^6}$ Now divide. To divide by a fraction, multiply by the reciprocal.

$\phantom{\left(\dfrac{4}{p^3}\right)^{-2}} = 1 \cdot \dfrac{p^6}{16}$

$\phantom{\left(\dfrac{4}{p^3}\right)^{-2}} = \dfrac{p^6}{16}$

Approach 2: In this approach, the reciprocal of $\left(\frac{4}{p^3}\right)^2$ is interpreted as $\left(\frac{p^3}{4}\right)^2$.

$$\left(\frac{4}{p^3}\right)^{-2} = \left(\frac{p^3}{4}\right)^2 \quad \text{Write the reciprocal of } \left(\frac{4}{p^3}\right)^2. \text{ Then apply the power rule for quotients.}$$

$$= \frac{p^6}{16}$$

Both approaches in Example 4 gave the correct final answer. However, the second approach got us there more quickly. In the future, try to make use of the following idea:

If A and B are nonzero real numbers and n is a natural number, then $\left(\frac{A}{B}\right)^{-n} = \left(\frac{B}{A}\right)^n$.

Example 5

Simplify: $\left(\frac{d^5}{2c^7}\right)^{-4}$

Solution

$$\left(\frac{d^5}{2c^7}\right)^{-4} = \left(\frac{2c^7}{d^5}\right)^4 \quad \text{Flip the fraction inside of the parentheses and change the exponent}$$
$$\text{outside of the parentheses to a positive number. Then apply the power rule}$$
$$= \frac{2^4 c^{28}}{d^{20}} \quad \text{for quotients.}$$

$$= \frac{16 c^{28}}{d^{20}}$$

Remember, as we have seen in previous sections, if an exponent is put onto a set of parentheses containing a sum or a difference, then the power rules for exponents do not apply. However, the rule for negative exponents can still be applied.

Example 6

Simplify: $(5x - 1)^{-24}$

Solution

$$(5x - 1)^{-24} = \frac{1}{(5x - 1)^{24}} \quad \text{Apply the rule for negative exponents, and then stop.}$$

When the exponent involves a negative sign, we sometimes aren't required to apply the rule for negative exponents.

Example 7

Simplify the following expressions.

1. $(k + 2z)^{-(-8)}$

2. $(p^{-7})^{-5}$

Solutions

1. $(k + 2z)^{-(-8)}$

$(k + 2z)^{-(-8)} = (k + 2z)^{8}$ The opposite of negative 8 is positive 8. Rewrite the expression with a positive exponent, and then stop.

2. $(p^{-7})^{-5}$

$(p^{-7})^{-5} = p^{35}$ Apply the power rule for powers. The resulting exponent is positive, and so we are done.

Practice A

Using the ideas we've studied, simplify each expression. Numerical answers should be written without any exponents — for example, 7^2 should be written as 49. When you are finished, turn the page and check your work.

1. y^{-5}

2. m^{-2}

3. 3^{-2}

4. 5^{-1}

5. $(xy)^{-4}$

6. $\left(\dfrac{5m}{g^4}\right)^{-3}$

7. $(a + 2b)^{-12}$

8. $(m - n)^{-(-4)}$

B. Working with Negative Exponents

Let's explore the second form of the rule for negative exponents:

$$\frac{1}{A^{-n}} = A^{n}$$

To see why this is true, consider the following process with the expression $\dfrac{1}{x^{-5}}$.

$\dfrac{1}{x^{-5}} = 1 \div x^{-5}$ A fraction bar denotes the operation of division.

$= 1 \div \dfrac{1}{x^5}$ Apply the negative exponent rule: $x^{-5} = \dfrac{1}{x^5}$

$= 1 \cdot \dfrac{x^5}{1}$ To divide by a fraction, invert it and multiply.

$= x^5$ Simplify the answer.

This example suggests the following rule for working with exponents:

> In a fraction, a factor can be moved from the numerator to the denominator or from the denominator to the numerator. We do this by changing the sign of the exponent.

This rule only applies to *factors* in the numerator and denominator. We can use this rule to move the variable x to the denominator in the expression $\dfrac{x^{-3}y^4}{5}$. However, we can't use this rule to move the variable x to the denominator in $\dfrac{x^{-3}+y^4}{5}$.

Let's look at some examples of using this rule with expressions involving negative exponents.

Example 8

Simplify the following expressions.

1. $x^{-2}y^5$ 2. $\dfrac{a^4 b^2}{c^{-6}}$ 3. $\dfrac{1}{x^{-3}y^{-2}z^{-1}}$

Solutions

1. $x^{-2}y^5$

   We begin by writing this expression as a fraction, with 1 as the denominator.

   $$\frac{x^{-2}y^5}{1} = \frac{y^5}{1\,x^2}$$ Move the factor x^{-2} from the numerator to the denominator by changing the exponent -2 to $+2$.

   $$= \frac{y^5}{x^2}$$

2. $\dfrac{a^4 b^2}{c^{-6}}$

   $$\frac{a^4 b^2}{c^{-6}} = \frac{a^4 b^2 c^6}{1}$$ Move the factor c^{-6} into the numerator by changing the exponent -6 to $+6$. With this rule, if nothing is left in a numerator or a denominator, we leave a 1, so the denominator of this fraction is written as 1.

   $$= a^4 b^2 c^6$$ Finish by simplifying the expression.

   When doing problems like this in the future, we can skip the step in which a fraction is written with a denominator of 1.

3. $\dfrac{1}{x^{-3}y^{-2}z^{-1}}$

   $$\frac{1}{x^{-3}y^{-2}z^{-1}} = x^3 y^2 z^1$$ Move all of the variables into the numerator and change the signs of all of the exponents. The denominator of 1 is not shown.

   $$= x^3 y^2 z$$ It's not necessary to write an exponent of 1, so the expression has been simplified by writing the variable z without an exponent.

Sometimes we have to simplify and reduce numerical coefficients as part of the process of simplifying a fraction.

Example 9

Simplify: $\dfrac{24\,a^7\,b^9}{2^3\,a^4\,b^{-6}}$

Solution

$$\dfrac{24\,a^7\,b^9}{2^3\,a^4\,b^{-6}} = \dfrac{24\,a^7\,b^9}{8\,a^4\,b^{-6}}$$

Begin by simplifying the numerical coefficient in the denominator: $2^3 = 8$

$$= 3\,a^{7-4}\,b^{9-(-6)}$$

Next, reduce the numerical coefficients, $\dfrac{24}{8} = 3$, and use the quotient rule for exponents to combine variables.

$$= 3\,a^3\,b^{15}$$

Finish by simplifying the expression.

In this example, the exponents on the variables in the denominator are less than the exponents on the variables in the numerator. Because of this, the quotient rule for exponents gives us positive exponents on the variables in our answer.

If the exponents on variables in the denominator are *not* less than exponents on the corresponding variables in the numerator, then the quotient rule for exponents leads to a negative exponent in the answer. In cases like that, the rule for moving factors can shorten the process somewhat.

Example 10

Simplify: $\dfrac{x\,y^6\,z^{-8}}{x^5\,y^2\,z^{-3}}$

Solution

Approach 1: Use the quotient rule for exponents.

$$\dfrac{x\,y^6\,z^{-8}}{x^5\,y^2\,z^{-3}} = x^{1-5}\,y^{6-2}\,z^{-8-(-3)}$$

Begin by applying the quotient rule for exponents. Then simplify the exponents.

$$= x^{-4}\,y^4\,z^{-5}$$

Apply the rule for negative exponents.

$$= \dfrac{1}{x^4} \cdot y^4 \cdot \dfrac{1}{z^5}$$

Finish by multiplying the expressions together.

$$= \dfrac{y^4}{x^4\,z^5}$$

Approach 2: Use the rule for moving factors.

$$\dfrac{x\,y^6\,z^{-8}}{x^5\,y^2\,z^{-3}} = \dfrac{y^6\,y^{-2}}{x^5\,x^{-1}\,z^{-3}\,z^8}$$

For each variable, move the one with the lowest exponent. The x in the numerator has the lowest exponent, so move it to the denominator. The y in the denominator has the lowest exponent, so move it to the numerator. The z in the numerator has the lowest exponent, so move it to the denominator.

$$= \dfrac{y^4}{x^4\,z^5}$$

Use the product rule for exponents to combine variables.

When using Approach 2, we make the process even shorter with mental math:

1. For x, the larger exponent, 5, was in the denominator and $5 - 1 = 4$. Therefore, we end up with x^4 in the denominator.

2. For y, the larger exponent, 6, was in the numerator and $6 - 2 = 4$. Therefore, we end up with y^4 in the numerator.

3. For z, the larger exponent, -3, was in the denominator and $-3 + 8 = 5$. Therefore, we end up with z^5 in the denominator.

By doing these computations mentally, the rule for moving factors allows us to arrive at the result in just one step of writing:

$$\frac{xy^6z^{-8}}{x^5y^2z^{-3}} = \frac{y^4}{x^4z^5}$$

Example 11

Write $\frac{9a^5b^3}{5x^3y^2}$ so that no denominator appears.

Solution

$$\frac{9a^5b^3}{5x^3y^2} = 9a^5b^35^{-1}x^{-3}y^{-2}$$

We're instructed to write the expression without a denominator, so *every* factor in the denominator must be moved to the numerator.

Because of the instructions in this example, our final answer included negative exponents. We typically don't want negative exponents in our answers. However, as we'll see in the next section, when we perform operations with numbers in scientific notation, we often end up with negative exponents in our answer.

Practice A — Answers

1. $\frac{1}{y^5}$

2. $\frac{1}{m^2}$

3. $\frac{1}{9}$

4. $\frac{1}{5}$

5. $\frac{1}{x^4y^4}$

6. $\frac{g^{12}}{125\,m^3}$

7. $\frac{1}{(a+2b)^{12}}$

8. $(m-n)^4$

Let's finish by looking at the problem from the beginning of this section.

Example 12

Evaluate: $\dfrac{1}{10^{-2}} + \dfrac{3}{4^{-3}}$

Solution

$\dfrac{1}{10^{-2}} + \dfrac{3}{4^{-3}} = 1 \cdot 10^2 + 3 \cdot 4^3$ Move the exponential expressions from the denominators into the numerators, and switch the signs of the exponents. Then follow the correct order of operations to simplify the number.

$$= 1 \cdot 100 + 3 \cdot 64$$

$$= 100 + 192$$

$$= 292$$

Practice B

Now it's your turn. Using the techniques we have discussed in this section, simplify each expression. Unless otherwise directed, make sure your final answer does not include any negative exponents. When you are finished, turn the page and check your answers.

9. Simplify: $x^{-4}y^7$

10. Simplify: $\dfrac{a^2}{b^{-4}}$

11. Simplify: $\dfrac{x^3 y^4}{z^{-8}}$

12. Simplify: $\dfrac{6 m^{-3} n^{-2}}{7 k^{-1}}$

13. Simplify: $\dfrac{1}{a^{-2} b^{-6} c^{-8}}$

14. Simplify: $\dfrac{3a(a-5b)^{-2}}{5b(a-4b)^5}$

15. Simplify: $\dfrac{36 x^8 b^3}{3^2 x^{-2} b^{-5}}$

16. Write $\dfrac{2^4 m^{-3} n^7}{4^{-1} x^5}$ as an expression in which no denominator appears.

17. Evaluate: $\dfrac{2}{5^{-2}} + 6^{-2} \cdot 2^3 \cdot 3^2$

Exercises 4.5

Simplify each of the following expressions. Assume all variables are nonzero. Numerical answers should be written without any exponents — for example, 2^3 should be written as 8.

1. x^{-2}

2. a^{-10}

3. b^{-12}

4. y^{-1}

5. n^{-5}

6. 8^{-2}

7. 2^{-7}

8. 6^{-3}

9. -7^{-2}

10. $(-3)^{-4}$

11. $(-5)^{-2}$

12. -10^{-4}

13. $(k^2 m)^{-1}$

14. $(q^2 r^4)^{-3}$

15. $(xy^5)^{-7}$

16. $\left(\frac{2}{c^3}\right)^{-5}$

17. $\left(\frac{9x^4}{y^{11}}\right)^{-2}$

18. $\left(\frac{4a^2}{b^{10}}\right)^{-3}$

19. $\left(\frac{n^5 m^8}{3}\right)^{-4}$

20. $(x+1)^{-2}$

21. $(x-5)^{-13}$

22. $(a-1)^{-12}$

23. $x^3 y^{-2}$

24. $x^7 y^{-5}$

25. $a^4 b^{-1}$

26. $x^3 y^2 z^{-6}$

27. $x^3 y^{-4} z^2 w$

28. $x^5 y^{-5} z^{-2}$

29. $x^9 y^{-6} z^{-1} w^{-5} r^{-2}$

30. $4x^{-6} y^2$

31. $5x^2 y^2 z^{-5}$

32. $4x^3 (x+1)^2 y^{-4} z^{-1}$

33. $7a^2 (a-4)^3 b^{-6} c^{-7}$

34. $18b^{-6} (b^2-3)^{-5} c^{-4} d^5 e^{-1}$

35. $7(w+2)^{-2} (w+1)^3$

36. $2(a-8)^{-3} (a-2)^5$

37. $(x^2+3)^3 (x^2-1)^{-4}$

38. $(x^4+2x-1)^{-6} (x+5)^4$

39. $5x^3 (2x^{-7})$

40. $3y^{-3} (9x)$

41. $6a^{-4} (2a^{-6})$

42. $4a^2 b^2 a^{-5} b^{-2}$

43. $5^{-1} a^{-2} b^{-6} b^{-11} c^{-3} c^9$

44. $2^3 x^2 2^{-3} x^{-2}$

45. $7a^{-3} b^{-9} \cdot 5a^6 b c^{-2} c^4$

46. $(x+5)^2 (x+5)^{-6}$

47. $(a-4)^3 (a-4)^{-10}$

48. $-4a^3 b^{-5} (2a^2 b^7 c^{-2})$

49. $-2x^{-2} y^{-4} z^4 (-6x^3 y^{-3} z)$

50. $(2x^{-2} y^7)^4 \cdot (3xy^{-3})^2$

51. $(5ab^{-6})^2 \cdot (2a^{-7} b^2)^3$

52. $\frac{1}{a^{-4}}$

53. $\frac{1}{a^{-1}}$

54. $\frac{6}{a^2 b^{-4}}$

55. $\dfrac{3c^5}{a^3 b^{-3}}$

56. $\dfrac{16 a^{-2} b^{-6} c}{2yz^{-5} w^{-4}}$

57. $\dfrac{24 y^2 z^{-8}}{6 a^2 b^{-1} c^{-9} d^3}$

58. $\dfrac{3^{-1} b^5 (b+7)^{-4}}{9^{-1} a^{-4} (a+7)^2}$

59. $\dfrac{36 a^6 b^5 c^8}{3^2 a^3 b^7 c^9}$

60. $\dfrac{45 a^4 b^2 c^6}{15 a^2 b^7 c^8}$

61. $\dfrac{3^3 x^4 y^3 z}{3^2 x y^5 z^5}$

62. $\dfrac{21 x^2 y^2 z^5 w^4}{7xy z^{12} w^{14}}$

63. $\dfrac{33 a^{-4} b^{-7}}{11 a^3 b^{-2}}$

64. $\dfrac{2^6 x^{-5} y^{-2} a^{-7} b^5}{2^{-1} x^{-4} y^{-2} b^6}$

65. $\dfrac{(x+3)^3 (y-6)^4}{(x+3)^5 (y-6)^{-8}}$

66. $\left(\dfrac{5^{-1} a^3 b^{-6}}{x^{-2} y^9}\right)^2$

67. $\left(\dfrac{4 m^{-3} n^6}{2 m^{-5} n}\right)^3$

68. $\left(\dfrac{r^5 s^{-4}}{m^{-8} n^7}\right)^{-4}$

69. $\left(\dfrac{h^{-2} j^{-6}}{k^{-4} p}\right)^{-5}$

70. $8^2 \cdot 2^{-4} + \dfrac{1}{6^{-3}}$

71. $\dfrac{3}{5^{-1}} + 10^4 \cdot 4^{-2}$

72. $\left(\dfrac{7}{8}\right)^{-2} \cdot \left(\dfrac{1}{14}\right)^{-1}$

73. $9^2 \cdot 3^{-3} - \dfrac{6}{8^{-2}}$

For each of the following exercises, rewrite the expression so that no denominator appears.

74. $\dfrac{4 x^3}{y^7}$

75. $\dfrac{5 x^4 y^3}{a^3}$

76. $\dfrac{23 a^4 b^5 c^{-2}}{x^{-6} y^5}$

77. $\dfrac{2^3 b^5 c^2 d^{-9}}{4 b^4 c d^2}$

78. $\dfrac{18 x^3 y^{-7}}{3 x^5 z^2}$

Practice B — Answers

9. $\dfrac{y^7}{x^4}$

10. $a^2 b^4$

11. $x^3 y^4 z^8$

12. $\dfrac{6k}{7 m^3 n^2}$

13. $a^2 b^6 c^8$

14. $\dfrac{3a}{5b(a-5b)^2(a-4b)^5}$

15. $4 x^{10} b^8$

16. $64 m^{-3} n^7 x^{-5}$

17. 52

4.6 Applications of Exponent Rules

Overview

Now it's time to explore two important applications that will require you to apply what you know about exponents — unit conversions and scientific notation.

Here's a related problem:

> Kenyatta is planning to put 3 inches (0.25 feet) of course mulch in a rectangular garden that measures 40 feet by 32 feet. He calculates the volume of mulch needed to be: $0.25(40)(32) = 320$ ft^3. However, the nursery sells mulch in cubic yards, not cubic feet. How many cubic yards of mulch should Kenyatta order?

In order to answer this question, we need to convert cubic feet into cubic yards. We'll learn about how the rules of exponents apply when we convert units, and then return to this problem.

This section will show you how to:

- ◆ Use fractions and rules of exponents to convert units of measure
- ◆ Convert numbers between standard decimal notation and scientific notation
- ◆ Multiply and divide numbers that are written in scientific notation

A. Converting Units of Measure

Many real-life applications require us to convert measurements from one type of unit into another. One method for doing this involves the use of conversion fractions. A **conversion fraction** is a fraction whose numerator and denominator represent the same quantity but are written with different units. We know that 12 inches is equal to 1 foot, for example. This fact allows us to write the following conversion fractions:

$$\frac{12 \text{ inches}}{1 \text{ foot}} \quad \text{and} \quad \frac{1 \text{ foot}}{12 \text{ inches}}$$

When converting units of measure, we can multiply a given measurement by a conversion fraction and cancel out the given type of unit. This will leave us with an equivalent measurement that has a different type of unit. Doing this is often referred to as *dimensional analysis*.

Example 1

Convert 96 inches into feet.

Solution

Because we are multiplying by a fraction, we begin by rewriting the given measure as a fraction with a denominator of 1.

$$96 \text{ inches } = \frac{96 \text{ in}}{1}$$ We need a conversion fraction that will cancel out inches and leave us with feet.

$$= \frac{96 \text{ in}}{1} \cdot \frac{1 \text{ ft}}{12 \text{ in}}$$ We use the conversion fraction that has inches in the denominator and feet in the numerator.

$$= \frac{96 \text{ i\!n\!}}{1} \cdot \frac{1 \text{ ft}}{12 \text{ i\!n\!}}$$ Cancel out the inches, leaving feet as the only unit of measure

$$= \frac{96 \text{ ft}}{12}$$ Simplify the result.

$$= 8 \text{ ft}$$

In this example, we treated the units of measure like variables and showed the cancellation taking place. We could have used the quotient rule and zero exponent rule in this process, and we would have gotten the same result, but with more steps:

$$\frac{96}{12} \text{ in}^0 \text{ft} = 8 \cdot 1 \cdot \text{ft} = 8 \text{ ft}$$

Sometimes, there are exponents with units of measure. When reporting area, we use square units such as in^2, cm^2, and ft^2. When reporting volume, we use cubic units such as mm^3, yd^3, and m^3. When we use conversion fractions to change units of measure, we have to take this into account.

Example 2

There are 100 cm in 1 m. Convert 3 m^2 into cm^2.

Solution

We'll use the fact that there are 100 cm in 1 m to help us write a conversion fraction for this problem.

$$3 \text{ m}^2 = \frac{3 \text{ m}^2}{1}$$ Write the measure as a fraction with a denominator of 1.

$$= \frac{3 \text{ m}^2}{1} \cdot \frac{100 \text{ cm}}{1 \text{ m}}$$ Write a conversion fraction with meters in the denominator. The m^2 in the numerator is not completely canceled by the m in the denominator: $\left(\frac{\text{m}^2}{\text{m}} = \text{m} \right)$.

$$= \frac{3 \text{ m}^2}{1} \cdot \frac{100 \text{ cm}}{1 \text{ m}} \cdot \frac{100 \text{ cm}}{1 \text{ m}}$$ Insert a second conversion fraction.

$$\frac{3 \text{ m}^2}{1} \cdot \left(\frac{100 \text{ cm}}{1 \text{ m}} \cdot \frac{100 \text{ cm}}{1 \text{ m}} \right)$$ Because all of the fractions are being multiplied, we multiply the two conversion fractions together first.

$$\frac{3 \text{ m}^2}{1} \cdot \left(\frac{10,00 \text{ cm}^2}{1 \text{ m}^2} \right)$$

Multiply the numbers, and use the product rule for exponents to combine units.

$$\frac{3 \cancel{\text{m}^2}}{1} \cdot \frac{10,000 \text{ cm}^2}{1 \cancel{\text{m}^2}}$$

Now we can **cancel the m².**

$$= \frac{30,000 \text{ cm}^2}{1}$$

Finish simplifying the result.

$$= 30,000 \text{ cm}^2$$

When converting square units or cubic units, we can shorten the process slightly by writing a single conversion fraction and then squaring (or cubing) all of the numbers and units in that fraction. By doing this, we avoid writing multiple conversion fractions. In the previous example, we could have written a single conversion fraction, $\frac{10^2 \text{ cm}^2}{1^2 \text{ m}^2}$, instead of the two conversion fractions that we wrote.

Example 3

There are 2.54 cm in 1 in. Convert 100 cubic centimeters into cubic inches. Round your answer to the nearest hundredth.

Solution

$$100 \text{ cm}^3 = \frac{100 \text{ cm}^3}{1}$$

$$= \frac{100 \text{ cm}^3}{1} \cdot \frac{1^3 \text{ in}^3}{2.54^3 \text{ cm}^3}$$

To convert cubic units, **write one conversion fraction** with numbers and units being cubed.

$$= \frac{100 \cancel{\text{cm}^3}}{1} \cdot \frac{1 \text{ in}^3}{16.387064 \cancel{\text{cm}^3}}$$

Cancel the cm³ and use a calculator to evaluate 2.54^3. We don't round off yet because that could cause a round-off error.

$$= \frac{100}{16.387064} \text{ in}^3$$

Divide 100 by 16.387064, and round the result to the nearest hundredth.

$$\approx 6.10 \text{ in}^3$$

The symbol "≈" means "is approximately equal." We use it when reporting an answer that is rounded off.

Sometimes, multiple conversion fractions are unavoidable, as in the next example.

Example 4

There are 2.54 cm in 1 in, and 1,000 cm in 1 m. Convert 70,000 square inches into square meters. Round your answer off to the nearest hundredth.

Solution

We aren't given enough information to convert square inches directly into square meters. However, we can use one conversion fraction to convert in² into cm², and another conversion fraction to convert cm² into m².

$$70,000 \text{ in}^2 = \frac{70,000 \text{ in}^2}{1}$$

$$= \frac{70,000 \text{ in}^2}{1} \cdot \frac{2.54^2 \text{ cm}^2}{1^2 \text{ in}^2} \cdot \frac{1^2 \text{ m}^2}{1,000^2 \text{ cm}^2}$$

$$= \frac{70,000 \text{ in}^2}{1} \cdot \frac{6.4516 \text{ cm}^2}{1 \text{ in}^2} \cdot \frac{1 \text{ m}^2}{1,000,000 \text{ cm}^2}$$

$$= \frac{451,612}{1,000,000} \text{ m}^2$$

$$\approx 0.45 \text{ m}^2$$

Now that we understand how conversion fractions work with square and cubic units, we're ready to revisit the problem from the beginning of this section.

Example 5

Kenyatta plans to put 3 inches (0.25 feet) of course mulch in a rectangular garden that measures 40 feet by 32 feet. He calculates the volume of mulch needed to be: $0.25(40)(32) = 320 \text{ ft}^3$. However, the nursery sells mulch in cubic yards, not cubic feet. There are 3 feet in 1 yard. How many cubic yards of mulch should Kenyatta order?

Solution

In order to solve this problem, we need to convert 320 cubic feet into cubic yards:

$$320 \text{ ft}^3 = \frac{320 \text{ ft}^3}{1}$$

$$= \frac{320 \text{ ft}^3}{1} \cdot \frac{1^3 \text{ yd}^3}{3^3 \text{ ft}^3}$$

$$= \frac{320 \text{ ft}^3}{1} \cdot \frac{1 \text{ yd}^3}{27 \text{ ft}^3}$$

$$= \frac{320}{27} \text{ yd}^3$$

$$\approx 11.85185185 \text{ yd}^3$$

Rounding off our answer off to the nearest hundredth, we can say that Kenyatta will need about 11.85 cubic yards of mulch.

Practice A

Convert the given measurements into equivalent measurements with different units. For each problem, show how the expression is set up with conversion fraction(s). Then write your final answer, rounding off to the nearest hundredth if necessary. Turn the page to check your work.

1. 3 ft = 1 yd. Convert 2 cubic yards into cubic feet.

2. 1,000 mm = 1 m. Convert 13,726,500 square millimeters into square meters.

3. 2.54 cm = 1 in, and 12 in = 1 ft. Convert 150 square centimeters into square feet.

B. Standard Decimal Notation and Scientific Notation

There are over 43,000,000,000,000,000,000 possible configurations of a Rubik's cube. One molecule of tryptophan has a mass of about 0.00000000000000000000339 grams. When numbers are very large or very small, they can be cumbersome to read and write. Fortunately, these kinds of numbers can be better expressed as a product involving a power of 10.

Let's see how this works with a more manageable number: 2,480.

Written in Standard Decimal Notation		*Written as a Product Involving a Power of 10*
2480	$=$	248.0×10^1
	$=$	24.80×10^2
	$=$	2.480×10^3

The final form in the table above is called *scientific notation*. You should notice two things about the number written in scientific notation:

- The first factor, 2.480, has one nonzero digit to the left of the decimal point. In other words, the first factor is between 1 and 10.

- The second factor, 10^3, has an exponent that is equal to the number of spaces that the decimal point in the original number was moved.

Also notice that the symbol "×" is used for multiplication. This is something mathematicians try to avoid, due to possible confusion involving the variable x. However, this symbol is preferred when writing numbers in scientific notation.

Here's another example illustrating how standard decimal notation and scientific notation are related.

$$0.00059 \quad = \quad \frac{0.0059}{10} \quad = \quad \frac{0.0059}{10^1} \quad = \quad 0.0059 \times 10^{-1}$$

$$= \quad \frac{0.059}{100} \quad = \quad \frac{0.059}{10^2} \quad = \quad 0.059 \times 10^{-2}$$

$$= \quad \frac{0.59}{1000} \quad = \quad \frac{0.59}{10^3} \quad = \quad 0.59 \times 10^{-3}$$

$$= \quad \frac{5.9}{10,000} \quad = \quad \frac{5.9}{10^4} \quad = \quad 5.9 \times 10^{-4}$$

In its final form, the number is again written in scientific notation. Let's analyze the factors of this expression.

- The first factor, 5.9, is between 1 and 10.

- The second factor, 10^{-4}, has an exponent that is the *opposite* of the number of spaces that the decimal point in the original number was moved.

Now that we've seen two examples of numbers written in scientific notation, let's look at the formal definition. A number is in **scientific notation** if it is written as: $A \times 10^n$, where $1 \le A < 10$ and n is an integer. Notice that 1 is an acceptable value for the first factor but 10 is not.

We can use the following procedure to convert a number from standard decimal notation into scientific notation.

Converting from Standard Decimal to Scientific Notation

1. To form the first factor, move the decimal point until there is one nonzero digit to its left. In other words, move the decimal in order to form a number that is between 1 and 10.

2. To form the second factor, write a power of 10. Determine the exponent as follows:
   - If the decimal point was moved *left* to form the first factor, then the exponent will be the same as the number of spaces that the decimal point moved.
   - If the decimal point was moved *right* to form the first factor, then the exponent will be the *opposite* of the number of spaces that the decimal point moved.

We can ensure that the exponent on the 10 is correct by thinking of this fact: If the number in standard decimal notation is greater than 10, then the number in scientific notation has a *positive* power of 10. If the number in standard decimal notation is less than 1, then the number in scientific notation has a *negative* power of 10.

Example 6

Write the following numbers in scientific notation.

1. 981
2. 54.066
3. 0.000000000004632
4. 0.027

Solutions

1. 981

$981 = 981.0$	When writing integers, the decimal point is typically not shown. However, the number 981 is equal to 981.0. In this form, we see that the decimal point comes after the 1.
$= 9.81 \times 10^2$	In scientific notation, the decimal point is now two places to the left of its original position. Because we moved the decimal to the left, the exponent on the 10 is positive 2.

We can double check our answer by verifying that, because 981 is greater than 10, the scientific notation involves a positive power of 10.

2. 54.066

$54.066 = 5.4066 \times 10^1$ The decimal point moves one place to the left of its original position, so the power of 10 is 1.

$= 5.4066 \times 10$ When the power of 10 is one, we can omit the exponent.

3. 0.000000000004632

$0.000000000004632 = 4.632 \times 10^{-12}$ The decimal point moves twelve places to the right of its original position. Because we move the decimal point to the right, the power of 10 is –12.

Again, we can check to make sure our answer is reasonable. Because 0.000000000004632 is less than 1, the scientific notation involves a negative power of 10.

4. 0.027

$0.027 = 2.7 \times 10^{-2}$ The decimal point is moved two places to the right of its original position, so the power of 10 is –2.

Now, let's talk about how to convert a number from scientific notation into standard decimal notation. This process is essentially the reverse of the process we've been using to convert numbers *into* scientific notation.

Converting from Scientific to Standard Decimal Notation

1. Locate the decimal point in the first factor, and move it the number of spaces prescribed by the exponent on second factor:
   ◆ Move the decimal point to the right when the exponent is positive.
   ◆ Move the decimal point to the left when the exponent is negative.

2. Insert zeros as necessary to ensure that the number in standard form has the correct number of digits.

Again, if you get confused about which direction to move the decimal point, remember the fact that numbers less than 1 have *negative* powers of 10 when written in scientific notation and numbers greater than 10 have *positive* powers of 10 when written in scientific notation.

Practice A — Answers

1. $\dfrac{2 \text{ yd}^3}{1} \cdot \dfrac{3^3 \text{ ft}^3}{1^3 \text{ yd}^3}; 54 \text{ ft}^3$

2. $\dfrac{13,726,500 \text{ mm}^2}{1} \cdot \dfrac{1^2 \text{ m}^2}{1,000^2 \text{ mm}^2};$

   About 13.73 m²

3. $\dfrac{150 \text{ cm}^2}{1} \cdot \dfrac{1^2 \text{ in}^2}{2.54^2 \text{ cm}^2} \cdot \dfrac{1^2 \text{ ft}^2}{12^2 \text{ in}^2};$

   About 0.16 ft²

Example 7

Write the following numbers in standard decimal notation.

1. 4.673×10^4

2. 2.9×10^7

3. 4.21×10^{-5}

4. 1.006×10^{-18}

Solutions

1. 4.673×10^4

$4.673 \times 10^4 = 46730.$ The exponent of 10 is +4, so we move the decimal point to the *right* 4 places. This requires us to insert one zero on the left side of the decimal point.

$= 46,730$ We then omit the decimal point and insert a comma between the digits in the hundreds and thousands place.

Because the scientific notation involves a positive power of 10, the number in standard decimal notation ends up being greater than 10.

2. 2.9×10^7

$2.9 \times 10^7 = 29000000$ The exponent of 10 is +7. Move the decimal point 7 places to the right and insert the necessary zeros.

$= 29,000,000$ We also insert commas in the customary positions when writing our final answer.

3. 4.21×10^{-5}

$4.21 \times 10^{-5} = 0.0000421$ This time, the exponent on the 10 is −5. Move the decimal point 5 places to the left, inserting the necessary zeros between the decimal point and the first nonzero digit.

The scientific notation involves a negative power of 10, so the number in standard decimal notation ends up being less than 1.

4. 1.006×10^{-18}

$1.006 \times 10^{-18} = 0.000000000000000001006$ The exponent of 10 is −18, so we move the decimal point 18 places to the left and insert the necessary zeros to the right of the decimal point.

In part 3 of the the previous example, the number in scientific notation includes an exponent of negative *five*, but the number in standard decimal notation only has *four* zeros inserted between the decimal point and the first nonzero digit. We see something similar in part 4. In scientific notation, the exponent on the 10 is negative *eighteen*, but in standard decimal notation there are *seventeen* zeros inserted between the decimal point and the first nonzero digit.

Practice B

Now it's your turn. Write the following numbers using scientific notation. Turn the page when you are finished and check your answers there.

4. 346

5. 72.33

6. 5387.7965

7. 179,000,000,000,000,000,000

8. 0.0086

9. 0.000098001

10. 0.00000000000000000054

Convert the following numbers into standard decimal notation.

11. 9.25×10^2 12. 4.01×10^5 13. 1.2×10^{-1} 14. 8.88×10^{-5}

C. Multiplying and Dividing Numbers in Scientific Notation

There are many times — particularly in the sciences — when it's necessary to find the product or quotient of two numbers written using scientific notation. We will do this by using the same approach we used when multiplying or dividing monomials. Remember that when multiplying monomials, we simply multiply the coefficients and use the product rule for exponents to combine the variables.

Example 8

Multiply: $2x^3 \cdot 4x^8$

Solution

$$2x^3 \cdot 4x^8 = 8x^{11}$$ We multiply the coefficients, $2 \cdot 4 = 8$, and use the product rule for exponents to combine the variables, $x^3 x^8 = x^{11}$.

When multiplying numbers in scientific notation, we treat the starting numbers like coefficients and multiply them together. We treat the 10s like variables and combine them using the product rule for exponents.

Example 9

Multiply: $(2 \times 10^3)(4 \times 10^8)$

Solution

$(2 \times 10^3)(4 \times 10^8) = 8 \times 10^{11}$ This problem is similar to Example 8. 2 and 4 are the coefficients and the 10s are the variables. We multiply coefficients, $2 \cdot 4 = 8$, and use the product rule for exponents to combine the 10s: $10^3 \cdot 10^8 = 10^{11}$.

When multiplying numbers in scientific notation, we sometimes end up with a result that is not in scientific notation. When this happens, we must take an additional step to convert our answer back into scientific notation. Here is an example of how that works.

Example 10

Multiply: $(5 \times 10^{17})(8.1 \times 10^{-22})$

Solution

$$(5 \times 10^{17})(8.1 \times 10^{-22}) = 40.5 \times 10^{-5}$$ We multiply 5 and 8.1, and use the product rule for exponents to combine the 10s.

$$= 4.05 \times 10^1 \times 10^{-5}$$ Because 40.5 is greater than 10, we must move the decimal point one place to the left to convert it into scientific notation: $40.5 = 4.05 \times 10^1$.

$$= 4.05 \times 10^{-4}$$ We finish by using the product rule for exponents to combine the 10s.

Now let's take a look at division. When dividing monomials, we divide the coefficients and use the quotient rule for exponents to combine the variables.

Example 11

Divide: $\dfrac{9 n^{18}}{3 n^6}$

Solution

$$\dfrac{9 n^{18}}{3 n^6} = 3 n^{12}$$ Divide the coefficients, $\dfrac{9}{3} = 3$, and then use the quotient rule for exponents to combine the variables: $\dfrac{n^{18}}{n^6} = n^{12}$.

Once again, when dealing with numbers in scientific notation, we treat the starting numbers like coefficients and the 10s like variables.

Example 12

Divide: $\dfrac{9 \times 10^{18}}{3 \times 10^6}$

Solution

$$\dfrac{9 \times 10^{18}}{3 \times 10^6} = 3 \times 10^{12}$$ Divide the coefficients, 9 and 3, and use the quotient rule to combine the 10s.

When dividing numbers in scientific notation, some people are tempted to write the answer with a positive power of 10 in the denominator. Resist that temptation. If you do that, the number is *not* in scientific notation. Consider the following example.

Example 13

Divide: $\dfrac{8.6 \times 10^3}{4.3 \times 10^{11}}$

Solution

$\dfrac{8.6 \times 10^3}{4.3 \times 10^{11}} = \dfrac{2}{10^8}$ Mathematically, this is correct, but it's not written in scientific notation, so it's not the correct answer.

$= 2 \times 10^{-8}$ This is mathematically correct, and it *is* in scientific notation, so this is the correct answer.

The previous example shows how the strategy for dividing numbers in scientific notation differs from the strategy for simplifying fractions involving monomials. When simplifying fractions involving monomials, we're not finished until every variable has a positive exponent. When dividing numbers in scientific notation, everything must end up in the numerator, even if the final result has a negative power of 10.

As we saw with multiplication, it's possible to end up with a starting number that's not between 1 and 10 when we divide numbers in scientific notation. When this happens, we take an additional step to convert our answer back into scientific notation.

Example 14

Divide: $\dfrac{2 \times 10^7}{8 \times 10^{-16}}$

Solution

$\dfrac{2 \times 10^7}{8 \times 10^{-16}} = 0.25 \times 10^{23}$ When dividing the coefficients, 2 and 8, we end up with $\frac{1}{4}$ or 0.25. Fractions and scientific notation don't mix, so we use 0.25.

$= 2.5 \times 10^{-1} \times 10^{23}$ 0.25 is less than 1, so we move the decimal point one place to the right to convert it into scientific notation: $0.25 = 2.5 \times 10^{-1}$.

$= 2.5 \times 10^{22}$ Finish by using the product rule for exponents to combine the 10s.

Practice B — Answers

4. 3.46×10^2	8. 8.6×10^{-3}	12. 401,000
5. 7.233×10	9. 9.8001×10^{-5}	13. 0.12
6. 5.3877965×10^3	10. 5.4×10^{-17}	14. 0.0000888
7. 1.79×10^{20}	11. 925	

Now that we are familiar with the mechanics of multiplying and dividing numbers in scientific notation, let's take a look at how to use a calculator to perform these operations.

Example 15

Use a graphing calculator to perform each operation.

1. Multiply: $(1.4 \times 10^{23})(2 \times 10^{12})$

2. Divide: $\frac{7.5 \times 10^{-10}}{2.5 \times 10^{-48}}$

Solutions

1. Multiply: $(1.4 \times 10^{23})(2 \times 10^{12})$

Step 1: One way to enter a number in scientific notation is with the [EE] key. This can be accessed by pressing [2nd] and [,]. The comma key is above the number 7. Doing this causes the letter E to appear on the screen instead of "×10". Begin by entering the first number in parentheses.

```
(1.4E23)
```

Step 2: Next, open another set of parentheses and enter the second number. Again, use the [EE] key to insert an E in place of "×10". After closing the parentheses, press [ENTER] to display the answer.

The screen reads 2.8E35. This means that our answer is: 2.8×10^{35}

```
(1.4E23)(2E12)
            2.8E35
```

2. Divide: $\frac{7.5 \times 10^{-10}}{2.5 \times 10^{-48}}$

Step 1: This time, we'll enter *10^ to indicate that we're multiplying by a power of 10. We do this by pressing the [×] key, then entering the number 10, and then pressing the [^] key.
Begin by entering the first number in parentheses.
   If you have a newer calculator, the display may show a nice exponential expression, 10^{-10}, as part of the number in scientific notation. If this happens, use the arrow key to move the cursor out of the exponent before closing the parentheses. If you have an older calculator, the expression will look like this: (7.5*10^-10)

```
(7.5*10⁻¹⁰)
```

Step 2: Now press the [÷] key, and a division slash will appear. After the slash, enter the second number in parentheses. Once again, we'll use *10^ to enter the power of 10. Finish by pressing [ENTER].

The calculator has converted the power of 10 into the letter E. Our answer is: 3×10^{38}

```
(7.5*10⁻¹⁰)/(2.5▸
            3E38
```

The problems in the previous example both contain coefficients that are fairly easy to deal with, so it would probably be faster to do them without a calculator. In many cases, however, we encounter problems with coefficients that are not nice at all. In those cases, it is important to be able to use a calculator to perform the operation.

Practice Set C

Perform each multiplication or division on your own. Make sure your final answer is written in scientific notation. When you are finished, turn the page to check your answers.

15. Multiply: $(3 \times 10^5)(2 \times 10^{12})$

16. Multiply: $(5 \times 10^{18})(3 \times 10^6)$

17. Multiply: $(2.1 \times 10^{-9})(3 \times 10^{-11})$

18. Divide: $\dfrac{7.2 \times 10^{-12}}{2.4 \times 10^{-21}}$

19. Divide: $\dfrac{3.6 \times 10^5}{9 \times 10^{19}}$

20. Divide: $\dfrac{1.0192 \times 10^3}{9.8 \times 10^{-10}}$

Exercises 4.6

Use the conversion table on this page to convert each measurement, rounding off to the nearest hundredth.

1.   Convert 2 cubic yards into cubic feet.

2.   Convert 4 square meters into square millimeters.

3.   Convert 4 cubic yards into cubic meters.

4.   Convert 10 square miles into square kilometers.

5.   Convert 30,000 cubic inches into cubic yards.

6.   Convert 100,000 square centimeters into square kilometers.

7.   Convert 1 square mile into square yards.

8.   Convert 14,000 cubic inches into cubic meters.

9.   Convert 254,000,000 cubic centimeters into cubic inches.

10. Convert 313,000 square millimeters into square inches.

11. Convert 313,000 square centimeters into square yards.

12. Convert 313,000 cubic kilometers into cubic miles.

1 foot	=	12 inches
1 yard	=	3 feet
1 mile	=	5280 feet
1 inch	=	2.54 centimeters
1 meter	=	100 centimeters
1 meter	=	1000 millimeters
1 kilometer	=	1000 meters
1 mile	≈	1.609 kilometers

For the following exercises, perform the necessary measurement conversions and answer the questions.

13. Veronica is buying carpet to put on the floor of a rectangular room measuring 243 inches by 330 inches. Carpet is sold by the square foot. How many square feet of carpet will Veronica need to buy? If necessary, round answer to the nearest hundredth.

14. A soccer field measures 115 yards by 74 yards. The groundskeeper is ordering new sod for the field, but sod is sold by the square foot. How many square feet of sod should be purchased in order to cover the entire soccer field? If necessary, round answer to the nearest whole number.

15. The Open Sea exhibit at the Oregon Coast Aquarium contains about 107,000 cubic feet of water. How many cubic meters is this? If necessary, round answer to the nearest whole number.

16. Marta wants to put 1.5 inches of fine mulch in a rectangular flower bed measuring 22 feet by 18 feet. The nursery sells mulch in cubic yards. How many cubic yards of mulch will Marta need to buy? If necessary, round answer to the nearest hundredth.

Convert the numbers used in the following problems to scientific notation.

17. Mount Kilimanjaro is the highest mountain in Africa. It is 5,890 meters high.

18. The planet Mars is about 222,900,000,000 meters from the sun.

19. An irregularly shaped galaxy named NGC 4449 is about 250,000,000,000,000,000,000,000 meters from earth.

20. The smallest known insects are about the size of a typical grain of sand. They are about 0.0002 meter in length.

21. Atoms such as hydrogen, carbon, nitrogen, and oxygen are about 0.0000000001 meter across.

22. A Cells in the human liver has a mass of about 0.000000008 gram.

23. The human sperm cell has a mass of about 0.000000000017 gram.

24. The are of the island of Manhattan in New York is about 59,000,000 square meters.

25. The second largest moon of Saturn is Rhea. Rhea has a surface area of about 7,350,000,000,000 square meters — roughly the same surface area as Australia.

26. The volume of the planet Venus is 927,590,000,000,000,000,000 cubic meters.

27. The average mass of a newborn American female is about 3360 grams.

28. The mass of the Eiffel tower in Paris, France, is 8,000,000 grams.

29. In 1981, a Japanese company built the largest oil tanker to date. The ship has a mass of about 510,000,000,000 grams. This oil tanker is more than 6 times as massive as the U.S. aircraft carrier Nimitz.

30. The mass of an amoeba is about 0.000004 grams.

31. The principal protein of muscle is myosin, which has a mass of 0.00000000000000000103 gram.

32. Amino acids are molecules that combine to make up protein molecules. The amino acid tryptophan has a mass of 0.00000000000000000000340 gram.

33. The approximate time it takes for a human being to die of asphyxiation is 316 seconds.

34. On the average, the male housefly lives 1,468,800 seconds — 17 days.

35. Aluminum-26 has a half-life of 740,000 years.

36. In its orbit around the sun, the earth moves a distance one and one half feet in about 0.0000316 second.

37. A pi-meson is a subatomic particle that has a half-life of about 0.0000000261 second.

Practice C — Answers

15. 6×10^{17}

16. 1.5×10^{25}

17. 6.3×10^{-20}

18. 3×10^9

19. 4×10^{-15}

20. 1.04×10^{12}

For the following problems, convert the numbers from scientific notation to standard decimal notation.

38. The sun is about 1×10^{11} meters from earth.

39. The mass of the earth is about 5.98×10^{27} grams.

40. Light travels about 5.866×10^{12} miles in a year.

41. One year is about 3×10^{7} seconds.

42. Rubik's cube has about 4.3×10^{19} different configurations.

43. A photon is a particle of light. A 100-watt light bulb emits 1×10^{20} photons every second.

44. There are about 6×10^{7} cells in the retina of the human eye.

45. A car traveling at an average speed will travel a distance about equal to the length of the smallest fingernail in 3.16×10^{-4} seconds.

46. A ribosome of *E. coli* has a mass of about 4.7×10^{-19} grams.

47. A mitochondrion, the energy-producing element of a cell, is about 1.5×10^{-6} meters in diameter.

48. There is a species of frogs in Cuba that attain a length of at most 1.25×10^{-2} meters.

49. There are 6.022×10^{23} atoms in a mole.

50. Super Bowl Fifty was viewed by 1.119×10^{8} people on television.

51. If the number of albinos in the world is divided by the total number of people in the world, the result is 5.4054×10^{-5}.

52. A snowflake weighs approximately 2.205×10^{-9} tons.

For the following problems, perform the indicated operation. Make sure your final answer is written in scientific notation.

53. $(2 \times 10^{4})(3 \times 10^{5})$

54. $(4 \times 10^{2})(8 \times 10^{6})$

55. $(6 \times 10^{14})(6 \times 10^{-10})$

56. $(3 \times 10^{-5})(8 \times 10^{7})$

57. $(2 \times 10^{-1})(3 \times 10^{-5})$

58. $(9 \times 10^{-5})(1 \times 10^{-11})$

59. $(3.1 \times 10^{4})(3.1 \times 10^{-6})$

60. $(4.2 \times 10^{-12})(3.6 \times 10^{-20})$

61. $(1.1 \times 10^{6})^2$

62. $(5.9 \times 10^{14})^2$

63. $(1.02 \times 10^{-17})^2$

64. $(8.8 \times 10^{-50})^2$

65. $\dfrac{8.8 \times 10^{29}}{1.1 \times 10^{12}}$

66. $\dfrac{5.1 \times 10^{75}}{1.7 \times 10^{81}}$

67. $\dfrac{7.92 \times 10^{-31}}{1.056 \times 10^{17}}$

68. $\dfrac{9.956 \times 10^{-3}}{2.096 \times 10^{-24}}$

69. $\dfrac{3.0625 \times 10^{-56}}{9.8 \times 10^{-69}}$

70. $\dfrac{1.04594 \times 10^{25}}{9.64 \times 10^{-47}}$

71. $\dfrac{2.173248 \times 10^{4}}{7.056 \times 10^{-12}}$

72. $\dfrac{(8.2 \times 10^{13})(3.7 \times 10^{-42})}{7.585 \times 10^{-18}}$

73. $\dfrac{(4.15 \times 10^{5})(9.6 \times 10^{19})}{6.64 \times 10^{31}}$

74. $\dfrac{2.0412 \times 10^{-5}}{(8.4 \times 10^{-27})(2.7 \times 10^{4})}$

75. $\dfrac{9.03 \times 10^{35}}{(1.05 \times 10^{12})(2.15 \times 10^{-7})}$

CHAPTER 5
Factoring

When multiplying, the result is known as a product, and the numbers being multiplied are known as factors. In the equation $3 \cdot 2 = 6$, for example, the product is 6, and the factors are 3 and 2. **Factoring** is the process of breaking a number or algebraic expression down into its factors. In previous math courses, you may have used this process to create prime factorizations. For example, the prime factorization of 24 is $2 \cdot 2 \cdot 2 \cdot 3$. In this chapter, we will be factoring polynomials.

Our primary focus will be on second-degree polynomials, but we will also encounter higher-degree expressions. After learning the various methods and techniques that can be used when factoring polynomials, we'll use factoring to help us simplify rational expressions.

This chapter is composed of the following sections:

5.1 Factoring Out the Greatest Common Factor

Overview

At this point, you should be comfortable with operations — addition, subtraction, multiplication, division, and exponents — and how they pertain to polynomials. Now it's time to learn how to reverse the multiplication process. Here's a problem that does this:

Factor out the GCF: $9x^3 + 18x^2 + 27x$

GCF stands for *Greatest Common Factor*, but what does the GCF look like when variables are involved? And how does one go about factoring out a GCF? We need answers for these questions before we are ready to tackle this problem.

This section will introduce you to factoring and teach you how to:

♦ Determine the other factor when given a polynomial and one monomial factor

♦ Identify and factor out the greatest common factor (GCF)

♦ Use grouping to factor polynomials

A. Finding a Missing Factor

In Chapter 4, we studied multiplication of polynomials. We were given factors and asked to find their product. To review, consider this example.

Example 1

For each pair of factors given, find the product.

1. 4 and 8

2. $6x^2$ and $2x - 7$

3. $x - 2y$ and $3x + y$

4. $a + 8$ and $a + 8$

Solutions

Factors are numbers, or algebraic quantities being multiplied. A product is the result of multiplying. Our job is to multiply the given numbers or expressions together. Of course, we are also expected to simplify each result, if possible.

1. 4 and 8

$4 \cdot 8 = 32$

2. $6x^2$ and $2x - 7$

$6x^2(2x - 7) = 12x^3 - 42x^2$

3. $x - 2y$ and $3x + y$

$(x - 2y)(3x + y) = 3x^2 + xy - 6xy - 2y^2$
$$= 3x^2 - 5xy - 2y^2$$

4. $a + 8$ and $a + 8$

$(a + 8)^2 = (a + 8)(a + 8)$
$$= a^2 + 8a + 8a + 64$$
$$= a^2 + 16a + 64$$

Now we'll reverse the situation. Instead of finding the product, we will start with the product and will try to find the factors. Below are examples of how this works when the product is a monomial.

Example 2

The number 24 is the product, and one factor is 6. What is the other factor?

Solution

We want a number we can multiply by 6 to get 24, so $6n = 24$. We know from experience that $n = 4$.

This problem can be solved with mental math, but as problems become progressively more complex, we may not find the solutions as easily. We need to have a method for finding factors. This method doesn't need to be too complicated, though. We can use the relatively simple problem $6n = 24$ as a guide.

To find the value of n, we divide 24 by 6:

$$\frac{24}{6} = 4$$

Now we know the other factor is 4. In general, any time we know the product and one of the factors, we can divide the product by the known factor in order to find the other factor.

Example 3

For each problem, find the missing factor.

1. The product is $18x^3y^4z^2$, and one factor is $9xy^2z$.

2. The product is $-21a^5b^{11}$ and one factor is $3ab^4$.

Solutions

1. We know the product and one of the factors. We find the other factor by dividing.

$18x^3y^4z^2 \div 9xy^2z = \dfrac{18x^3y^4z^2}{9xy^2z}$
$$= 2x^2y^2z$$

We can check our work by multiplying both factors to verify that the product is $18x^3y^4z^2$.

$(2x^2y^2z)(9xy^2z) = 18x^{2+1}y^{2+2}z^{1+1}$
$$= 18x^3y^4z^2 \checkmark$$

2. $-21\,a^5\,b^{11} \div 3a\,b^4 = \dfrac{-21\,a^5\,b^{11}}{3a\,b^4}$

$\qquad\qquad\qquad\quad = -7\,a^{5-1}\,b^{11-4}$

$\qquad\qquad\qquad\quad = -7\,a^4\,b^7$

Once again, we use multiplication to check our answer.

$\quad -7\,a^4\,b^7 \cdot 3a\,b^4 = -21\,a^{4+1}\,b^{7+4}$

$\qquad\qquad\qquad\quad = -21\,a^5\,b^{11}\ \checkmark$

To understand the process of factoring out a monomial from a polynomial with more than one term, let's consider the case in which $12\,x^2 + 20x$ is the product, and one of the factors is $4x$. To find the other factor, we divide $12\,x^2 + 20x$ by $4x$, making use of the distributive property for division:

$$\frac{A+B}{C} = \frac{A}{C} + \frac{B}{C}$$

In other words, if we divide a polynomial with multiple terms by a monomial, then *each term* of the polynomial must be divided by the monomial. We have seen this before, most notably when converting linear equations into slope intercept form. Here's how it looks with the polynomial product and monomial factor given above:

$$\frac{12\,x^2 + 20x}{4x} = \frac{12\,x^2}{4x} + \frac{20x}{4x}$$

$$= 3x + 5$$

It's important to note that the distributive property for division *only* applies when dividing an expression with multiple terms by another expression. The reverse is not true:

$$\frac{A}{B+C} \text{ is } not \text{ equal to } \frac{A}{B} + \frac{A}{C}.$$

Example 4

The product is $3\,x^7 - 2\,x^6 + 4\,x^5 - 3\,x^4$, and one factor is x^4. Find the other factor.

Solution

We again use division to find the other factor.

$(3\,x^7 - 2\,x^6 + 4\,x^5 - 3\,x^4) \div x^4 = \dfrac{3\,x^7 - 2\,x^6 + 4\,x^5 - 3\,x^4}{x^4}$ Write the problem using a fraction bar.

$\qquad\qquad\qquad\qquad\quad = \dfrac{3\,x^7}{x^4} - \dfrac{2\,x^6}{x^4} + \dfrac{4\,x^5}{x^4} - \dfrac{3\,x^4}{x^4}$ Apply the distributive property for division.

$\qquad\qquad\qquad\qquad\quad = 3\,x^3 - 2\,x^2 + 4x - 3$ Finish by simplifying the expression.

The other factor is $3\,x^3 - 2\,x^2 + 4x - 3$. We use multiplication to check our work.

$\quad x^4 \cdot (3\,x^3 - 2\,x^2 + 4x - 3) = 3\,x^7 - 2\,x^6 + 4\,x^5 - 3\,x^4\ \checkmark$

In the future, we can shorten the process slightly by writing the initial division problem with a fraction bar instead of the "÷" symbol.

Example 5

The product is $10x^3y^6 + 15x^3y^4 - 5x^2y^4$, and one factor is $5x^2y^4$. Find the other factor.

Solution

$$\frac{10x^3y^6 + 15x^3y^4 - 5x^2y^4}{5x^2y^4} = \frac{10x^3y^6}{5x^2y^4} + \frac{15x^3y^4}{5x^2y^4} - \frac{5x^2y^4}{5x^2y^4}$$ Write the original problem with a fraction bar and apply the distributive property of division.

$$= 2xy^2 + 3x - 1$$ Finish by writing the simplified expression.

Use multiplication to check.

$$5x^2y^4 \cdot (2xy^2 + 3x - 1) = 10x^3y^6 + 15x^3y^4 - 5x^2y^4 ✓$$

Sometimes we have to divide out a negative factor.

Example 6

For each problem, find the missing factor.

1. The product is $-4a^2 - b^3 + 2c$, and one factor is -1.

2. The product is $-3a^2b^5 - 15a^3b^2 + 9a^2b^2$, and one factor is $-3a^2b^2$.

Solutions

1. $-4a^2 - b^3 + 2c$

$$\frac{-4a^2 - b^3 + 2c}{-1} = \frac{-4a^2}{-1} - \frac{b^3}{-1} + \frac{2c}{-1}$$

$$= 4a^2 + b^3 - 2c$$

Now check the result.

$$-1(4a^2 + b^3 - 2c) = -4a^2 - b^3 + 2c ✓$$

2. $-3a^2b^5 - 15a^3b^2 + 9a^2b^2$

$$\frac{-3a^2b^5 - 15a^3b^2 + 9a^2b^2}{-3a^2b^2} = \frac{-3a^2b^5}{-3a^2b^2} - \frac{15a^3b^2}{-3a^2b^2} + \frac{9a^2b^2}{-3a^2b^2}$$

$$= b^3 + 5a - 3$$

And here's the verification of our answer:

$$-3a^2b^2(b^3 + 5a - 3) = -3a^2b^5 - 15a^3b^2 + 9a^2b^2 ✓$$

Practice A

Now it's your turn. Find the missing factors in each of these problems. Then check your results by writing a multiplication equation. When you're done, turn the page to check your work.

1. The product is 84, and one factor is 6.

2. The product is $14x^3y^2z^5$, and one factor is $7xyz$.

3. The product is $3x^2 - 6x$, and one factor is $3x$.

4. The product is $5y^4 + 10y^3 - 15y^2$, and one factor is $5y^2$.

5. The product is $4x^5y^3 - 8x^4y^4 + 16x^3y^5 + 24xy^7$, and one factor is $4xy^3$.

6. The product is $-25a^5 - 35a^3 + 5a^2$, and one factor is $-5a^2$.

7. The product is $-a^2 + b^2$, and one factor is -1.

B. Factoring Out the Greatest Common Factor

In the previous examples, we knew the product and one of the factors, so we simply used division to figure out the other factor. But suppose we're given a polynomial product without any factors. We need to be able to determine a monomial that is a factor of each of the terms in the polynomial. If such a monomial exists, it's called a **common factor**. In the expression $8x + 12$, for example, we know that 2 is a factor of $8x$ and of 12. We can thus use 2 as one factor and use division to find the other factor: $\frac{8x}{2} + \frac{12}{2}$ = $4x + 6$. We now rewrite $8x + 12$ as $2(4x + 6)$. This is called *factoring out* a common factor.

The Greatest Common Factor

Whenever we factor out a monomial from a polynomial, we try to find a monomial that is more than just a common factor of each term of the polynomial. We are looking for the *greatest* common factor of the terms in the polynomial.

The **greatest common factor** (GCF) of two numbers is the largest number that is a factor of each. For example, consider the numbers 8 and 12. The factors of 8 are 1, 2, 4, and 8. The factors of 12 are 1, 2, 3, 4, 6, and 12. The common factors are 1, 2, and 4. The greatest of these, 4, is the GCF of 8 and 12.

When dealing with a polynomial that has multiple terms, the GCF of theses terms has the following qualities:

1. The numerical coefficient is the GCF of the numerical coefficients of the terms of the polynomial.

2. The variables represent the *lowest* powers of the variables in the terms of the polynomial.

The rule pertaining to variables might seem counterintuitive, so let's consider x^4 and x^7. The factors of x^4 are 1, x, x^2, x^3, and x^4. The factors of x^7 are 1, x, x^2, x^3, x^4, x^5, x^6, and x^7. When we compare these lists of factors, we see that x^4 is the GCF of x^4 and x^7.

To factor out the GCF, we first must determine what the GCF is and write it. Then we divide each term of the original polynomial by the GCF and write the resulting polynomial in parentheses.

Example 7

Fctor out the GCF for the following expressions.

1. $3x - 18$ **2.** $9x^3 + 18x^2 + 27x$ **3.** $10x^2y^3 - 20xy^4 - 35y^5$

Solutions

1. $3x - 18$

$3x - 18$ GCF is 3   The first term contains x^1, and the second term contains no variable — we can think of it as containing x^0. The *lowest* power of the variable is zero, so no variable at all is included with the GCF.

$3x - 18 = 3(x - 6)$ We write 3 and determine the other factor by dividing each term of the original expression by 3. We do this with mental math: $\frac{3x}{3} = x$ and $\frac{-18}{3} = -6$.

We've factored out the GCF. The factored expression is $3(x - 6)$.

2. $9x^3 + 18x^2 + 27x$

This is the problem from the beginning of this section.

$9x^3 + 18x^2 + 27x$ GCF is $9x$.   The first term contains x^3, the second contains x^2, and the third contains x^1. The *lowest* power of the variable is one, so the GCF contains x^1.

$9x^3 + 18x^2 + 27x = 9x(x^2 + 2x + 3)$ We write $9x$ and use mental math to fill in the parentheses: $\frac{9x^3}{9x} = x^2$, $\frac{18x^2}{9x} = 2x$, and $\frac{27x}{9x} = 3$.

3. $10x^2y^3 - 20xy^4 - 35y^5$

$10x^2y^3 - 20xy^4 - 35y^5$ GCF is $5y^3$.   Because this polynomial has two variables, we determine the lowest power for each. The lowest power of x is zero because the third term does not contain x. The lowest power of y is 3, from the first term. The GCF contains y^3.

$10x^2y^3 - 20xy^4 - 35y^5 = 5y^3(2x^2 - 4xy - 7y^2)$ Write $5y^3$. Then fill in the parentheses by mentally dividing each term of the original polynomial by $5y^3$.

When a polynomial in one variable is written in standard form so that the variable powers are descending, the GCF should have the same sign as the first term. That means that if the first term of the polynomial is negative, then the GCF should be negative.

Example 8

Factor out the GCF: $-12x^5 + 8x^3 - 4x^2$

Solution

This polynomial is in standard form, and the first term is negative. We factor out a negative GCF.

$-12x^5 + 8x^3 - 4x^2$ GCF is $-4x^2$.   The GCF of 12, 8, and 4 is 4, but because the first term of the polynomial is negative, we use -4 in the GCF. The lowest power of the variable is 2, so we include x^2 in the GCF.

$-12x^5 + 8x^3 - 4x^2 = -4x^2(3x^3 - 2x + 1)$ Write $-4x^2$ and fill in the parentheses by mentally dividing each term in the polynomial by $-4x^2$.

Practice B

Now you can give it a try. For each polynomial given, factor out the GCF. When you're finished, turn the page and see how you did.

8. $4x - 48$

9. $6y^3 + 24y^2 + 36y$

10. $10a^5b^4 - 14a^4b^5 - 8b^6$

11. $-14m^4 + 28m^2 - 7m$

C. Using Grouping to Factor a Polynomial

Let's consider the product $Ax + Ay$. We know how to factor this expression:

$$Ax + Ay = A(x + y)$$

Sometimes in algebra, we see expressions with unexpected — or just plain weird — GCFs. Consider this product: $☺x + ☺y$. We don't typically see smiley faces in algebra, but we should treat them the same way we would treat any other variable. In this case, the smiley face is the GCF:

$$☺x + ☺y = ☺(x + y)$$

Now let's consider an expression that's more common in algebra: $(a + b)x + (a + b)y$. The smiley faces from the previous expression have been replaced by $(a + b)$. We treat the $(a + b)$ in this problem exactly the same way we treated the smiley faces in the last problem. In this case, $(a + b)$ is the GCF:

$$(a + b)x + (a + b)y = (a + b)(x + y)$$

Notice that we only write the GCF once when writing the factored expression. In other words, ☺x + ☺y is not the same as ☺☺$(x + y)$, and $(a + b)x + (a + b)y$ is not the same as $(a + b)(a + b)(x + y)$. Let's look at some expressions involving GCFs that are not monomials.

Example 9

Factor out the GCF for the following expressions.

1. $(x - 7)a + (x - 7)b$

2. $3x^2(x + 1) - 5x(x + 1)$

Solutions

1. $(x - 7)a + (x - 7)b$

$(x - 7)a + (x - 7)b$ GCF is $(x - 7)$.   We treat $(x - 7)$ as a single entity. Both parts of the product contain $(x - 7)$, so the GCF is $(x - 7)$.

$(x - 7)a + (x - 7)b = (x - 7)(a + b)$ We write $(x - 7)$. Then we divide $(x - 7)$ out of both parts of the original product: $\frac{(x - 7)a}{(x - 7)} = a$ and $\frac{(x - 7)b}{(x - 7)} = b$.

2. $3x^2(x + 1) - 5x(x + 1)$

$3x^2(x + 1) - 5x(x + 1)$ GCF is $x(x + 1)$ x and $(x + 1)$ are common factors in both parts of the product, so the GCF contains both x and $(x + 1)$.

$3x^2(x + 1) - 5x(x + 1) = x(x + 1)(3x - 5)$ We write $x(x + 1)$ and then find the other factor by dividing both parts of the original product by $x(x + 1)$: $\frac{3x^2(x + 1)}{x(x + 1)} = 3x$ and $\frac{-5x(x + 1)}{x(x + 1)} = -5$.

Practice A — Answers

1. 14; Check: $6 \cdot 14 = 84$

2. $2x^2yz^4$; Check: $2x^2yz^4 \cdot 7xyz = 14x^3y^2z^5$

3. $x - 2$; Check: $3x(x - 2) = 3x^2 - 6x$

4. $y^2 + 2y - 3$; Check: $5y^2(y^2 + 2y - 3) = 5y^4 + 10y^3 - 15y^2$

5. $x^4 - 2x^3y + 4x^2y^2 + 6y^4$; Check: $4xy^3(x^4 - 2x^3y + 4x^2y^2 + 6y^4) = 4x^5y^3 - 8x^4y^4 + 16x^3y^5 + 24xy^7$

6. $5a^3 + 7a - 1$; Check: $-5a2(5a^3 + 7a - 1) = -25a^5 - 35a^3 + 5a^2$

7. $a^2 - b^2$; Check: $-(a^2 - b^2) = -a^2 + b^2$

If you have a hard time following the process in the rpvi, you might find it helpful to start by rewriting the common parenthetical expression as a smiley face or some other shape. The factoring is then a bit easier to see.

$$3x^2(x+1) - 5x(x+1) \;=\; 3x^2\text{☺} - 5x\text{☺} \qquad \text{Rewrite } (x+1) \text{ as a smiley face.}$$
$$=\; x\text{☺}(3x - 5) \qquad\qquad \text{Factor out the GCF.}$$
$$=\; x(x+1)(3x - 5) \qquad \text{Rewrite the smiley face as } (x+1).$$

Sometimes the only common factor of all the terms in a polynomial is 1, but there's no benefit to factoring 1 out of a polynomial. $4x + 5$ is the same as $1(4x - 5)$, and the new expression isn't any different than the original. It is thus trivial to write this polynomial in factored form. In some cases, however, it may be possible to produce a non-trivial factored form for the polynomial.

Consider the polynomial $x^3 + 3x^2 - 6x - 18$. This polynomial doesn't have a single factor, other than 1, that is common to all four terms. However, if we group the first pair of terms and group the second pair of terms, each resulting binomial does have a nontrivial common factor:

In the binomial $x^3 + 3x^2$, the GCF is x^2.

In the binomial $-6x - 18$, the GCF is -6.

If we factor x^2 out of the first pair of terms and factor -6 out of the second pair of terms, we have this expression: $x^2(x + 3) - 6(x + 3)$. Notice that the parenthetical expressions are identical. This means we can now factor out a GCF of $(x + 3)$. Our final factorization is $(x + 3)(x^2 - 6)$.

This method of splitting a polynomial into two groups prior to factoring out GCFs is called **factoring by grouping**. Try the grouping method when the original polynomial has these two qualities:

1. There is no factor that is common to all of the terms.

2. The number of terms is even, but more than 2.

The grouping method could be used on polynomials with 4 terms, 6 terms, 8 terms, etc. However, in this book, we will focus on polynomials that have four terms. Keep in mind that when we factor by grouping, the sign of the GCF for a pair of terms (+ or −) will be the same as the sign of the first term in that pair.

Practice B — Answers

8. $4(x - 12)$

9. $6y(y^2 + 4y + 6)$

10. $2b^4(5a^5 - 7a^4b - 4b^2)$

11. $-7m(2m^3 - 4m + 1)$

Example 10

Factor by grouping: $8a^2b^4 - 4b^4 + 14a^2 - 7$

Solution

Notice that there are four terms, and there is no factor that is common to all terms. This is why the grouping method is prescribed.

$8a^2b^4 - 4b^4 + 14a^2 - 7 = 8a^2b^4 - 4b^4 \mid +14a^2 - 7$	Begin by grouping the first pair of terms and the second pair of terms.
$= 4b^4(2a^2 - 1) \mid +14a^2 - 7$	Factor out the GCF from the first pair of terms.
$= 4b^4(2a^2 - 1) \mid +7(2a^2 - 1)$	Factor out the GCF from the second pair of terms. Keep the plus sign (+) in front of the 7. In factored form, both groups have the same parenthetical expression: $(2a^2 - 1)$.
$= (2a^2 - 1)(4b^4 + 7)$	Factor $(2a^2 - 1)$ out as a GCF.

If the parenthetical expressions aren't the same after factoring both pairs of terms, we can't continue. This means that the original expression is not factorable by using the grouping method.

Sometimes in the process of factoring by grouping, it's helpful to factor out a GCF of 1.

Example 11

Factor by grouping: $14x^3 + 21x^2 + 2x + 3$

Solution

$14x^3 + 21x^2 + 2x + 3 = 14x^3 + 21x^2 \mid +2x + 3$	Group the first and second pairs of terms.
$= 7x^2(2x + 3) \mid +2x + 3$	Factor out the GCF from the first pair of terms.
$= 7x^2(2x + 3) \mid +1(2x + 3)$	The second pair of terms doesn't have a common factor other than 1, but we factor out the 1 and find the same parenthetical expression, $(2x + 3)$, in both groups.
$= (2x + 3)(7x^2 + 1)$	Factor out the parenthetical GCF.

Don't forget — if a group begins with a negative term, then we need to factor out a negative GCF.

Example 12

Factor by grouping: $4y^6 - 3y^5 - 8y + 6$

Solution

$$4y^6 - 3y^5 - 8y + 6 = 4y^6 - 3y^5 \mid -8y + 6$$
$$= y^5(4y - 3) \mid -2(4y - 3) \qquad \text{The second group begins with a negative term,}$$
$$\text{so we factor out a negative GCF.}$$
$$= (4y - 3)(y^5 - 2)$$

Practice C

It's your turn. Factor each expression. Check your work on the next page.

12. Factor out the GCF: $(y + 4)a + (y + 4)b$

13. Factor out the GCF: $8m^3(n - 4) - 6m^2(n - 4)$

14. Factor by grouping: $ax + ay + bx + by$

15. Factor by grouping: $2am + 8m + 5an + 20n$

16. Factor by grouping: $15mx + 10nx - 6my - 4ny$

17. Look at the polynomial from Example 10: $8a^2b^4 - 4b^4 + 14a^2 - 7$. Switch the second and third terms in this polynomial to write: $8a^2b^4 + 14a^2 - 4b^4 - 7$. Now use the grouping method to factor the new expression. Do you get the same result?

Exercises 5.1

For the following problems, find the missing factor.

1. Product = 30; one factor is 6.

2. Product = 45; one factor is 9.

3. Product = $21b$; one factor is $7b$.

4. Product = $15a$; one factor is $5a$.

5. Product = $6x^2y$; one factor is $3x$.

6. Product = $9a^4b^5$; one factor is $9a^4$.

7. Product = $22b^8c^6d^3$; one factor is $-11b^8c^4$.

8. Product = $-60x^5b^3f^9$; one factor is $-15x^2b^2f^2$.

9. Product = $4x + 10$; one factor is 2.

10. Product = $6y + 18$; one factor is 6.

11. Product = $21x^2 + 28x$; one factor is $7x$.

12. Product = $45y^2 + 50y$; one factor is $5y$.

13. Product = $12b^7 + 16b^5 + 20b^4$; one factor is $4b^4$.

14. Product = $9a^6 + 6a^5 - 18a^4 + 24a^2$; one factor is $3a^2$.

15. Product = $-10x^3 - 35x^2$; one factor is $-5x^2$.

16. Product = $-12x^3y^5 + 20x^3y^2$; one factor is $-4x^3y^2$.

17. Product = $-a + b$; one factor is -1.

18. Product = $-n - m$; one factor is -1.

19. Product = $-2x + 4y - z$; one factor is -1.

For the following problems, factor out the GCF.

20. $9a + 18$

21. $16x + 12$

22. $21y - 28$

23. $18f - 36$

24. $12x^2 + 18x$

25. $30y^2 + 10y$

26. $8y^2 - 18$

27. $-7x^2 - 21$

28. $-3y^2 + 6$

29. $5a^2x^2 + 10x$

30. $24ax^2 + 28a$

31. $16x^2 + 8x - 24$

32. $-12x^2 - 8x - 16$

33. $-15y^5 - 24y^4 + 9y^3$

34. $-30a^2b^2 + 40a^2b^2 + 50a^2b^2$

35. $13x^2y^5c^5 - 26x^2y^5c^3 - 39x^2y^5$

36. $(x + 4)b + (x + 4)c$

37. $(x - 9)a + (x - 9)b$

38. $(2x + 7)11a + (2x + 7)13b$

39. $2w(9a - b) - 7x(9a - b)$

40. $7p(p - 1) - 11(p - 1)$

41. $-5x(4x + 3) + 3(4x + 3)$

42. $-7n(n^2 + 10) - 8(n^2 + 10)$

43. $21q(7q^3 - 15) + 19(7q^3 - 15)$

44. $12x^3(5 - v) + 15x^2(5 - v)$

45. $24n^5(5x + 1) - 12n^3(5x + 1)$

For the following problems, use the grouping method to factor the polynomials. Keep in mind that some polynomials may not be factorable using the grouping method.

46. $5x^3 + 30x^2 + 8x + 48$

47. $6n^5 - 15n^4 + 14n - 35$

48. $xy - 7x + 4y - 28$

49. $xy + x + 3y + 3$

50. $mp + 3mq - np - 3nq$

51. $ar + 4as - 5br - 20bs$

52. $14ax - 6bx + 21ay - 9by$

53. $36ak - 8ah - 27bk + 6bh$

54. $a^2b^2 + 2a^2 + 3b^2 + 6$

55. $3n^3 + 6n^2 + 11n + 33$

56. $8y^4 - 5y^3 + 12z^2 - 10z$

57. $x^2 + 4x - 3y^2 + y$

58. $x^2 - 3x + xy - 3y$

59. $2n^2 + 12n - 5mn - 30m$

60. $4pq - 7p + 3q^2 - 21$

61. $8x^2 + 16xy - x - 2y$

Practice C — Answers

12. $(y + 4)(a + b)$

13. $2m^2(n - 4)(4m - 3)$

14. $(a + b)(x + y)$

15. $(2m + 5n)(a + 4)$

16. $(5x - 2y)(3m + 2n)$

17. Yes

5.2 Factoring Trinomials of the Form $x^2 + bx + c$

Overview

We have learned how to multiply a monomial by a polynomial containing multiple terms and how to factor a monomial out of a polynomial containing multiple terms. We've also learned how to multiply binomials together using the FOIL method. Wouldn't it be great if there was a factoring method that could undo this? Fortunately, there is such a method.

Consider this problem:

Factor: $y^2 - 2y - 24$

The expression $y^2 - 2y - 24$ can't be generated by multiplying a monomial by a polynomial. However, it can be generated by multiplying two binomials together. Because of this, the expression $y^2 - 2y - 24$ is factorable. In this section, we will develop a strategy for factoring expressions like this, and then return to this problem.

By studying this section, you will learn how to:

◆   Factor trinomials of the form $x^2 + bx + c$

◆   Factor out the GCF prior to factoring trinomials of the form $x^2 + bx + c$

A. Factoring Trinomials of the Form $x^2 + bx + c$

In this section, we focus on second-degree trinomials that have a leading coefficient of 1.

There are a lot of key terms in that last sentence, so let's do a little review. First, a polynomial in one variable has a *degree* of 2 if the highest power of the variable is 2. Second, a polynomial with three terms is called a *trinomial*. Third, the *coefficient* of a term is the term's constant factor.

We can thus say, for example, that the expression $5x^2 - 7x + 3$ is a second-degree trinomial whose coefficients are 5, –7, and 3. If a polynomial is written in standard form, with descending variable powers, then the coefficient of the first term is called the **leading coefficient**. Because $5x^2 - 7x + 3$ is written in standard form, the *leading coefficient* is 5. If a polynomial is written in standard form and the coefficient of the first term is not written, we safely assume that the leading coefficient is 1.

Now let's see what happens when we multiply the binomials $(x + 4)$ and $(x + 7)$:

$$(x + 4)(x + 7) = x^2 + 7x + 4x + 28 \qquad \text{We begin by using the FOIL method.}$$
$$= x^2 + 11x + 28 \qquad \text{We then simplify by combining like terms. Our simplified answer is a trinomial.}$$

The first term of that trinomial is the product of the first terms of the binomials: $x \cdot x = x^2$ The last term in the trinomial comes from the product of the last terms from the binomials: $4 \cdot 7 = 28$. The middle term comes from adding the outer and inner products: $7x + 4x = 11x$.

But look how the coefficient of the middle term of the simplified product is the sum of the last terms in the original binomials: $4 + 7 = 11$. This relationship is important to us as we think about factoring. Let's look at a few more products.

Binomial Factors		Simplified Product	Comment
$(x + 3)(x + 9)$	=	$x^2 + 12x + 27$	The middle coefficient of the product is $3 + 9 = 12$.
$(x + 7)(x - 2)$	=	$x^2 + 5x - 14$	The middle coefficient of the product is $7 + (-2) = 5$.
$(x - 5)(x - 4)$	=	$x^2 - 9x + 20$	The middle coefficient of product is $(-5) + (-4) = -9$.

In each case, the middle coefficient in the simplified product is the sum of the last terms in the original binomials. This relationship will always hold if both of the original binomials are first degree expressions with a leading coefficient of 1. Below are cases in which the relationship does not hold.

Binomial Factors		Simplified Product	Comment
$(2x + 4)(x + 6)$	=	$2x^2 + 16x + 24$	The sum of the last terms is $4 + 6 = 10$, but the middle coefficient of the product is 16.
$(x^2 + 3)(x + 10)$	=	$x^3 + 10x^2 + 3x + 30$	The simplified product isn't even a trinomial.

In the first case, the original binomials both have a degree of one, but the leading coefficient of the first is 2. In the second case, only one of the factors is a first-degree binomial.

So if we multiply two first-degree binomials with a leading coefficient of 1, the simplified product will have the following characteristics:

◆ It is a second-degree trinomial with a leading coefficient of 1.

◆ The last term is the product of the last terms in the binomials.

◆ The middle coefficient is the *sum* of the last terms in the binomials.

Using this information, we now develop a factoring procedure.

Factoring a Second-Degree Trinomial with a Leading Coefficient of 1

1. Write two sets of parentheses to contain the binomial factors: (   )(   )

2. The first term of both factors is the variable being squared in the original trinomial.

3. Determine two numbers whose product is equal to the last term in the original trinomial *and* whose sum is equal to the middle coefficient in the original trinomial. These numbers represent the last terms of the binomial factors.

Let's take a closer look at how this procedure works in the following examples.

Example 1

Factor: $x^2 + 5x + 6$

Solution

Step 1: Write two sets of parentheses: ( )( )

Step 2: The variable x is being squared in the first term, so each factor begins with x: $(x \quad)(x \quad)$

Step 3: The third term of the trinomial is 6, and the middle coefficient is 5. We must find two numbers whose product is 6 and whose sum is 5. The numbers are 3 and 2. We complete the factors by placing +3 and +2 in the parentheses:

$$(x + 3)(x + 2)$$

The factorization is complete, but we'll check our work to be sure:

$$(x + 3)(x + 2) = x^2 + 2x + 3x + 6$$
$$= x^2 + 5x + 6 \checkmark$$

Now it's time to revisit the problem from the beginning of this section.

Example 2

Factor: $y^2 - 2y - 24$

Solution

Step 1: Write two sets of parentheses: ( )( )

Step 2: The variable y is being squared in the first term, so each factor begins with y: $(y \quad)(y \quad)$

Step 3: The third term of the trinomial is −24, and the middle coefficient is −2. We find two numbers whose product is −24 and whose sum is −2. The required numbers are −6 and 4:

$$(y - 6)(y + 4)$$

The factorization is complete, but of course, we check our work just to be sure:

$$(y - 6)(y + 4) = y^2 + 4y - 6y - 24$$
$$= y^2 - 2y - 24 \checkmark$$

Note that −6 and 4 are the only factors of −24 that work in the factorization for this example. The other pairs of factors, such as −2 and 12, for example, do not work:

$$(y - 2)(y + 12) = y^2 + 10y - 24$$
$$(y + 3)(y - 8) = y^2 - 5y - 24$$
$$(y - 4)(y + 6) = y^2 + 2y - 24$$

We know these other pairs of factors don't work because the middle term in each of these products is incorrect.

Example 3

Factor: $a^2 - 11a + 30$

Solution

Step 1: Write two sets of parentheses: ( )( )

Step 2: The variable a is being squared in the first term , so each factor begins with a: (a)(a)

Step 3: The third term of the trinomial is +30, and the middle coefficient is −11. We need to find two numbers whose product is +30 and whose sum is −11. The required numbers are −5 and −6:

$(a - 5)(a - 6)$

Use multiplication to check:

$$(a - 5)(a - 6) \; = \; a^2 - 6a - 5a + 30$$
$$= \; a^2 - 11a + 30 \; \checkmark$$

Here are tips that might help you figure out the numbers whose product is the last term of the trinomial and whose sum is the middle coefficient of the trinomial:

1. Think of pairs of factors that give you the last term in the trinomial. There are *unlimited* ways to find two numbers that will add up to the middle coefficient, but there are relatively few ways to find two numbers that multiply together to equal the last term.

2. Look at the *sign* of the last term of the trinomial. If the sign is positive, the two factors have the same sign because (+)(+) = (+) and (−)(−) = (+). If the sign of the last term is negative, the two factors have opposite signs because (+)(−) = (−) and (−)(+) = (−).

Let's use these factoring tips as we look at another example.

Example 4

Factor: $x^2 - 7x + 12$

Solution

Step 1: Write two sets of parentheses: ( )( )

Step 2: Fill in the first term of both binomials: (x)(x)

Step 3: Think of pairs of numbers whose product is 12. There are only three pairs:

1 and 12   2 and 6   3 and 4

Notice that the third term of the trinomial is +12. Because the sign is positive, the two factors of 12 that we are looking for must have the same sign. Because the numbers must add up to a negative value, −7, we know that both factors must be negative.

Our choices are now:

> –1 and –12   –2 and –6   –3 and –4

The last pair, –3 and –4, is only pair listed that adds up to –7. Now we know which numbers to place into the parentheses, and can complete our factorization:

$$x^2 - 7x + 12 = (x - 3)(x - 4)$$

Don't assume that every second-degree trinomial with a leading coefficient of 1 is factorable.

Example 5

Factor: $x^2 - 10x - 21$

Solution

Step 1: Write two sets of parentheses: ( )( )

Step 2: Fill in the first term of both binomials: $(x\quad)(x\quad)$

Step 3: We begin by thinking of pairs of numbers whose product is 21. There are only two pairs:

> 1 and 21   3 and 7

Notice that the third term of the trinomial is –21. The sign is negative, so the two factors of 21 that we are looking for have opposite signs. It seems like we should be able to use the 3 and the 7, but the only way they can add up to –10 is if they are both negative:

$$-3 + (-7) = -10.$$

If the 3 and the 7 have opposite signs, then the only possible sums are 4 and –4:

$$-3 + 7 = 4 \qquad \text{and} \qquad 3 + (-7) = -4$$

In fact, no pair of integers multiply to equal –21 and add up to equal –10. Because of this, the trinomial is not factorable. It is also known as a *prime* polynomial.

Practice A

Now that you've seen this method in action, use it yourself to factor the following trinomials, if possible. When you're finished, turn the page and check your answers.

1. $k^2 + 8k + 15$ 3. $m^2 - 10m + 24$ 5. $g^2 + 12g + 3$

2. $y^2 + 7y - 30$ 4. $x^2 - 13x - 30$

B. Using Multiple Factoring Techniques

As you learn various methods of factoring, keep in mind that you may have to use multiple techniques to factor a polynomial. You should always begin by checking to see if there is a greatest common factor other than 1 that can be factored out of an expression. If such a GCF exists, factor it out first. Then check the expression inside of parentheses to see if it can be factored any further.

Example 6

Factor completely: $3x^2 - 15x - 42$

Solution

3 is the GCF of the terms in the polynomial, so first factor out 3.

$3x^2 - 15x - 42 = 3(x^2 - 5x - 14)$ With the GCF factored out, look at the trinomial inside of the parentheses. It's factorable, so work through the process. In each step, keep the GCF, 3, in front of the parentheses.

$\quad\quad = 3(\quad)(\quad)$ Write two sets of parentheses.

$\quad\quad = 3(x\quad)(x\quad)$ Because x is being squared in the trinomial, each binomial factor begins with x.

$\quad\quad = 3(x - 7)(x + 2)$ The numbers -7 and $+2$ multiply together to make -14, and they add to make -5. Place -7 and $+2$ into the parentheses.

As before, check the result by multiplying:

$$3(x - 7)(x + 2) = 3(x^2 + 2x - 7x - 14)$$
$$= 3(x^2 - 5x - 14)$$
$$= 3x^2 - 15x - 42 \checkmark$$

In the previous example, the leading coefficient is *positive*, so we factored out a *positive* GCF. However, if we can factor a GCF out of a polynomial with a *negative* leading coefficient, we need to make sure to write a *negative* GCF. The next example illustrates this.

Example 7

Factor completely: $-7x^5 + 56x^4 - 84x^3$

Solution

$$
\begin{aligned}
-7x^5 + 56x^4 - 84x^3 &= -7x^3(x^2 - 8x + 12) && \text{Factor out the GCF.} \\
&= -7x^3(\quad)(\quad) && \text{Write two sets of parentheses.} \\
&= -7x^3(x\quad)(x\quad) && \text{Place the appropriate variable into the parentheses.} \\
&= -7x^3(x - 6)(x - 2) && \text{Determine the pair of numbers that multiply} \\
&&& \text{to make 12 and add to make -8. Place these} \\
&&& \text{numbers into the parentheses.}
\end{aligned}
$$

Check the result by multiplying.

$$
\begin{aligned}
-7x^3(x - 6)(x - 2) &= -7x^3(x^2 - 2x - 6x + 12) \\
&= -7x^3(x^2 - 8x + 12) \\
&= -7x^5 + 56x^4 - 84x^3 \checkmark
\end{aligned}
$$

Practice B

Factor the following expressions completely. Then check your work on the next page.

6. $2m^2 - 20m + 32$

7. $-5x^2 - 15x + 20$

8. $3y^4 + 24y^3 + 36y^2$

Exercises 5.2

For the following problems, factor the trinomials when possible. You may have to factor out a GCF as the first step.

1. $x^2 + 4x + 3$

2. $x^2 + 6x + 8$

3. $x^2 + 7x + 12$

4. $x^2 + 6x + 5$

5. $y^2 + 8y + 12$

6. $y^2 - 5y + 6$

7. $y^2 - 5y + 4$

8. $a^2 + a - 6$

9. $a^2 + 3a - 4$

10. $x^2 + 4x - 21$

11. $x^2 - 4x - 21$

12. $c^2 - 8c - 48$

13. $y^2 + 10y + 16$

14. $x^2 + 6x - 16$

15. $n^2 + 2n + 9$

16. $t^2 - 5t + 8y^2 - 8y + 7$

17. $y^2 - 5y - 24$

18. $a^2 + a - 30$

19. $a^2 - 3a + 2$

20. $a^2 - 12a + 20$

21. $y^2 - 4y - 32$

22. $x^2 + 13x + 42$

23. $x^2 + 2x - 35$

24. $x^2 + 13x + 40$

25. $y^2 + 6y - 27$

26. $b^2 - 5b + 24$

27. $x^2 + 7x + 18$

28. $b^2 + 15b + 56$

29. $3a^2 + 24a + 36$

30. $4x^2 + 12x + 8 \; 2a^2 - 18a + 40$

31. $5y^2 - 70y + 440$

32. $6x^2 - 54x + 48$

33. $x^3 + 6x^2 + 8x$

34. $x^3 - 8x^2 + 15x$

35. $x^2 + x - 10$

36. $y^2 - 4y - 35$

37. $x^4 + 9x^3 + 14x^2$

38. $-2a^3 - 12a^2 - 10a$

39. $-4a^3 + 40a^2 - 84a$

40. $-3xm^2 - 33xm - 54x$

41. $-2y^2n^2 + 10y^2n + 48y^2$

42. $4x^4 - 42x^3 + 144x^2$

43. $y^5 + 13y^4 + 42y^3$

44. $4x^2a^6 - 48x^2a^5 + 252x^2a^4$

Practice A — Answers

1. $(k + 3)(k + 5)$

2. $(y + 10)(y - 3)$

3. $(m - 6)(m - 4)$

4. $(x - 15)(x + 2)$

5. not factorable (prime)

Practice B — Answers

6. $2(m - 8)(m - 2)$

7. $-5(x + 4)(x - 1)$

8. $3y^2(y + 2)(y + 6)$

5.3 Special Cases for Factoring

Overview

We know how to factor out a GCF and how to factor a second-degree trinomial whose leading coefficient is 1. Now it's time to take a look at a couple of special cases involving factoring. Consider this problem, for example:

Factor: $x^2 - 16$

This isn't a trinomial, so there isn't a GCF, other than 1. However, this expression is still factorable. We'll do a little review and then come back to this problem.

This section will teach you how to:

- Factor a difference squares.
- Use multiple techniques to factor an expression involving a difference of squares
- Factor a perfect square trinomial

A. Factoring a Difference of Squares

In Chapter 4, you learned that conjugates are binomials with the same first term and opposite last terms. When we use the FOIL method to multiply conjugates, the product of the outer terms and the product of the inner terms are canceled out when simplifying the product, as in this example:

$$(x + 3)(x - 3) = x^2 - 3x + 3x - 9$$

The middle terms cancel each other out, leaving the simplified expression, $x^2 - 9$. This always happens when we multiply conjugates, so in general, we can say:

$$(a + b)(a - b) = a^2 - b^2$$

The terms a^2 and b^2 in the above equation are found by squaring a and b. A term that is found by squaring another term is called a perfect square, so a^2 and b^2 are perfect squares. The minus sign between a^2 and b^2 means that we are looking at a difference of two perfect squares. It's typical to refer to the expression $a^2 - b^2$ as a *difference of squares*.

We now use the relationship in reverse. When given a difference of squares, the factored form consists of conjugates. We find the terms in the conjugates by determining the values that can be squared to form each of the perfect squares in the original expression. Algebraically, it looks like this:

$$a^2 - b^2 = (a + b)(a - b)$$

With that in mind, let's look once again at the problem from the beginning of this section.

Example 1

Factor: $x^2 - 16$

Solution

Both x^2 and 16 are perfect squares, and there is a minus sign between them, so this is a difference of squares. The terms that produce x^2 and 16 when squared are x and 4. We use those terms to write conjugate factors:

$$x^2 - 16 = (x + 4)(x - 4)$$

We check our factorization by multiplying.

$$(x + 4)(x - 4) \quad x^2 - 4x + 4x - 16$$
$$x^2 - 16 \checkmark$$

Note that in the real number system, there is no factored form for the *sum* of two squares.

Example 2

Factor the following expressions.

1. $n^2 + 81$ 　　　　　　　　　　　　　　2. $49\,a^2 b^4 - 121$

Solutions

1. $n^2 + 81$

Both n^2 and 81 are perfect squares, but there is a plus sign between them. This is a sum of squares, so this expression is not factorable.

2. $49\,a^2 b^4 - 121$

Both $49\,a^2 b^4$ and 121 are perfect squares. The terms that produce $49\,a^2 b^4$ and 121 when squared are $7a b^2$ and 11. We use these terms to write conjugate factors:

$$49\,a^2 b^4 - 121 = (7ab^2 + 11)(7ab^2 - 11)$$

Let's again check our factorization by multiplying.

$$(7ab^2 + 11)(7ab^2 - 11) \quad 49\,a^2 b^4 - 11ab^2 + 11ab^2 - 121$$
$$49\,a^2 b^4 - 121 \checkmark$$

By the way, we know that the term $49\,a^2 b^4$ from Example 2 is a perfect square because the coefficient, 49, is a perfect square and all of the variables have even exponents. A variable with an odd exponent is not a perfect square.

Practice A

Now it's time for you to give it a try. Factor each expression, if possible. When you are finished, turn the page and check your work.

1. $m^2 - 25$ 2. $x^2 + 144$ 3. $49\,a^4 - b^2 c^2$ 4. $x^8 y^4 - 100\,w^{12}$

B. Using Multiple Factoring Techniques

In the last section, we saw some problems that required us to factor out the GCF prior to factoring a second-degree trinomial. That can also happen in cases involving a difference of squares.

Example 3

Factor: $3\,x^2 - 27$

Solution

This expression is not a difference of squares because the numbers 3 and 27 are not perfect squares. However, 3 is a common factor for both of the terms.

$3x^2 - 27 = 3(x^2 - 9)$ Factor out the GCF of 3. The other factor, $x^2 - 9$, is a difference of squares.

$= 3(x + 3)(x - 3)$ We complete the process by factoring $x^2 - 9$. We're careful to not drop the GCF of 3 from our final answer.

In general, when we are asked to factor a polynomial, we should follow this process:

1. Check to see if there is a GCF that can be factored out. If there is, factor it out.

2. Continue factoring by using other methods.

3. After each step of factoring, check the factors to see if they can be factored any further.

Let's look at some more examples of how this works.

Example 4

Factor the following expressions.

1. $4\,a^8 b - 36\,b^5$ 2. $x^{16} - y^8$

Solution

1. $4\,a^8 b - 36\,b^5$

$4\,a^8 b - 36\,b^5 = 4b(a^8 - 9\,b^4)$ Factor out the GCF, $4b$. We see a difference of squares.

$= 4b(a^4 + 3\,b^2)(a^4 - 3\,b^2)$ Now factor the difference of squares.

Neither the binomial factors is a difference of squares. This means we are done.

2. $x^{16} - y^8$

First, we factor the original difference of squares and check our binomial factors.

$x^{16} - y^8 = (x^8 + y^4)(x^8 - y^4)$ In the first set of parentheses, we have the sum of two squares, which is not factorable. In the second set of parentheses, we have another difference of squares, so we proceed to factor that.

$= (x^8 + y^4)(x^4 + y^2)(x^4 - y^2)$ In the third set of parentheses, we have yet another difference of squares, so we factor that.

$= (x^8 + y^4)(x^4 + y^2)(x^2 + y)(x^2 - y)$ Finally, the factorization is complete.

These types of products appear from time to time, so pay attention. You may have to factor more than once.

Example 5

Factor: $3x^3 + x^2 - 12x - 4$

Solution

This expression has four terms, and the only common factor is 1.

$3x^3 + x^2 - 12x - 4 = 3x^3 + x^2 \mid -12x - 4$ See if this expression can be factored by grouping.

$= x^2(3x + 1) \mid -4(3x + 1)$ The expressions in both sets of parentheses are identical. Factor out the parenthetical GCF.

$= (3x + 1)(x^2 - 4)$ Now the second set of parentheses contains a difference of squares, so factor that.

$= (3x + 1)(x + 2)(x - 2)$ The factorization is complete.

Practice B

Now you try it. Factor each expression completely. When you're finished, turn the page and check your answers.

5. $3x^2 - 75$ **7.** $a^3 b^4 m - a m^3 n^2$ **9.** $2p^3 + 5p^2 - 18p - 45$

6. $m^4 - n^4$ **8.** $16y^8 - 1$

C. Factoring a Perfect Square Trinomial

In Chapter 4, we learned the power rule for products: $(ab)^n = a^n b^n$. We also saw that a similar rule cannot be applied to sums or differences. For example, $(a + b)^2$ is not the same thing as $a^2 + b^2$. In order to expand $(a + b)^2$, we employed the FOIL method:

$$(a + b)^2 = (a + b)(a + b)$$
$$= a^2 + ab + ab + b^2$$
$$= a^2 + 2ab + b^2$$

We thus have two different forms: $(a + b)^2 = a^2 + 2ab + b^2$ and $(a - b)^2 = a^2 - 2ab + b^2$.

Let's look at these products more closely. Notice that they have the following characteristics:

♦ The first term of the product, a^2, is the square of the first term of the binomial, a.

♦ The middle term of the product is twice the product of the terms in the binomial. In the first case, $(a + b)^2$, $2(a)(+b)$ gives us a middle term of $+2ab$ in the product. In the second case, $(a - b)^2$, $2(a)(-b)$ gives us a middle term of $-2ab$ in the product.

♦ The last term of the product, b^2, is the square of the last term of the binomial. In the first case, $(a + b)^2$, $(b)(b) = b^2$. In the second case, $(a - b)^2$, $(-b)(-b) = b^2$. In both cases, we end up with a positive last term in the product.

If we understand these characteristics, we can instantly expand a squared binomial, as we can see in the following example.

Example 6

Multiply: $(x - 7)^2$

Solution

$(x - 7)^2 = x^2 - 14x + 49$ The first term is the square of x: $(x)(x) = x^2$

The middle term is twice the product of x and -7: $2(x)(-7) = -14x$

The last term is the square of -7: $(-7)(-7) = 49$

Let's now shift our focus to factoring. We sometimes need to factor trinomials that are perfect squares. When we do, the factored form consists of a squared binomial. The first step in this process is recognizing whether or not a trinomial is, in fact, a perfect square. A perfect square trinomial always has these characteristics:

◆ The first and last terms are perfect squares.

◆ The middle term is twice the product of the terms that, when squared, produce the first and last terms.

After we identify that a trinomial is a perfect square, we write the factored form as a squared binomial. The first term of the binomial, when squared, will give the first term of the product. The second term of the binomial, when squared, will give the last term of the product. When writing the second term of the binomial, the sign must match the sign of the middle term in the product:

$$a^2 + 2ab + b^2 = (a + b)^2 \text{ and } a^2 - 2ab + b^2 = (a - b)^2$$

Let's see how this works.

Example 7

Determine whether each expression is a perfect square trinomial. If the expression is a perfect square trinomial, factor it.

1. $x^2 + 6x + 9$
2. $x^2 + 10x + 16$
3. $x^4 - 10x^2y^3 + 25y^6$
4. $4a^4 + 32a^2b - 64b^2$

Solution

1. $x^2 + 6x + 9$

$x^2 = (x)(x)$ ✓   Begin by checking the first and last terms. Both are perfect squares.

$9 = (+3)(+3)$ ✓   Because the middle term of the product is positive, write 9 as the square of +3.

$6x = 2(x)(3)$ ✓   Check the middle term. $6x$ is twice the product of x and +3, so we know that $x^2 + 6x + 9$ is a perfect square trinomial.

$x^2 + 6x + 9 = (x + 3)^2$ Factor the expression by writing a squared binomial. The terms of the binomial are x and +3, from the first step of the problem.

Practice B — Answers

5. $3(x + 5)(x - 5)$
6. $(m^2 + n^2)(m - n)(m + n)$
7. $am(ab^2 + mn)(ab^2 - mn)$
8. $(4y^4 + 1)(2y^2 + 1)(2y^2 - 1)$
9. $(2p + 5)(p + 3)(p - 3)$

2. $x^2 + 10x + 16$

$$x^2 = (x)(x) \checkmark$$
$$16 = (+4)(+4) \checkmark$$

The first and last terms are perfect squares. Because the middle term of the product is positive, write 16 as the square of +4.

$$10x \neq (x)(4)$$

Check the middle term. $10x$ is not twice the product of x and 4, so the expression $x^2 + 10x + 16$ is *not* a perfect square trinomial.

The expression $x^2 + 10x + 16$ can be factored, but the factored form is not a squared binomial.

3. $x^4 - 10x^2y^3 + 25y^6$

$$x^4 = (x^2)(x^2) \checkmark$$
$$25y^6 = (-5y^3)(-5y^3) \checkmark$$

Begin by checking the first and last terms. Both are perfect squares. Because the middle term of the product is negative, write $25y^6$ as the square of $-5y^3$.

$$-10x^2y^3 = 2(x^2)(-5y^3) \checkmark$$

Check the middle term. $-10x^2y^3$ is twice the product of x^2 and $-5y^3$. Therefore, the expression $x^4 - 10x^2y^3 + 25y^6$ is a perfect square trinomial.

$$x^4 - 10x^2y^3 + 25y^6 = (x^2 - 5y^3)^2$$

Factor the expression by writing a squared binomial. The terms of the binomial are x^2 and $-5y^3$.

4. $4a^4 + 32a^2b - 64b^2$

$$4a^4 = (2a^2)(2a^2) \checkmark$$
$$-64b^2 \neq (8b)(8b)$$

Checking the first and last terms. The first term is a perfect square, but the last term is not.

$$-64b^2 \neq (-8b)(-8b)$$

We can't square a real-number quantity and get a negative result, so the first and last terms of a perfect square trinomial must both be positive.

The expression $4a^4 + 32a^2b - 64b^2$ is not a perfect square trinomial.

In number 4 of the previous example, notice that we can factor out a GCF:

$$4a^4 + 32a^2b - 64b^2 = 4(a^4 + 8a^2b - 16b^2)$$

When factoring polynomials, always begin by factoring out a GCF if possible. However, even after factoring out the GCF, the trinomial in parentheses, $a^4 + 8a^2b - 16b^2$, is still not a perfect square trinomial.

Practice C

Factor each trinomial, if possible. When you are finished, check your answers on the next page.

10. $m^2 - 8m + 16$

11. $k^2 + 10k + 25$

12. $4a^2 + 12a + 9$

13. $9x^2 - 24xy + 16y^2$

14. $2w^3z + 16w^2z^2 + 32wz^3$

15. $x^2 + 12x + 49$

Exercises 5.3

Factor each polynomial completely, if possible. You may have to begin by factoring out a GCF.

1. $a^2 - 9$

2. $a^2 - 25$

3. $x^2 - 16$

4. $a^2 - 100$

5. $h^2 + 100$

6. $b^2 - 36$

7. $s^2 + 25$

8. $4a^2 - 64$

9. $3x^2 - 27$

10. $36x^2 + y^2$

11. $4a^2 - 25$

12. $9x^2 - 100$

13. $36y^2 - 25$

14. $121a^2 - 9$

15. $12a^2 - 75$

16. $81n^2 + 4k^2$

17. $8y^2 - 50$

18. $a^2b^2 - 9$

19. $x^2y^2 - 25$

20. $x^4y^4 - 36$

21. $x^4y^4 - 9a^2$

22. $a^2b^4 - 16y^4$

23. $4a^2b^2 - 9b^2$

24. $a^2 - b^2$

25. $64x^4y^2 + 9z^6$

26. $a^4 - b^4$

27. $x^4 - y^4$

28. $x^8 - y^2$

29. $a^8 - y^2$

30. $b^6 - y^2$

31. $b^6 - x^4$

32. $2x^3 - x^2 - 50x + 25$

33. $25 - a^2$

34. $100 - 36b^4$

35. $128 - 32x^2$

36. $x^4 - 16$

37. $a^4 - b^4$

38. $a^{16} - b^4$

39. $x^{12} - y^{12}$

40. $a^2c - 9c$

41. $a^3c^2 - 25ac^2$

42. $49x^2y^4z^6 - 64a^4b^2c^8d^{10}$

43. $2x^3 - x^2 - 50x + 25$

44. $4x^3 + 5x^2 - 4x - 5$

45. $12x^3 + 28x^2 - 27x - 63$

46. $16x^3 - 48x^2 - 49x + 147$

47. $x^2 + 8x + 16$

48. $x^2 + 10x + 25$

49. $a^2 + 4a + 4$

50. $w^2 - 18w + 81$

51. $p^2 - 7p + 49$

52. $c^2 + 6c + 9$

53. $m^2 - 16m + 64$

54. $t^2 - t + 1$

55. $b^2 + 22b + 121$

56. $g^2 + 60g + 900$

57. $x^2 - 50x + 625$

58. $3vk^2 + 12vk + 12v$

59. $7an^2 - 42an + 63a$

60. $-13x^2 + 26x - 13$

61. $-5jp^2 - 40jp - 80j$

62. $4a^2 + 12a + 9$

63. $9x^2 + 6x + 1$

64. $4z^2 - 28z + 49$

65. $16a^2 - 24a + 9$

5.4 Simplifying Rational Expressions

Overview

There are a number of mathematical procedures that require factoring in order to complete. One such procedure is the simplification of rational expressions. Consider this problem:

Simplify: $\dfrac{x^2 - 25}{x^2 - 7x + 10}$

As it's written, there's nothing we can do to reduce this fraction. However, as you will learn, factoring skills can help us rewrite the fraction in a form that is reducible.

As you work through this section, you will learn how to:

- ◆ Use factoring to help you reduce rational expressions
- ◆ Identify acceptable variable values in rational expressions

A. Reducing Rational Expressions

The word "rational" comes from the word "ratio." In mathematics, ratios are typically expressed as fractions. Therefore, "rational" means roughly the same thing as "fractional."

A **rational number** is a number that can be expressed as a fraction of two integers. For example, 0.7 can be expressed as $\dfrac{7}{10}$, so 0.7 is a rational number.

A **rational expression** is a fraction in which the numerator and denominator are polynomials. For example, $\dfrac{x + 6}{x^2 + 4x}$ is a rational expression.

Before we simplify any rational expressions, let's briefly review what we know about simplifying numerical fractions. A numerical fraction is simplified if the numerator and denominator do not have any common factors other than 1. For example, the fraction $\dfrac{11}{18}$ is simplified because 1 is the only factor that is common to both 11 and 18. However, the fraction $\dfrac{12}{18}$ is not simplified because 12 and 18 have a number of common factors other than 1.

We can simplify tthe fraction $\dfrac{12}{18}$ by writing the numerator and denominator as factored expressions using the GCF, which is 6. Then we can cancel out the common factor.

$$\frac{12}{18} \;=\; \frac{6 \cdot 2}{6 \cdot 3} \;=\; \frac{\cancel{6} \cdot 2}{\cancel{6} \cdot 3} \;=\; \frac{2}{3}$$

It's important to remember that when simplifying a fraction, we can only cancel or reduce numbers in the numerator and denominator that are being *multiplied*. We cannot cancel or reduce numbers being added or subtracted.

Example 1

Simplify each expression, if possible.

1. $\dfrac{5x}{10y}$

2. $\dfrac{x+5}{y-10}$

Solutions

1. $\dfrac{5x}{10y}$

The numerator and denominator of the first fraction contain numbers and variables that are being multiplied. Therefore, this fraction can be simplified:

$$\dfrac{5x}{10y} = \dfrac{1(5)x}{2(5)y}$$
$$= \dfrac{x}{2y}$$

2. $\dfrac{x+5}{y-10}$

The numerator and denominator of the second fraction contain numbers and variables that are being added or subtracted. We can't cancel or reduce numbers or variables being added or subtracted. Therefore, this fraction *cannot* be simplified.

☠ Warning! Incorrect Approach! ☠

Some people mistakenly attempt to reduce a fraction like the one in number 2 of the previous example:

$$\dfrac{x+5}{y-10} = \dfrac{x+1(5)}{y-2(5)} = \dfrac{x+1(\cancel{5})}{y-2(\cancel{5})} = \dfrac{x+1}{y-2}$$

This is not correct! It's easy to show that the resulting fraction is *not* equivalent to the fraction we started with. Assign any nonzero values to x and y, and see what happens. For example, if we let $x = 3$ and $y = 4$, we get the following:

Original Fraction

$$\dfrac{x+5}{y-10} = \dfrac{3+5}{4-10} = \dfrac{8}{-6} = -\dfrac{4}{3}$$

Incorrectly Reduced Fraction

$$\dfrac{x+1}{y-2} = \dfrac{3+1}{4-2} = \dfrac{4}{2} = 2$$

The values $-\dfrac{4}{3}$ and 2 are not the same, so the fractions $\dfrac{x+5}{y-10}$ and $\dfrac{x+1}{y-2}$ are *not* equivalent.

Example 2

Simplify, if possible: $\frac{15x + n}{20y - m}$

Solution

The 15 and the x in the numerator are being multiplied together. The 20 and the y are being multiplied together in the denominator. It seems like we should be able to reduce this fraction.

However, in the numerator, the term $15x$ is being added to n. Similarly, in the denominator, m is being subtracted from the term $20y$. Because of this, we cannot reduce the 15 and the 20. The fraction $\frac{15x + n}{20y - m}$ cannot be reduced. It is simplified as much as it can be.

As we've seen before, if the fraction in Example 2 had been $\frac{15x \cdot n}{20y \cdot m}$, we would have been able to reduce it because a fraction can be reduced if the numerator and denominator are both monomials:

$$\frac{15x \cdot n}{20y \cdot m} = \frac{3(5)\,xn}{4(5)\,ym} = \frac{3xn}{4ym}$$

If the numerator of a fraction contains a polynomial with multiple terms, and if the denominator contains a monomial, then the fraction can be split up to facilitate reducing:

$$\frac{8x + 42}{6} = \frac{8x}{6} + \frac{42}{6} = \frac{4x}{3} + 7$$

A fraction like $\frac{9x}{3x + 7}$, with a monomial numerator and a multi-term denominator, *cannot* be split up. Some fractions like this can still be reduced, but we will need another plan of attack for doing that.

At this point, you may feel confused or frustrated. Fractions have a way of doing that to people. Let's put some happiness back into this topic.

Example 3

Simplify: $\frac{6x☺}{12☺}$

Solution

Other than the smiley faces, this is a typical fraction with monomials in the numerator and denominator. This fraction is reducible, so we simplify it. We reduce the 6 and 12 and cancel out the smiley faces. Our result is $\frac{x}{2}$.

The next example is almost identical to this one, though with a little less happiness.

Example 4

Simplify: $\frac{6x(2n + 1)}{12(2n + 1)}$

Solution

We replace the smiley faces with the expression $(2n + 1)$. By treating the expression $(2n + 1)$ as a single entity, this fraction behaves the same way as a fraction with monomials in the numerator and denominator. We reduce the 6 and 12, cancel the parenthetical factors, and get $\frac{x}{2}$.

The preceding examples suggest the following rule:

> If polynomials in the numerator and denominator of a fraction are written in factored form, and if a factor in the numerator matches exactly with a factor in the denominator, then the like factors can — and should — be canceled.

Now we have a plan of attack that can be used to simplify fractions involving polynomials with multiple terms in the denominator:

1. Factor any polynomials that contain multiple terms.

2. If a factor in the numerator and a factor in the denominator match up, cancel out the matching factors.

We'll look at how this works in the following example.

Example 5

Simplify the following rational expressions.

1. $\dfrac{8x}{8x^2 + 24x}$ 2. $\dfrac{x^2 - 25}{x^2 - 7x + 10}$ 3. $\dfrac{-2x^4 - 6x^3 + 20x^2}{-2x^3 - 18x^2}$

Solutions

1. $\dfrac{8x}{8x^2 + 24x}$

We can't split this fraction up. However, we can factor the denominator.

$\dfrac{8x}{8x^2 + 24x} = \dfrac{8x}{8x(x + 3)}$ In the denominator, factor out the GCF.

$\qquad = \dfrac{8x}{8x(x + 3)}$ A factor, $8x$, appears in the numerator and denominator. Cancel the matching factors.

$\qquad = \dfrac{1}{x + 3}$ Because the entire numerator was canceled out, we leave a 1 in the numerator of our answer.

Because the entire GCF is canceled out in the denominator, there's no more need for parentheses down there. That's why the final answer is written without parentheses.

2. $\dfrac{x^2 - 25}{x^2 - 7x + 10}$

This is the problem from the beginning of this section.

$\dfrac{x^2 - 25}{x^2 - 7x + 10} = \dfrac{(x + 5)(x - 5)}{(x - 2)(x - 5)}$ Factor the numerator and the denominator.

$\qquad = \dfrac{(x + 5)(x - 5)}{(x - 2)(x - 5)}$ Cancel the matching factors.

$\qquad = \dfrac{x + 5}{x - 2}$

3. $\dfrac{-2x^4 - 6x^3 + 20x^2}{-2x^3 - 18x^2}$

$\dfrac{-2x^4 - 6x^3 + 20x^2}{-2x^3 - 18x^2} = \dfrac{-2x^2(x^2 + 3x - 10)}{-2x^2(x + 9)}$ Factor the GCF out of the numerator and denominator.

$= \dfrac{-2x^2(x + 5)(x - 2)}{-2x^2(x + 9)}$ Factor the trinomial in the numerator.

$= \dfrac{\cancel{-2x^2}(x + 5)(x - 2)}{\cancel{-2x^2}(x + 9)}$ Cancel the matching factors.

$= \dfrac{(x + 5)(x - 2)}{x + 9}$

We could also write the final answer as $\dfrac{x^2 + 3x - 10}{x + 9}$, but it is not necessary to do so.

In number 3 of the previous example, factoring the trinomial in the second step didn't yield any factors that matched the $(x + 9)$ in the denominator. However, we should *always* factor polynomials completely when simplifying rational expressions. Sometimes the resulting factors match up. In that case, we cancel those factors to get an answer that is completely simplified.

Practice A

Now you get to simplify some rational expressions. When you are done, turn the page to check your work.

1. Simplify: $\dfrac{28x^4}{21x^4 - 84x^3}$

2. Simplify: $\dfrac{x^2 - 9x - 22}{x^2 - 15x + 44}$

3. Simplify: $\dfrac{x^2 + 3x + 2}{x^4 - 16}$

4. Simplify: $\dfrac{9x^2 + 36x - 108}{6x^2 - 24x + 24}$

5. Simplify: $\dfrac{2x^3 + 3x^2 - 2x - 3}{x^2 - 7x + 6}$

B. Acceptable Values for Variables in a Rational Expression

There are many applications of rational expressions in real life. For now, we will focus on identifying acceptable values for the variables in rational expressions.

When we were learning about slope, some of the fractions we ended up with had 0 in the denominator. When calculating the slope of the line passing through (8, 3) and (8, 14), for example, the calculation looks like this:

$$m = \frac{14 - 3}{8 - 8} = \frac{11}{0}$$

When the denominator of a fraction is 0, that fraction doesn't represent a number. In other words, the value is undefined.

We see this phenomenon with rational expressions as well. Anytime the denominator of a rational expression contains a variable, it's possible for the denominator to take on a value of 0. In other words, it is possible for the fraction to represent a value that is undefined.

In applications involving rational expressions, it's not acceptable for the rational expression to be undefined. If replacing a variable with a particular number causes the denominator of the expression to equal 0, then that number is not an acceptable value for the variable in that expression.

Here are some examples that illustrate this.

Example 6

Determine the acceptable values for the variable x: $\dfrac{x+7}{x-5}$

Solution

It's faster to determine any numbers that are *not* acceptable values of x. If we replace x with 5 in this expression, the denominator will equal 0:

$$\frac{x+7}{x-5} = \frac{(5)+7}{(5)-5} = \frac{12}{0}$$

Therefore, 5 is not an acceptable value for x. If we replace x with any real number except 5, the denominator of the fraction will not equal 0. We can now state our answer:

All real numbers except 5 are acceptable values for x.

In the previous example, we could have set up and solved an equation to determine the value of x that makes the denominator equal 0:

$$
\begin{array}{rcc}
x-5 &=& 0 \\
+5 & & +5 \\
\hline
x &=& 5
\end{array}
$$

However, we don't have to formally solve an equation like this. It's fine to use mental math.

Before we look at the next examples, let's review a familiar concept from basic math. We know that if we multiply any real number times zero, the product is zero. From experience, we also know that if we multiply any two nonzero real numbers, the product cannot possibly be zero. These ideas are formally summarized in the following rule.

The Zero-Product Property

If A and B represent real numbers, and if $A \cdot B = 0$, then $A = 0$ or $B = 0$.

Practice A — Answers

1. $\dfrac{4x}{3(x-4)}$ or $\dfrac{4x}{3x-12}$

2. $\dfrac{x+2}{x-4}$

3. $\dfrac{x+1}{(x^2+4)(x-2)}$ or $\dfrac{x+1}{x^3-2x^2+4x-8}$

4. $\dfrac{3(x+6)}{2(x-2)}$ or $\dfrac{3x+18}{2x-4}$

5. $\dfrac{(2x+3)(x+1)}{x-6}$ or $\dfrac{2x^2+5x+3}{x-6}$

The zero-product property will help us to identify unacceptable values of the variable in the following examples.

Example 7

Determine the acceptable values for the variable n: $\dfrac{n^2 + 8n}{n^2 + 3n - 54}$

Solution

We need to determine which value(s) of n will cause the denominator to equal 0. We can figure this out rather quickly if the denominator is factored:

$$\frac{n^2 + 8n}{n^2 + 3n - 54} = \frac{n^2 + 8n}{(n + 9)(n - 6)}$$

Based on the zero-product property, we know that the denominator will equal zero if either of the factors takes on a value of zero. Using mental math, we can quickly deduce that:

- If n is replaced with -9, the first factor in the denominator will equal 0
- If n is replaced with 6, the second factor in the denominator will equal 0

Therefore, all real numbers except -9 and 6 are acceptable values for the variable n.

Now let's look at a problem involving simplifying *and* acceptable variable values.

Example 8

For the expression: $\dfrac{x^2 + 2x - 120}{x^2 - 14x + 40}$, do the following.

  a. Simplify the expression.

  b. Determine the acceptable values for the variable x in the *original* expression.

  c. Determine the acceptable values for the variable x in the *simplified* expression.

Solution

a. First, we simplify the rational expression.

$$\frac{x^2 + 2x - 120}{x^2 - 14x + 40} = \frac{(x + 12)(x - 10)}{(x - 4)(x - 10)} \qquad \text{Begin by factoring the numerator and the denominator.}$$

$$= \frac{x + 12}{x - 4}$$

b. By looking at the factored form of the original denominator, $(x - 4)(x - 10)$, we deduce that *all real numbers except 4 and 10* are acceptable values for x in the original expression.

c. By looking at the denominator of the reduced fraction, we deduce that *all real numbers except 4* are acceptable values for x in the simplified expression.

Practice B

It's your turn to tackle some problems involving acceptable variable values. When you're ready, check your answers below.

6. Determine the acceptable values for the variable y: $\dfrac{2y + 7}{y + 13}$

7. Determine the acceptable values for the variable w: $\dfrac{w^2 + 8w}{w^2 - 3w - 28}$

8. Simplify the expression $\dfrac{x^2 + x - 72}{x^2 - 2x - 48}$ and then determine the acceptable values for x in the original expression and then the simplified expression.

Exercises 5.4

For each of the following problems, simplify the rational expression.

1. $\dfrac{5n^6}{5n^7 - 65n^6}$

2. $\dfrac{11q^5}{22q^4 + 55q^3}$

3. $\dfrac{7x^8}{35x^5 + 49x^4}$

4. $\dfrac{-10m^3}{2m^6 - 6m^5 - 36m^4}$

5. $\dfrac{-36p^4}{4p^9 + 16p^8 - 48p^7}$

6. $\dfrac{-6y^7}{-3y^4 - 24y^3 - 21y^2}$

7. $\dfrac{r^2 - 81}{r^2 - 13r + 36}$

8. $\dfrac{x^2 + 16x - 17}{x^2 - 6x + 5}$

9. $\dfrac{g^2 - g - 12}{g^2 - 10g + 24}$

10. $\dfrac{d^2 + 14d + 48}{d^2 - 36}$

11. $\dfrac{x^2 - 23x + 60}{x^2 + 8x - 33}$

12. $\dfrac{z^2 - 144}{z^2 + 7z - 60}$

13. $\dfrac{a^2 + 2a + 1}{a^2 - 1}$

14. $\dfrac{x^2 - 25}{x^2 - 10x + 25}$

15. $\dfrac{10n^3 - 70n^2 + 100n}{25n^3 - 75n^2 + 50n}$

16. $\dfrac{16t^4 - 64t^2}{12t^4 + 72t^3 + 96t^2}$

17. $\dfrac{x^4 - 81}{x^2 + 11x + 24}$

18. $\dfrac{y^2 - 15y + 14}{y^4 - 1}$

19. $\dfrac{3x^3 + x^2 - 12x - 4}{x^2 + 8x - 20}$

20. $\dfrac{m^3 + 5m^2 - 16m - 80}{m^2 + m - 20}$

For the following problems, determine which numbers are acceptable values for the variable in the expression.

21. $\dfrac{x + 9}{4x}$

22. $\dfrac{2v + 18}{-7v}$

23. $\dfrac{13x + 39}{x - 11}$

24. $\dfrac{12x^2}{x + 20}$

25. $\dfrac{-42}{k^2 - 3k - 108}$

26. $\dfrac{4h + 7}{h^2 - 9h + 8}$

27. $\dfrac{16p - 80}{p^2 - 121}$

28. $\dfrac{5c + 35}{c^2 - 100}$

29. $\dfrac{3d + 27}{11d^2 + 77d}$

30. $\dfrac{2 - y}{-6y^2 + 48y}$

31. $\dfrac{x^2 + 22x + 120}{2x^3 + 18x^2 - 44x}$

32. $\dfrac{7x^2 - 28}{6x^3 - 24x^2 + 18x}$

For the following problems, simplify the rational expression. Then determine which numbers are acceptable values for the variable in the original expression and the simplified expression.

33. $\dfrac{x^2 - 49}{x^2 + 2x - 35}$

34. $\dfrac{-14x^2 + 70x}{x^2 - 11x + 30}$

35. $\dfrac{-9x^2 - 99x}{x^2 + 5x - 66}$

36. $\dfrac{x^2 - 8x + 7}{x^2 + 14x - 15}$

37. $\dfrac{x^2 + 16x + 60}{x^2 + x - 30}$

38. $\dfrac{n^2 + 5n - 36}{n^2 - 15n + 44}$

39. $\dfrac{2x^3 + 16x^2 + 30x}{7x^2 + 35x}$

40. $\dfrac{-21x^3 + 189x}{11x^2 - 33x}$

Practice B — Answers

6. All real numbers except -13

7. All real numbers except -4 and 7

8. Simplified expression: $\dfrac{x + 9}{x + 6}$; original expression: all real numbers except -6 and 8; simplified expression: all real numbers except -6

CHAPTER 6
Quadratic Equations

For the final chapter in this book, our focus returns to solving equations. In the first three chapters , we focused on solving and graphing linear equations and linear systems. Based on what we learned about degree in Chapter 4, we know that linear expressions and equations can also be called first-degree expressions and equations.

Now it's time for us to learn how to solve second-degree equations. These equations can also be called *quadratic* equations. As we progress through the chapter, we'll see how skills such as factoring and graphing can help us in certain situations. We will also be introduced to the concept of *square root* and explore how square roots can be used to solve quadratic equations.

This chapter will cover the following topics:

6.1 Solving Quadratic Equations by Factoring

Overview

We have become familiar with several factoring techniques. Now we'll use that knowledge to help us solve equations and application problems. Here's an application problem:

> A contractor plans to pour a concrete walkway around a rectangular swimming pool that is 20 feet wide and 40 feet long. The area of the walkway should be 544 square feet. If the walkway has a uniform width, how wide should the contractor make it?

This problem can be modeled with a second-degree equation, which can also be called a *quadratic equation*. However, we haven't solved any quadratic equations yet. We need to learn how factoring can be used to solve a quadratic equation before we're ready to work through this problem.

In this section, you will learn how to:

◆ Use the zero-product property to solve equations

◆ Solve quadratic equations by factoring

◆ Solve application problems by factoring

A. The Zero-Product Property

A polynomial in one variable is quadratic if the highest power of the variable is 2. When writing a quadratic polynomial in standard form, we write $ax^2 + bx + c$, where a, b, and c represent numerical coefficients and where a cannot be equal to zero. A quadratic equation in one variable is an equation that can be written in the form, $ax^2 + bx + c = 0$, where $a \neq 0$. This form is the standard form of a quadratic equation.

Example 1

Write $3x^2 = -2x + 1$ in standard form.

Solution

Option 1: Here is one way to convert this equation into standard form.

$$3x^2 = -2x + 1 \quad \text{Leave the expression } -2x + 1 \text{ alone and get 0 on the left side.}$$

$$\underline{-3x^2} \qquad \underline{-3x^2} \quad \text{Subtract } 3x^2 \text{ from both sides.}$$

$$0 = -3x^2 - 2x + 1$$

$$-3x^2 - 2x + 1 = 0 \qquad \text{Switch the sides of this equation so that 0 is on the right side.}$$

Option 2: Here is another way to convert the equation to standard form.

$$3x^2 \;=\; -2x + 1 \quad \text{Leave the expression } 3x^2 \text{ alone and get 0 on the right side.}$$

$$\underline{+2x} \qquad \underline{+2x} \quad \text{Add } 2x \text{ to both sides.}$$

$$3x^2 + 2x \;=\; 1$$

$$\underline{-1} \qquad \underline{-1} \quad \text{Subtract 1 from both sides.}$$

$$3x^2 + 2x - 1 \;=\; 0$$

Both options give us a quadratic equation in standard form. If you examine the terms of the quadratic polynomials in both of these equations, you'll notice that the terms in the equation we got from Option 1 have opposite signs as the terms in the equation we got from Option 2.

Now that we know what a quadratic equation in one variable looks like, it's time to develop methods for solving quadratic equations. The method we will use in this section is based on the **zero-product property** that we learned about in Chapter 5:

If A and B represent real numbers, and if $A \cdot B = 0$, then $A = 0$ or $B = 0$.

This property, along with a little mental math, can be used to determine values of a variable that will cause a factored expression to take on a value of zero. In the following examples, we'll use the zero-product property to help us solve some equations.

Example 2

Solve the following equations.

1. $9x = 0$
2. $-2x^2 = 0$
3. $5(x - 1) = 0$
4. $x(x + 6) = 0$
5. $(x + 2)(x + 3) = 0$

Solutions

1. $9x = 0$

The factors are 9 and x. The first factor, 9, is obviously not zero, so x must be zero. Our solution is $x = 0$.

2. $-2x^2 = 0$

We can rewrite this equation as $(-2)(x)(x) = 0$. The first factor, -2, is obviously not zero. The second and third factors are both x. That means that x must be zero. Our solution is again $x = 0$.

3. $5(x - 1) = 0$

Yet again, we have a first factor that can't possibly equal zero. Using mental math, we know that the second factor, $x - 1$, can be made to equal zero if x is replaced with 1. Our solution is $x = 1$.

4. $x(x + 6) = 0$

Using mental math, we deduce the following:

- ♦ If x is replaced with 0, then the first factor becomes zero.
- ♦ If x is replaced with –6, then the second factor takes on a value of zero.

Our solutions are $x = 0$ and $x = -6$. We can also write the solutions as: $x = 0, -6$

5. $(x + 2)(x + 3) = 0$

Use mental math to find the solutions:

- ♦ If x is replaced with –2, then the first factor takes on a value of zero.
- ♦ If x is replaced with –3, then the second factor takes on a value of zero.

Our solutions are $x = -2, -3$

Sometimes we encounter factors that are more complicated. When that happens, it's helpful to use pencil and paper to set up and solve an equation. We will see such a case in the next example.

Example 3

Solve: $(x + 10)(4x - 5) = 0$

Solution

Using mental math, we know the first factor gives us the solution $x = -10$. Because the second factor is more complicated, we set up and solve an equation in order to determine the second solution:

$$
\begin{aligned}
4x - 5 &= 0 \\
\underline{+5} \quad &\quad \underline{+5} \\
4x &= 5 \\
\frac{4x}{4} &= \frac{5}{4} \\
x &= \frac{5}{4}
\end{aligned}
$$

Now we can write our solutions: $x = -10, \frac{5}{4}$

Practice A

Use the zero-product property to solve the equations. Then turn the page to check your solutions.

1. $6(a - 4) = 0$ 　　　　　2. $(y + 6)(y - 7) = 0$ 　　　　　3. $(x + 5)(3x - 4) = 0$

B. Solving Quadratic Equations by Factoring

It's time to learn how to use factoring to help us solve quadratic equations. To do this, we need to make use of the zero-product property, and in order to use the zero-product property, one side of the equation must be 0. With that in mind, we have the following three-step process.

Solving Quadratic Equations by Factoring

1. To use the zero-product property, one side of the equation must be 0. If necessary, use algebra to put all nonzero terms on one side of the equal sign and 0 on the other side, thus rewriting the equation in standard form: $ax^2 + bx + c = 0$.

2. Factor the quadratic expression: $(\quad)(\quad) = 0$

3. Use the zero-product property to solve the equation.

We will use this process to work through the following examples.

Example 4

Solve $x^2 - 7x + 12 = 0$

Solution

$$x^2 - 7x + 12 \;=\; 0 \qquad \text{The equation already has 0 on one side, so we skip step 1.}$$
$$(x - 3)(x - 4) \;=\; 0 \qquad \text{Now factor the expression on the left side of the equation.}$$

Using mental math, we write the solutions: $x = 3$, and $x = 4$.
We check our work by substituting each value back into the original equation.

First, if $x = 3$, then:

$$x^2 - 7x + 12 \;=\; 0$$
$$(3)^2 - 7(3) + 12 \;=\; 0$$
$$9 - 21 + 12 \;=\; 0$$
$$0 \;=\; 0 \checkmark$$

If $x = 4$, then:

$$x^2 - 7x + 12 \;=\; 0$$
$$(4)^2 - 7(4) + 12 \;=\; 0$$
$$16 - 28 + 12 \;=\; 0$$
$$0 \;=\; 0 \checkmark$$

In the examples that follow, our focus will be on solving equations. Therefore, we won't show the work that goes in to checking the solutions. However, in practice, it is always a good idea to check the solutions you find when solving an equation.

Example 5

Solve the following equations.

1. $x^2 = 25$
2. $x^2 = 2x$

Solutions

1. $x^2 = 25$

$$
\begin{aligned}
x^2 &= 25 \qquad &&\text{Before we can factor, we need to have 0 on one side of the equation.}\\
\underline{-25} \quad &\underline{-25}\\
x^2 - 25 &= 0 \qquad &&\text{Factor the left side of the equation.}\\
(x+5)(x-5) &= 0 \qquad &&\text{Finish by using mental math to determine the solutions.}\\
x &= -5, 5
\end{aligned}
$$

2. $x^2 = 2x$

$$
\begin{aligned}
x^2 &= 2x \qquad &&\text{Begin by getting 0 on one side of the equation.}\\
\underline{-2x} \quad &\underline{-2x}\\
x^2 - 2x &= 0 \qquad &&\text{Now factor and solve.}\\
x(x-2) &= 0\\
x &= 0, 2
\end{aligned}
$$

In part 2 of the previous example, the first factor, x, was a monomial. We need to be careful, however, to not write a solution of 0 every time we see a monomial factor. If a monomial factor does not include a variable, then we cannot write 0 as a solution.

Example 6

Solve each equation.

1. $8x^2 = 16x + 24$
2. $7x^3 + 63x^2 = 154x$

Solutions

1. $8x^2 = 16x + 24$

It's typically easier to solve quadratic equations in which the quadratic polynomial has a *positive* leading coefficient. The term $8x^2$ has a positive coefficient, so we'll leave it alone.

$$
\begin{aligned}
8x^2 &= 16x + 24 \qquad &&\text{Leave } 8x^2 \text{ where it is, and get 0 on the other side.}\\
\underline{-16x - 24} \quad &\underline{-16x - 24} \qquad &&\text{We can show both subtractions on the same line.}\\
8x^2 - 16x - 24 &= 0 \qquad &&\text{Factor out the GCF.}\\
8(x^2 - 2x - 3) &= 0 \qquad &&\text{Then factor the trinomial.}\\
8(x-3)(x+1) &= 0 \qquad &&\text{Now, write the solutions.}\\
x &= 3, -1 \qquad &&\text{The first factor, 8, doesn't include a variable, so 0 isn't a solution.}
\end{aligned}
$$

2. $7x^3 + 63x^2 = 154x$

This is not a quadratic equation. The highest power of the variable is 3, so this is a third-degree or *cubic* equation. However, we can still use factoring to help us solve it.

$$7x^3 + 63x^2 \ \ = \ \ \ \ \ \ \ 154x \quad \text{Get 0 on the right side of the equation.}$$
$$\underline{\ \ \ \ \ \ \ \ \ -154x \ \ \ \ \ \ \ \ \ -154x\ \ \ \ \ }$$
$$7x^3 + 63x^3 - 154x \ \ = \ \ \ \ \ \ \ \ 0 \quad \text{Then, factor out the GCF.}$$
$$7x(x^2 + 9x - 22) \ \ = \ \ 0 \quad \ \ \ \ \ \ \text{Next, factor the trinomial.}$$
$$7x(x + 11)(x - 2) \ \ = \ \ 0 \quad \ \ \ \ \ \ \text{Finish by writing the solutions.}$$
$$x \ \ = \ \ 0, -11, 2 \quad \text{The monomial factore, } 7x, \text{ includes a variable, so 0 is}$$
$$\text{included as a solution.}$$

In this chapter, our main focus is on solving *quadratic* equations, so most of the problems you see will involve those. However, if you come across a higher-degree equation, as in number 2 of the previous example, try using factoring to help you solve it.

Practice B

Now it's your turn. Use factoring to help you solve the following equations. You may turn the page when you're ready to check your solutions.

4. $x^2 - x - 42 = 0$

5. $r^2 - 49 = 0$

6. $x^2 + 2x - 24 = 0$

7. $n^2 = -13n - 40$

8. $v^2 - 11v = -28$

9. $10c^2 + 30c + 20 = 0$

10. $-7x^3 + 35x^2 - 42x = 0$

Practice A — Answers

1. $a = 4$

2. $y = -6, 7$

3. $x = -5, \frac{4}{3}$

C. Solving Application Problems

It's time to study some application problems involving quadratic equations. We find quadratic equations in a variety of situations that arise within the study of geometry, physics, economics, and other disciplines. There are also some purely mathematical application problems that can only be solved with quadratic equations.

To solve these problems, we again follow a five-step process:

1. Define a variable to represent the unknown quantity. It may be necessary to define two variables.

2. Using the variable defined in Step 1, translate the word problem into mathematical symbols and form an equation. If more than one variable is defined in Step 1, write a system of equations. Then reduce the system of equations to a single equation containing just one variable.

3. Solve this equation.

4. Check the solution by substituting the results into the equation found in Step 2.

5. Translate the mathematical solution into a written statement.

Remember, Step 1 is important! We need to begin the process by defining a variable. Then, after we've developed a quadratic equation in Step 2, we can try to solve it by factoring.

Example 7

Calina owns a small business, Calina's Creative Cases, that sells cell phone cases. She has determined that the number of cases sold, N, is related to the price of a case, x, by the equation $N = 35x - x^2$. What price or prices would result in Calina selling 216 cases?

Solution

Step 1: In this problem, Step 1 has been done for us: x represents the price of a cell phone case.

Step 2: Calina wants to sell 216 cases, so we replace N with 216 in the equation: $216 = 35x - x^2$.

Step 3: Now we solve the equation.

$$216 = 35x - x^2$$

The second-degree term, $-x^2$, is negative, so we don't want to leave it on the right side.

$$\underline{-35x + x^2 \qquad -35x + x^2}$$

$$x^2 - 35x + 216 = 0$$

Write the equation in standard form. Then factor and solve.

$$(x - 8)(x - 27) = 0$$

$$x = 8, 27$$

Practice B — Answers

4. $x = 7, -6$
6. $x = 4, -6$
8. $v = 4, 7$
10. $x = 0, 2, 3$

5. $r = -7, 7$
7. $n = -5, -8$
9. $c = -2, -1$

Step 4: We check our solutions, starting with 8 and then moving to 27.

$$216 = 35x - x^2 \qquad\qquad 216 = 35x - x^2$$
$$216 = 35(8) - (8)^2 \qquad\qquad 216 = 35(27) - (27)^2$$
$$216 = 280 - 64 \qquad\qquad 216 = 945 - 729$$
$$216 = 216 \checkmark \qquad\qquad 216 = 216 \checkmark$$

Step 5: We translate our answer into a written statement:

In order to sell 216 cases, Calina can price the cases at \$8 or \$27.

We can also use factoring when solving application problems that require us to find numbers that match a given description.

Example 8

The product of two consecutive integers is 156. Find them.

Solution

Step 1: As we have seen earlier, if x = the smaller integer, then $x + 1$ = the next consecutive integer.

Step 2: Because we know the product of these two integers is 156, our equation is $x(x +1) = 156$

Step 3: Now solve the equation.

$$x(x + 1) = 156 \quad \text{The left side of the equation is factored, but the right side isn't 0.}$$
$$\text{Use the distributive property to unfactor the left side.}$$
$$x^2 + x = 156 \quad \text{Now proceed to get 0 on the right side of the equation.}$$
$$\underline{\ -156 \quad\ -156}$$
$$x^2 + x - 156 = 0 \quad \text{Finish by factoring and solving.}$$
$$(x - 12)(x + 13) = 0$$
$$x = 12, -13$$

Step 4: Check both solutions, starting with 12 and moving to –13.

$$x(x + 1) = 156 \qquad\qquad x(x + 1) = 156$$
$$12(12 + 1) = 156 \qquad\qquad -13(-13 + 1) = 156$$
$$12(13) = 156 \qquad\qquad -13(-12) = 156$$
$$156 = 156 \checkmark \qquad\qquad 156 = 156 \checkmark$$

Step 5: 12 and 13 are consecutive integers whose product is 156, and so are –13 and –12.

We can also use quadratic equations to help us find the dimensions of objects.

Example 9

The length of a rectangle is 4 inches more than twice its width. The area is 30 square inches. Find the dimensions of the rectangle. Hint: the area of a rectangle is given by $A = LW$.

Solution

Steps 1 and 2: We need to find the length and the width of this rectangle, so we let L represent length and W represent width. Using these variables, we can set up a system of equations to get us started. We will then need to reduce our system down to a single equation containing one variable before proceeding to Step 3.

The length of the rectangle is 4 inches more than twice the width. That gives us this equation:

$$L = 2W + 4$$

We are also given the formula for the area of a rectangle and told that the area of this particular rectangle is 30 square inches. That gives us this equation:

$$30 = LW$$

The first equation allows us to substitute for L in the second equation:

$$30 = (2W + 4)W$$

We now have a single equation containing just one variable.

Step 3: Solve the equation.

$$30 = (2W + 4)W \quad \text{Use the distributive property to expand the right side.}$$

$$30 = 2W^2 + 4W \quad \text{The second-degree term is positive, so leave it on the right side of}$$

$$\underline{-30 \qquad\qquad -30} \quad \text{the equation. Get 0 on the left side.}$$

$$0 = 2W^2 + 4W - 30 \quad \text{Now factor and solve.}$$

$$0 = 2(W^2 + 2W - 15)$$

$$0 = 2(W + 5)(W - 3)$$

$$W = -5, 3 \qquad\qquad \text{Distance can't be negative, so we discard the negative solution.}$$

$$W = 3$$

If the width of the rectangle, W, is three inches, we can use substitution to find the length. We do this by using the first equation from our system in Step 2.

$$L = 2W + 4 = 2(3) + 4 = 10$$

Step 4: Check the values of L and W in both of the equations in Step 2.

$L = 2W + 4$ $\qquad\qquad$ $30 = LW$

$10 = 2(3) + 4$ $\qquad\qquad$ $30 = (3)(10)$

$10 = 6 + 4$ $\qquad\qquad\quad$ $30 = 30\ \checkmark$

$10 = 10\ \checkmark$

Step 5: The length of the rectangle is 10 inches, and the width is 3 inches.

We'll finish by looking at the problem from the beginning of this section.

Example 10

A contractor plans to pour a concrete walkway around a rectangular swimming pool that is 20 feet wide and 40 feet long. The area of the walkway should be 544 square feet. If the walkway has a uniform width, how wide should the contractor make it?

Solution

Step 1: Let x = the width of the walkway.

Step 2: Sketching a diagram is a good idea with a problem like this. It helps us to figure out the equation.

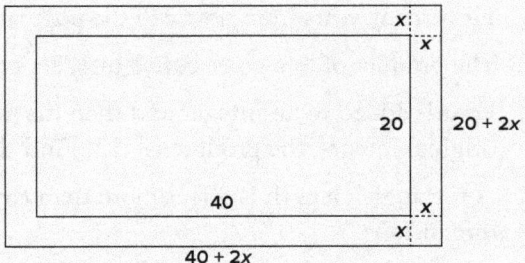

From our diagram, we see that [area of pool and walkway] – [area of pool] = [area of walkway]. So this is our equation: $[(20 + 2x)(40 + 2x)] - [20 \cdot 40] = [544]$

We can simplify this somewhat, and write $(20 + 2x)(40 + 2x) - 800 = 544$

Step 3: Solve the equation.

$(20 + 2x)(40 + 2x) - 800 = 544$ $\quad$ We must get rid of the parentheses before we can get 0

$800 + 40x + 80x + 4x^2 - 800 = 544$ $\quad$ on one side of the equation.

$4x^2 + 120x = 544$

$\underline{\qquad -544 \qquad -544\qquad}$

$4x^2 + 120x - 544 = 0$

$4(x^2 + 30x - 136) = 0$

$4(x + 34)(x - 4) = 0$

$x = -34, 4$ $\quad$ We discard the negative solution because in this

$x = 4$ $\qquad$ problem, only a positive solution makes sense.

Step 4: Check the solution.

$$(20 + 2x)(40 + 2x) - 800 = 544$$
$$(20 + 2(4))(40 + 2(4)) - 800 = 544$$
$$(28)(48) - 800 = 544$$
$$1344 - 800 = 544$$
$$544 = 544 \checkmark$$

Step 5: The contractor should make the walkway 4 feet wide.

Practice C

Now you get to solve a few application problems. Use the five-step process and factoring to help you. The turn the page to check your answers.

11. In t years from now, it's estimated that P, the population Yo City will be $P = t^2 - 24t + 96,000$. How many years from now will the population of Yo City be 95,865?

12. The product of two consecutive integers is 210. Find them both.

13. Four is added to an integer and then that sum is tripled. When this result is multiplied by the original integer, the product is −12. Find the integer.

14. A rectangle's length is 3 feet more than twice its width. The area is 14 square feet. Find the dimensions.

15. A contractor will pour a concrete walkway around a shed that is 15 feet wide and 25 feet long. The area of the walkway will be 276 square feet. If the walkway has uniform width, how wide must it be?

Exercises 6.1

For the following problems, solve the equation.

1. $9(a - 4) = 0$

2. $-2(m + 11) = 0$

3. $x(x + 7) = 0$

4. $n(n - 10) = 0$

5. $(y - 4)(y - 8) = 0$

6. $(k - 1)(k - 6) = 0$

7. $(x + 5)(x + 4) = 0$

8. $(j - 2)(j + 11) = 0$

9. $(w + 27)(w - 1) = 0$

10. $(y + 6)(2y + 1) = 0$

11. $(x - 3)(5x - 6) = 0$

12. $(5a + 1)(2a - 3) = 0$

13. $(6m + 5)(11m - 6) = 0$

14. $(2m - 1)(3m + 8) = 0$

15. $(4x + 5)(2x - 7) = 0$

16. $(3y + 1)(2y + 1) = 0$

17. $(x - 8)^2 = 0$

18. $(x - 2)^2 = 0$

19. $x(x - 4)^2 = 0$

20. $y(y + 9)^2 = 0$

21. $x^2 - 4 = 0$

22. $x^2 - 25 = 0$

23. $a^2 - 100 = 0$

24. $a^2 - 81 = 0$

25. $3a^2 - 75 = 0$

26. $5b^2 - 20 = 0$

27. $y^3 - y = 0$

28. $a^2 = 9$

29. $b^2 = 64$

30. $b^2 = 1$

31. $a^2 = 36$

32. $3a^2 = 12$

33. $-7b^2 = -63$

34. $3b^2 = 48$

35. $a^2 - 8a + 16 = 0$

36. $y^2 + 10y + 25 = 0$

37. $n^2 + 14n + 33 = 0$

38. $a^2 + 2a - 3 = 0$

39. $a^2 + 3a - 10 = 0$

40. $x^2 + 9x + 14 = 0$

41. $x^2 - 7x + 12 = 0$

42. $b^2 + 12b + 27 = 0$

43. $b^2 - 3b + 2 = 0$

44. $x^2 - 13x = -42$

45. $k^2 = 6k + 40$

46. $23p - 132 = p^2$

47. $48 = f^2 + 8f$

48. $c - 72 = -c^2$

49. $5x = -4 - x^2$

50. $a^3 = -8a^2 - 15a$

51. $y^3 = 9y^2 - 18y$

52. $7x^3 + 42x^2 = 112x$

53. $-2n^3 = 40n - 18n^2$

54. $-12h^4 - 12h^2 = -24h^3$

For the following exercises, use the five-step process and factoring to help you solve the problem.

55. The manufacturer of electronic fuel injectors determines that N, the number of injectors sold, is related to x, the price per injector, by $N = 22x - x^2$. At what price should the manufacturer price the injectors so that 112 of them are sold?

56. The owner of a stained-glass shop determines that N, the number of pieces of a particular type of glass sold in a month, is related to x, the price per piec, by $N = 21x - x^2$. At what price should the shop buyer price the glass so that 162 sell?

57. In t years from now, it's estimated that P, the population of Falls City, will be $P = t^2 - 15t + 12{,}036$.

   a. What is the population now?

   b. How many years from now will the population be 12,000?

58. In t years from now, it's estimated that P, the population of Silverton, will be $P = t^2 - 16t + 24{,}060$.

   a. What is the population now?

   b. How many years from now will the population be 24,000?

59. The height of a ball thrown into the air can be modeled by the equation $h = 48 + 32t + 16t^2$, here h represents the ball's height in feet, and t represents the number of seconds that elapse after the ball is thrown.

   a. How high is the ball at the instant it is thrown? Hint: let t = 0

   b. How high is the ball 1 second after it is thrown?

   c. When does the ball hit the ground? Hint: Determine the appropriate value for h first. Then solve for t.

60. The height of the glasses can be modeled by the equation $h = 64 - 16t^2$, where h represents the glasses' height above the ground in feet, and t represents the number of seconds that the glasses have been falling.

   a. How high above the ground was the woman's face when her glasses fell off?

   b. How many seconds after the glasses fell did they hit the ground?

61.   The length of a rectangle is 6 feet more than twice its width. The area is 8 square feet. Find the dimensions.

62.   The length of a rectangle is 18 inches more than three times its width. The area is 81 square inches. Find the dimensions.

63.   The length of a rectangle is two thirds its width. The area is 96 square feet. Find the dimensions.

64.   The length of a rectangle is four ninths its width. The area is 144 square feet. Find the dimensions.

65.   The area of a triangle is 14 square inches. The base is 3 inches longer than the height. Find both the length of the base and height. Hint: for a triangle, $A = \frac{1}{2}BH$.

66.   The area of a triangle is 34 square centimeters. The base is 1 cm longer than twice the height. Find both the length of the base and the height.

67.   The product of two consecutive integers is 72. Find them.

68.   The product of two consecutive integers is 42. Find them.

69.   The product of two consecutive odd integers is 143. Find them.

70.   The product of two consecutive even integers is 168. Find them.

71.   Three is added to an integer and that sum is doubled. When this result is multiplied by the original integer the product is 20. Find the integer.

72.   Four is added to three times an integer. When this sum and the original integer are multiplied, the product is –1. Find the integer.

73.   A contractor plans to pour a concrete walkway around a wading pool that is 4 feet wide and 8 feet long. The area of the walkway and pool is to be 96 square feet. If the walkway is to be of uniform width, how wide should it be?

Practice C — Answers

11.   The population of Yo City will be 95,865 in 9 or 15 years.

12.   14 and 15 are both consecutive integers whose product is 210, and so are –14 and –15.

13.   The integer is –2.

14.   The width of the rectangle is 2 feet, and the length is 7 feet.

15.   The contractor should make walkway 3 feet wide.

6.2 Understanding and Simplifying Square Roots

Overview

We've seen how factoring skills can be used in solving quadratic equations. However, as you know, not all quadratic expressions are factorable. This means that factoring isn't always a viable method for solving quadratic equations. Fortunately, there are other methods to use, and we will be learning them. First, though, we need to add another operation to our mathematical arsenal.

You are already comfortable performing the operations of addition, subtraction, multiplication, division, and even exponents on numbers and on polynomials. Now, it's time to explore *roots*. Here's a problem about *square roots*:

Simplify $\sqrt{45}$. Give an exact answer, and *not* a decimal approximation.

You may be able to locate the square root button C on a calculator and attempt to simplify the expression that way. If so, the number that shows up on your calculator screen certainly isn't very nice. It also isn't the exact value we're looking for because it's a rounded off decimal. We need to be able to simplify this expression and give an answer that is exact. So let's learn about square roots first and then return to this problem.

This section will show you how to:

◆ Understand square roots, and determine principal and secondary square roots

◆ Know when a square root represents a meaningful expression in the real number system

◆ Use product and quotient properties when simplifying rational and irrational square roots

A. Introduction to Square Roots

We've already seen perfect squares in our study of algebra. Consider the number 16. Because $4^2 = 16$, we know that 16 is a perfect square. We can thus refer to 16 as the square of 4. We also know that $(-4)^2 = 16$, so 16 is also the square of -4.

A square root of a number is a value that, when it is squared, will result in the number. Algebraically, we state it this way:

If $x^2 = A$, then x is a *square root* of A.

We have seen that $4^2 = 16$ and that $(-4)^2 = 16$. That means that both 4 and -4 are square roots of 16. We can generalize what we've seen here as follows:

- ◆ Every positive number has two square roots
- ◆ The square roots of a positive number are opposites of each other

So, what can be said about the square roots of numbers that are not positive?

As an extension of the zero-product property, we know that the number zero can have only one square root: The square root of zero is 0. Negative numbers don't have *any* square roots in the real number system because it's impossible to multiply a real number by itself and get a negative result.

Let's consider square roots of a couple more positive numbers.

▶ **Example 1**

Determine the square roots of each number.

1. 49

2. $\frac{49}{64}$

Solutions

1. The two square roots of 49 are 7 and -7 because $7^2 = 49$ and $(-7)^2 = 49$.

2. The two square roots of $\frac{49}{64}$ are $\frac{7}{8}$ and $\frac{-7}{8}$ because $\left(\frac{7}{8}\right)^2 = \frac{7}{8} \cdot \frac{7}{8} = \frac{49}{64}$ and $\left(\frac{-7}{8}\right)^2 = \frac{-7}{8} \cdot \frac{-7}{8} = \frac{49}{64}$.

We use a particular notation for distinguishing the positive square root of a number from the negative square root of the number. If x is a positive real number, then $\sqrt{x}$ represents the positive square root of x. The positive square root of a number is called the **principal square root**. The negative square root of x is indicated by $-\sqrt{x}$. The negative square root of a number is called the **secondary square root**.

In the expression $\sqrt{x}$, the root symbol, C is called a **radical sign**. The value within the radical sign, x, is called the **radicand**. The horizontal bar of the radical sign acts as a grouping symbol that shows what the radicand is. When we put the radical and radicand together, the expression, $\sqrt{x}$, is called a **radical expression**. These terms are derived from the Latin word *radix*, which means "root."

Using this notation, we can summarize how square roots work in this way:

Because $\sqrt{x}$ and $-\sqrt{x}$ are the two square roots of x,
$(\sqrt{x})(\sqrt{x}) = x$ and $(-\sqrt{x})(-\sqrt{x}) = x$.

In the following examples, we'll use this notation to consider the principal and secondary square roots of each number.

Example 2

Determine the principal and secondary square root for each number.

1. 9   **2.** 15

Solutions

1. The number 9 is a perfect square. We know that 9 = (3)(3) and 9 = (−3)(−3). Therefore, the principal square root is $\sqrt{9} = 3$. The secondary square root is $-\sqrt{9} = -3$.

2. Because 15 is not a perfect square, factorizations will not help us write the answer without a radical sign. The principal square root is $\sqrt{15}$. The secondary square root is $-\sqrt{15}$.

The square roots of 15 are called **irrational numbers** because these values cannot be written as a fraction of two integers. They also cannot be written as terminating or repeating decimals.

Example 3

Give decimal approximations for the principal and secondary square root of 34. Round answers to the nearest hundredth.

Solution

We again have a number that is not a perfect square. We will use a calculator to obtain a decimal approximation for the square roots of this number. We will round to two decimal places.

In order to evaluate $\sqrt{34}$, some calculators require the user to enter the square root sign, [√], and then type in ③④. Graphing calculators tend to use this format. Some scientific calculators, however, require the user to begin by entering ③④ and then pressing the square root key, [√]. You should become familiar with how your calculator works when evaluating square roots.

After we enter the expression into a calculator, the display reads 5.830951895. Note that the actual value of $\sqrt{34}$ is a never-ending, never-repeating decimal. Our calculator only shows us as much of the number as it can fit on the display. We round our answer to 5.83, which is an approximation the principle square root.

Our calculator does not have a −[√] key. However, once we have found the principal square root, all we need to do to in order to find the secondary square root is add a negative sign. Therefore, we can write the principal root, $\sqrt{34} \approx 5.83$, and the secondary square root, $-\sqrt{34} \approx -5.83$.

You will recall that the symbol "≈" is how we show that two expressions are "approximately equal to" each other. In this case, we need to use this symbol instead of an equals sign because our answers are approximations and not exact values.

Example 4

The number $\sqrt{50}$ is between what two whole numbers?

Solution

We can often use mental math to give us an idea of the value of a principal square root. This can be helpful in double-checking the accuracy of the approximation we get when using our calculators.

We know that $7^2 = 49$, so $\sqrt{49} = 7$. Similarly, we know that $8^2 = 64$, so $\sqrt{64} = 8$. Because 50 is between 49 and 64, we know that $\sqrt{50}$ must be a number between the whole numbers 7 and 8. If our calculators do not display a number between 7 and 8 when approximating $\sqrt{50}$, then we must have made a mistake when entering the expression.

Radical signs are often used in problems that ask us to evaluate a square root. When this happens, remember that the *principal* square root is denoted with "$\sqrt{x}$" while the *secondary* square root is denoted with "$-\sqrt{x}$." In these situations, do not give both square roots.

Example 5

Evaluate each expression.

1. $\sqrt{144}$ 2. $-\sqrt{81}$

Solutions

1. $\sqrt{144}$

   We are being asked for the *principal* square root of 144, so the answer is 12.

2. $-\sqrt{81}$

   In this case, we're asked for the *secondary* square root of 81. The answer is −9.

Practice A

Now you can give it a try. Write the principal and secondary square root of each number.

1. 36
2. $\frac{9}{16}$
3. 0.09
4. 35: Give an exact answer.
5. 35: Give a decimal approximation. Round to two decimal places.

Evaluate each expression

6. $-\sqrt{64}$ 7. $\sqrt{121}$

When you're done, turn the page and check your solutions.

B. Meaningful Expressions

We can't square a real number and get a negative result. This means that an expression like $\sqrt{-16}$ doesn't describe a real number. In more advanced algebra classes, you will examine a number system that allows you to evaluate such an expression, but for now, we simply state that $\sqrt{-16}$ is not a meaningful expression. We can also say that $\sqrt{-16}$ is "not real."

To state this concept algebraically, we can say that for $\sqrt{x}$ to be a real number and a meaningful expression, $x \geq 0$.

In the next example, we'll see how variable values can be restricted in order to ensure that a radical expression represents a real number.

Example 6

Place a restriction on each variable so that the radical expression represents a real number.

1. $\sqrt{x - 3}$ 2. $\sqrt{2m + 7}$

Solutions

1. $\sqrt{x - 3}$

The radicand, $x - 3$, cannot be negative if this expression is going to represent a real number. In other words, $x - 3$ must be greater than or equal to zero.

$$
\begin{array}{rcr}
x - 3 & \geq & 0 \\
+3 & & +3 \\
\hline
x & \geq & 3
\end{array}
$$

The solution, $x \geq 3$, is the variable restriction.

2. $\sqrt{2m + 7}$

We set the radicand greater than or equal to zero, and then solve.

$$
\begin{array}{rcr}
2m + 7 & \geq & 0 \\
-7 & & -7 \\
\hline
2m & \geq & -7 \\
\dfrac{2m}{2} & \geq & \dfrac{-7}{2} \\
m & \geq & -\dfrac{7}{2}
\end{array}
$$

Practice B

For each problem, write a restriction that must be placed on the variable so that the radical expression represents a real number. When you are done, turn the page and check your solutions.

8. $\sqrt{x + 5}$ 9. $\sqrt{y - 8}$ 10. $\sqrt{3a + 2}$ 11. $\sqrt{5m - 6}$

C. More Complicated Square Root Expressions

Now it's time to look at two properties of square roots: the product property and the quotient property.

The **product property of square roots** states that, in general, if x and y are positive real numbers, then $\sqrt{xy} = \sqrt{x}\,\sqrt{y}$. For example, notice that $\sqrt{9 \cdot 4} = \sqrt{36} = 6$ and $\sqrt{9}\,\sqrt{4} = 3 \cdot 2 = 6$. Because both $\sqrt{9 \cdot 4}$ and $\sqrt{9}\,\sqrt{4}$ equal 6, it must be that $\sqrt{9 \cdot 4} = \sqrt{9}\,\sqrt{4}$.

The **quotient property of square roots** states that, in general, if x and y are positive real numbers, then $\sqrt{\frac{x}{y}} = \frac{\sqrt{x}}{\sqrt{y}}$. Notice that $\sqrt{\frac{36}{4}} = \sqrt{9} = 3$ and $\frac{\sqrt{36}}{\sqrt{4}} = \frac{6}{2} = 3$. Because both $\sqrt{\frac{36}{4}}$ and $\frac{\sqrt{36}}{\sqrt{4}}$ equal 3, it must be that $\sqrt{\frac{36}{4}} = \frac{\sqrt{36}}{\sqrt{4}}$.

We will now look at some examples of how the product property and quotient property can be used to simplify square root expressions.

Example 7

Simplify the following expressions.

1. $\sqrt{25 \cdot 36 \cdot 81 \cdot 144}$ 　　　2. $-\sqrt{\frac{9}{64}}$ 　　　3. $\sqrt{\frac{27}{75}}$

Solutions

1. $\sqrt{25 \cdot 36 \cdot 81 \cdot 144}$

If we multiply all of the numbers inside of the radical together, we end up with the expression $\sqrt{10{,}497{,}600}$. This expression would not be particularly easy to evaluate. However, we can use the product property of square roots to help us:

$$\sqrt{25 \cdot 36 \cdot 81 \cdot 144} = \sqrt{25} \cdot \sqrt{36} \cdot \sqrt{81} \cdot \sqrt{144}$$

Rewrite the expression as the product of square roots. Then simplify each square root separately.

$$= 5 \cdot 6 \cdot 9 \cdot 12$$
$$= 3{,}240$$

2. $-\sqrt{\frac{9}{64}}$

$$-\sqrt{\frac{9}{64}} = -\frac{\sqrt{9}}{\sqrt{64}}$$

Use the quotient rule to rewrite the original expression as the quotient of two square roots. Then simplify each square root separately.

$$= -\frac{3}{8}$$

Practice A — Answers

1. 6 and −6 　　　3. 0.3 and −0.3 　　　5. 5.92 and −5.92 　　　7. 11
2. $\frac{3}{4}$ and $-\frac{3}{4}$ 　　　4. $\sqrt{35}$ and $-\sqrt{35}$ 　　　6. −8

3. $\sqrt{\dfrac{27}{75}}$

$$\sqrt{\dfrac{27}{75}} = \sqrt{\dfrac{9}{25}}$$ Begin by reducing the fraction inside the radical.

$$= \dfrac{\sqrt{9}}{\sqrt{25}}$$ Apply the quotient property of square roots. Then simplify each square root separately.

$$= \dfrac{3}{5}$$

☠ Warning! Incorrect Approach! ☠

The product and quotient properties of square roots allow us to split a single square root into multiple square roots. However, do *not* attempt to split square roots involving sums or differences. Some people mistakenly attempt to do this, and it typically ends badly:

$$\sqrt{16 + 9} = \sqrt{16} + \sqrt{9} = 4 + 3 = 7$$

This is incorrect! We know that 16 + 9 = 25, so the simplification process *should* look like:

$$\sqrt{16 + 9} = \sqrt{25} = 5$$

It is important to know that, in general, $\sqrt{x + y} \neq \sqrt{x} + \sqrt{y}$ and $\sqrt{x - y} \neq \sqrt{x} - \sqrt{y}$.

Now that you're familiar with the product and quotient properties of square roots, it's time to explore irrational square roots further. At the beginning of this section, we reviewed the concept of perfect squares. Now we'll give a more restrictive definition for perfect squares.

Perfect squares are real numbers that are the squares of *rational numbers*. The numbers 25 and $\frac{1}{4}$ are examples of perfect squares because $25 = 5^2$ and $\frac{1}{4} = \left(\frac{1}{2}\right)^2$ and because 5 and $\frac{1}{2}$ are both rational numbers. The number 2 is not a perfect square because $2 = \left(\sqrt{2}\right)^2$, and $\sqrt{2}$ is not a rational number. We saw other irrational square roots earlier in Examples 2 through 4. Any square root whose radicand is not a perfect square is an irrational number.

When simplifying irrational square roots with a calculator, the closest we can come to the value of the square root is a decimal approximation, as in Example 3. However, it's often better to give an exact value. When a square root is irrational, we can't write an exact value without a radical sign, but in many cases, we can simplify the square root.

Practice B — Answers

8. $x \geq -5$ 9. $y \geq 8$ 10. $a \geq -\dfrac{2}{3}$ 11. $m \geq \dfrac{6}{5}$

A square root is written in **simplified form** if the radicand does not have any perfect square factors other than 1. For example, $\sqrt{14}$ is in simplified form because the factors of 14 are 1, 2, 7, and 14. The only factor that is a perfect square is 1. However, $\sqrt{8}$ is not in simplified form because the factors of 8 are 1, 2, 4, and 8, and 4 is a perfect square. If an irrational square root is not in simplified form, we simplify it by rewriting the radicand as a product involving a perfect square and then applying the product rule for square roots.

Here's an example that illustrates this procedure.

Example 8

Simplify $\sqrt{8}$. Give an exact answer.

Solution

$$\sqrt{8} = \sqrt{4 \cdot 2}$$ We know that 4 is a perfect square and that it is also a factor of 8. We rewrite the radicand as $4 \cdot 2$.

$$= \sqrt{4}\sqrt{2}$$ Next we apply the product rule for square roots.

$$= 2\sqrt{2}$$ We finish by simplifying the rational square root, $\sqrt{4} = 2$, and leaving the irrational square root alone.

Now that you've seen how the product property of square roots can be used to simplify irrational square roots, let's take another look at the problem from the beginning of this section.

Example 9

Simplify $\sqrt{45}$. Give an exact answer.

Solution

$$\sqrt{45} = \sqrt{9 \cdot 5}$$ We know that 9 is a perfect square and that it is also a factor of 45. We rewrite the radicand as $9 \cdot 5$.

$$= \sqrt{9}\sqrt{5}$$ Next we apply the product rule for square roots.

$$= 3\sqrt{5}$$ We finish by simplifying the rational square root, $\sqrt{9} = 3$, and leaving the irrational square root alone.

Sometimes a radicand may contain more than one perfect square factor.

Example 10

Simplify $\sqrt{180}$.

Solution

$\sqrt{180} = \sqrt{9 \cdot 20}$ We know that 9 is a perfect square and that it is also a factor of 180. Rewrite the radicand as $9 \cdot 20$.

$= \sqrt{9 \cdot 4 \cdot 5}$ We know that 4 is a perfect square and that it is also a factor of 20. Rewrite the radicand as $9 \cdot 4 \cdot 5$.

$= \sqrt{9}\,\sqrt{4}\,\sqrt{5}$ Next, apply the product rule for square roots.

$= 3 \cdot 2 \cdot \sqrt{5}$ Then simplify the rational square roots, $\sqrt{9} = 3$ and $\sqrt{4} = 2$, and leave the irrational square root alone.

$= 6\sqrt{5}$ Finish by simplifying the expression.

If we realize that 36 is a factor of 180, the process is faster:

$\sqrt{180} = \sqrt{36 \cdot 5}$

$= \sqrt{36}\,\sqrt{5}$

$= 6\sqrt{5}$

Practice C

Now it's your turn to simplify each square root. Give exact answers, not decimal approximations. If a square root is irrational, then the simplified expression will still contain a radical. When you are finished, check your work on the next page.

12. $\sqrt{144 \cdot 49}$

13. $\sqrt{\dfrac{81}{121}}$

14. $\sqrt{\dfrac{72}{128}}$

15. $\sqrt{125}$

16. $\sqrt{98}$

17. $\sqrt{1300}$

18. $\sqrt{588}$

Exercises 6.2

For the following problems, write the principal and secondary square root of each number. If a number has irrational square roots, give decimal approximations rounded off to the nearest hundredth.

1. 64
2. 81
3. 25
4. 121
5. 144

6. 225
7. 10,000
8. 20
9. 71
10. 195

11. 47
12. $\frac{1}{16}$
13. $\frac{1}{49}$
14. $\frac{25}{36}$
15. $\frac{121}{225}$

16. 0.04
17. 0.16
18. 1.21

For the following problems, evaluate each expression. If the expression does not represent a real number, write "not real."

19. $\sqrt{49}$
20. $\sqrt{64}$
21. $-\sqrt{36}$
22. $-\sqrt{100}$

23. $-\sqrt{169}$
24. $-\sqrt{\frac{36}{81}}$
25. $-\sqrt{\frac{121}{169}}$
26. $\sqrt{-225}$

27. $\sqrt{-36}$
28. $-\sqrt{-1}$
29. $-\sqrt{-5}$
30. $-(-\sqrt{9})$

31. $-(-\sqrt{0.81})$

For the following problems, write the proper restrictions that must be placed on the variable so that the expression represents a real number.

32. $\sqrt{y + 10}$
33. $\sqrt{x + 4}$

34. $\sqrt{a - 16}$
35. $\sqrt{h - 11}$

36. $\sqrt{2k - 1}$
37. $\sqrt{7x + 8}$

38. $\sqrt{-2x - 8}$
39. $\sqrt{-5y + 15}$

For the following problems, simplify each irrational square root. Give exact answers, not decimal approximations.

40. $\sqrt{48}$
41. $\sqrt{75}$
42. $\sqrt{242}$
43. $\sqrt{24}$

44. $-\sqrt{56}$
45. $\sqrt{135}$
46. $\sqrt{162}$
47. $\sqrt{108}$

48. $\sqrt{490}$
49. $\sqrt{252}$
50. $\sqrt{320}$
51. $\sqrt{1000}$

52. $\sqrt{2450}$
53. $\sqrt{1125}$

Practice C — Answers

12. 84
13. $\frac{9}{11}$

14. $\frac{3}{4}$
15. $5\sqrt{5}$

16. $7\sqrt{2}$
17. $10\sqrt{13}$

18. $4\sqrt{3}$

6.3 Solving Quadratic Equations by Using Square Roots

Overview

Throughout this book, we've used inverse operations to help us solve equations. Now that we can simplify square roots, we have yet another tool at our disposal. Consider this problem:

Solve: $(a + 3)^2 = 5$

We will solve this equation with an approach that's similar to what we did in Chapter 1. However, because there's an exponent involved, we need to use a root as part of the process. We'll learn some basics about using roots to solve equations before we tackle this problem.

This section will teach you how to:

- ◆ Use square roots to solve quadratic equations
- ◆ Use square roots to solve equations that arise as a result of the Pythagorean Theorem
- ◆ Find the distance between two points on a coordinate plane

A. Using Square Roots to Solve Equations

In Chapter 1, we solved linear equations by using inverse operations to isolate the variable. For review, let's look at the following example.

Example 1

Solve: $3x - 18 = -12$

Solution

$$
\begin{aligned}
3x - 18 &= -12 \quad \text{18 is being subtracted from } 3x. \\
\underline{+\,18} \quad &\underline{+\,18} \quad \text{Undo the subtraction by } \textit{adding 18 to both sides.} \\
3x &= 6 \quad \text{Now } x \text{ is being multiplied by 3.} \\
\frac{3x}{3} &= \frac{6}{3} \quad \text{Undo the multiplication by } \textit{dividing both sides by 3.} \\
x &= 2
\end{aligned}
$$

Example 1 illustrates the fact that addition and subtraction are inverse operations, as are multiplication and division. Another pair of inverse operations are exponents and roots. In this chapter, we focus on the inverse relationship between an exponent of 2 and a second root.

When an exponent of 2 is put on an expression, we refer to this as *squaring* the expression. For example, x^2 is commonly called "the square of x" or "x squared." The inverse operation is to apply the second or "square" root. If we use square roots to undo operations while solving an equation, we must use both the principal and secondary square roots because we want to find all of the solutions for the equation.

Example 2

Solve: $x^2 = 25$

Solution

$$x^2 = 25$$

The variable x has an exponent of 2. We undo this operation by taking square roots of both sides.

$$x = \sqrt{25}$$
$$x = -\sqrt{25}$$

Because every positive number has *two* square roots, we find both the principal and secondary square root of 25.

$$x = 5, -5$$

We finish by simplifying our solutions.

Check the first solution: Then the second:

$$(5)^2 = 25 \qquad\qquad (-5)^2 = 25$$
$$25 = 25 \checkmark \qquad\qquad 25 = 25 \checkmark$$

We can use the symbol "$\pm$" to denote both the principal and secondary square roots. For this example, we can write the solutions as $x = \pm 5$.

In the previous example, we found two real-number solutions for the equation. In some cases, we find only one solution and in still other cases, we don't find any real-number solutions. Based on what we learned in the previous section, we can summarize these cases as follows:

Solving Equations of the Form $x^2 = K$

1. If K is positive, then there are two real number solutions: $x = \pm\sqrt{K}$.
   This fact is commonly referred to as the *square root property*.

2. If K is zero, then there is one real number solution: $x = 0$.

3. If K is negative, then no real number solutions exist.

Recall that a quadratic equation can be written in the form: $ax^2 + bx + c = 0$. If $b = 0$, then the equation doesn't have a first-degree term: $ax^2 + c = 0$. Whenever that happens, we can follow this procedure when solving:

1. Use inverse operations to isolate the squared variable

2. Use square roots to find the real-number solutions — if they exist.

Example 3

Solve the following equations.

1. $x^2 - 49 = 0$ **2.** $25\,a^2 = 36$ **3.** $4\,m^2 - 32 = 0$

Solutions

1. $x^2 - 49 = 0$

$$x^2 - 49 = 0$$

$$\underline{+49 \qquad +49}$$ Isolate x^2 by **adding 49 to both sides**.

$$x^2 = 49$$

$$x = \pm\sqrt{49}$$ Undo the exponent of 2 **by taking square roots of both sides**.

$$x = \pm 7$$ Simplify the solutions.

Check the first solution: Then the second:

$(7)^2 - 49 = 0$ $(-7)^2 - 49 = 0$

$49 - 49 = 0$ $49 - 49 = 0$

$0 = 0 \checkmark$ $0 = 0 \checkmark$

2. $25\,a^2 = 36$

$$\frac{25\,a^2}{25} = \frac{36}{25}$$ Isolate a^2 by **dividing both sides by 25**.

$$a^2 = \frac{36}{25}$$

$$a = \pm\sqrt{\frac{36}{25}}$$ Undo the exponent of 2 **by taking square roots of both sides**.

$$a = \pm\frac{\sqrt{36}}{\sqrt{25}}$$ Finish by simplifying the solutions.

$$a = \pm\frac{6}{5}$$

Check the first solution: Then the second:

$25\left(\frac{6}{5}\right)^2 = 36$ $25\left(\frac{-6}{5}\right)^2 = 36$

$25\left(\frac{36}{25}\right) = 36$ $25\left(\frac{36}{25}\right) = 36$

$36 = 36 \checkmark$ $36 = 36 \checkmark$

3. $4m^2 - 32 = 0$

$$4m^2 - 32 = 0$$

$$\underline{+32+32}$$ Isolate the term containing m^2 by adding 32 to both sides.

$$4m^2 = 32$$

$$\frac{4m^2}{4} = \frac{32}{4}$$ Isolate m^2 by dividing both sides by 4.

$$m^2 = 8$$

$$m = \pm\sqrt{8}$$ Undo the exponent of 2 by taking square roots of both sides.

$$m = \pm 2\sqrt{2}$$ Finish by simplifying the solutions. Because this is an irrational square root, our final answer includes a radical sign.

Check the first solution:

$$4\left(2\sqrt{2}\right)^2 = 32$$
$$4\left[2^2\left(\sqrt{2}\right)^2\right] = 32$$
$$4[4 \cdot 2] = 32$$
$$4 \cdot 8 = 32$$
$$32 = 32 \checkmark$$

Then the second:

$$4\left(-2\sqrt{2}\right)^2 = 32$$
$$4\left[(-2)^2\left(\sqrt{2}\right)^2\right] = 32$$
$$4[4 \cdot 2] = 32$$
$$4 \cdot 8 = 32$$
$$32 = 32 \checkmark$$

In number 3 of the previous example, the solution is an irrational square root, and we give an exact answer, not a decimal approximation. In general, we give an exact answer when solving square root equations unless the instructions indicate that we should give a rounded off decimal as the final answer.

Example 4

Solve: $14a^2 - 235 = 0$. If the solution is irrational, give a decimal approximation rounded to the nearest hundredth.

Solution

The instructions say that we should give a rounded off decimal if the final answer is an irrational square root. We'll work through this problem without explanation. See if you can follow along.

$$14a^2 - 235 = 0$$

$$\underline{+235+235}$$

$$14a^2 = 235$$

$$\frac{14a^2}{14} = \frac{235}{14}$$

$$a^2 = \frac{235}{14}$$

$$a = \pm\sqrt{\frac{235}{14}}$$

$$a \approx \pm 4.10$$

When we check the positive solution, we end up with the following:

$$14(4.10)^2 - 235 = 0$$
$$14(16.81) - 235 = 0$$
$$235.34 - 235 = 0$$
$$0.34 = 0$$

The left side of the equation is slightly different than the right because the answer we're using is a rounded-off decimal. There's a small amount of round-off error in this case, but because 0.34 is close to 0, we can be confident that our answer is reasonable.

Even though a rounded-off decimal doesn't represent an *exact* solution, there are many application problems in which a decimal approximation is preferable to a simplified radical.

Now let's look at an equation that cannot be solved with real numbers.

Example 5

Solve: $k^2 + 89 = 25$

Solution

$$
\begin{array}{rcr}
k^2 + 89 &=& 25 \\
-89 & & -89 \\
\hline
k^2 &=& -64 \\
k & & \pm\sqrt{-64}
\end{array}
$$

The solution is $k = \pm\sqrt{-64}$. The radicand is negative, so no real-number solutions exist.

Practice A

Now you try it. Use square roots to help you solve each quadratic equation, if possible. Unless otherwise indicated, give exact answers. When you are done, turn the page and check your solutions.

1. $x^2 - 144 = 0$

2. $9y^2 - 121 = 0$

3. $6a^2 = 108$

4. Solve $16m^2 - 2206 = 0$. Round the answer to the nearest hundredth.

5. $h^2 + 100 = 0$

B. Using Square Roots
to Solve More Complicated Quadratic Equations

We can also use square roots to help solve more complicated quadratic equations. If the variable appears inside of a set of parentheses being squared — and nowhere else in the equation — then we can use square roots to get rid of the exponent on the parentheses. After this, we will have to take additional steps in order to isolate the variable.

Example 6

Solve: $(x + 2)^2 = 81$

Solution

$(x + 2)^2 = 81$ The variable appears inside of a set of squared parentheses and nowhere else.

$x + 2 = \pm\sqrt{81}$ Undo the exponent of 2 by taking square roots of both sides.

$x + 2 = \pm 9$ Simplify the radical expression

At this point, we see that the expression "$x + 2$" takes on two different values. We will thus write and solve two different equations.

$$x + 2 = 9 \qquad\qquad x + 2 = -9$$
$$\underline{-2 \quad -2} \qquad\qquad \underline{-2 \quad -2}$$
$$x = 7 \qquad\qquad x = -11$$

We can write the solution as $x = 7, -11$. We won't check the solutions here, but rest assured that both are correct.

We can shorten the process slightly if we don't write two separate equations.

$$x + 2 = \pm 9$$
$$\underline{-2 \qquad -2}$$
$$x = -2 \pm 9$$

We can use mental math to find the solutions: $-2 + 9 = 7$, and $-2 - 9 = -11$. However, be careful when taking this approach.

☠ Warning! Incorrect Approach! ☠

Some people try to take the shorter approach by applying the "±" symbol after the subtraction step:

$$
\begin{array}{rcc}
x + 2 &=& \pm\, 9 \\
\underline{-2} & & \underline{-2} \\
x &=& \pm\,(9 - 2) \\[4pt]
x &=& \pm\, 7
\end{array}
$$

This is not correct! The "±" symbol *only* applies to the 9. It doesn't apply to the −2. Be careful when attempting to use the shortened approach in these types of problems.

When an irrational square root is involved, the shortened approach is slightly safer to use. Let's see how that works by revisiting the problem from the beginning of this section.

Example 7

Solve: $(a + 3)^2 = 5$

Solution

$$(a + 3)^2 = 5$$

$a + 3 = \pm\sqrt{5}$ Undo the exponent of 2 by taking square roots of both sides.

$a + 3 = \pm\sqrt{5}$ The square roots are irrational and cannot be simplified, so we proceed

$\quad\underline{-3}\qquad\underline{-3}\qquad$ to the next step and subtract 3 from both sides.

$$a = -3 \pm \sqrt{5}$$

After performing the subtraction in the final step of solving, there are two acceptable expressions that can be written: $\pm\sqrt{5} - 3$ and $-3 \pm \sqrt{5}$. The second is the preferred way.

Sometimes it's nicer to see both solutions in their entirety. We can write the solutions like this: $a = -3 + \sqrt{5}\,,\ -3 - \sqrt{5}$.

In the previous example, we gave exact solutions. If decimal approximations had been requested, then we could have used a calculator to help us find the following: $a \approx -0.76\,,\ -5.24$

Practice B

Use square roots to help you solve each of the following quadratic equations. Give exact answers, not decimal approximations. When you are done, turn the page and check your solutions.

6. $(a + 6)^2 = 64$ 8. $(y - 7)^2 = 49$ 10. $(x - 11)^2 = 0$

7. $(m - 4)^2 = 15$ 9. $(k - 1)^2 = 12$

C. The Pythagorean Theorem

A common application of square roots is solving right triangles. In more advanced math classes, "solving a triangle" means determining the measures of all of the sides *and* all of the angles. For now, we will think of "solving a triangle" as determining the lengths of all the sides.

If one of the angles of a triangle is a right angle — that is, an angle measuring $90°$ — then the triangle is a **right triangle**. In a right triangle, both of the sides that touch the right angle are called **legs**. The side of a right triangle that does *not* touch the right angle is called the **hypotenuse**, as you can see in Figure 1.

The sides of a right triangle have a special mathematical relationship. This relationship is given by the Pythagorean Theorem.

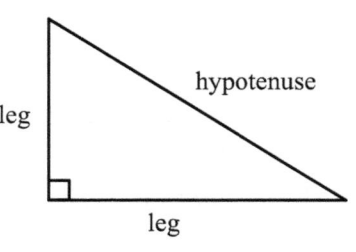

Figure 1.

The Pythagorean Theorem

If A and B represent the lengths of the legs of a right triangle, and if C represents the length of the hypotenuse, then $A^2 + B^2 = C^2$.

Because the Pythagorean Theorem involves squaring the side lengths of a right triangle, it follows that we will have to use square roots when solving equations given by this theorem.

Example 8

Find the value of n in Figure 2.

Solution

In this problem, we're given the lengths of both legs, 6 inches and 8 inches. That means that we can put number values in for A and B in the formula given by the Pythagorean Theorem.

$$A^2 + B^2 = C^2 \qquad \text{Replace } A \text{ with 6, } B \text{ with 8, and } C \text{ with } n.$$

$$6^2 + 8^2 = n^2 \qquad \text{Solve for } n.$$

$$36 + 64 = n^2$$

$$100 = n^2$$

$$\sqrt{100} = n \qquad \text{A geometric distance cannot be negative.}$$

$$10 = n \qquad \text{In this case, we are only interested in the principal square root.}$$

The length of the hypotenuse, n, is 10 inches.

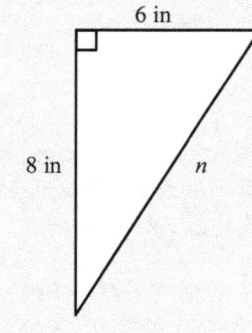

Figure 2.

When using the Pythagorean Theorem, it's important to identify whether a given number value is the length of a leg or the length of the hypotenuse. In Example 8, both of the given number values were lengths of legs. However, this isn't always the case.

Example 9

Find the value of x in Figure 3.

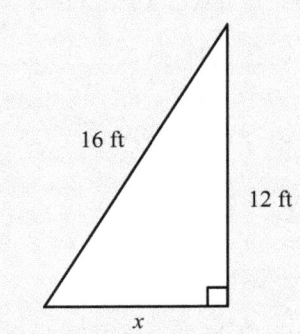

Figure 3.

Solution

In this problem, we're given the length of one leg and the length of the hypotenuse. We can plug in number values for A and C in the formula given by the Pythagorean Theorem.

$A^2 + B^2 \ = \ \ \ \ C^2$ Replace A with 12, C with 16,

$12^2 + x^2 \ = \ \ \ \ 16^2$ and B with x. Then solve for x.

$144 + x^2 \ = \ \ \ 256$

$\underline{\ \ -144 \ \ \ \ \ \ \ \ -144\ \ }$

$x^2 \ = \ \ \ \ 112$

$x \ = \ \ \sqrt{112}$ Only take the principal square root.

$x \ = \ \ 4\sqrt{7}$

The instructions for this problem didn't indicate that a decimal approximation would be okay, so we give an exact answer: The length of the other leg, x, is $4\sqrt{7}$ feet.

Example 10

A work crew needs to install a cable over the busy road shown in Figure 4. The cable will stretch diagonally across the road, from point A to point B. The workers know that the road is 9 meters wide and that point B is 30 meters down the road from point A. Based on this information, determine how long the cable must be to stretch from point A to point B. Round your answer to the nearest hundredth.

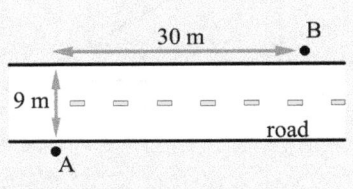

Figure 4.

Practice B — Answers

6. $a = 2, -14$

7. $m = 4 \pm \sqrt{15}$

8. $y = 0, 14$

9. $k = 1 \pm 2\sqrt{3}$

10. $x = 11$

Solution

As you can see, the width of the road, 9 m, and the distance down the road from point A to point B, 30 m, are the lengths of the legs of a right triangle. When the cable is stretched from point A to point B, it will form the hypotenuse. We will use the Pythagorean Theorem to solve this problem.

$$A^2 + B^2 = C^2 \quad \text{Replace } A \text{ and } B \text{ with 9 and 30.}$$
$$9^2 + 30^2 = C^2 \quad \text{Solve for } C.$$
$$81 + 900 = C^2$$
$$981 = C^2$$
$$\sqrt{981} = C$$
$$31.32 \approx C$$

The cable needs to be about 31.32 meters long.

Practice C

Now that you've seen the Pythagorean Theorem in action, try a few problems. Unless you're directed to do otherwise, give exact answers and not decimal approximations. When you are done, turn the page and check your solutions.

11. Find the value of w.

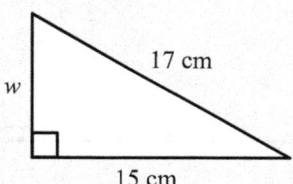

12. Find the value of y.

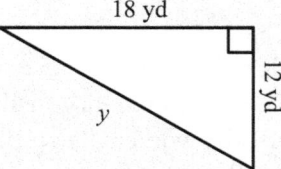

13. A 14-foot ladder leans against the wall of a building. The bottom of the ladder is 3 feet away from the base of the wall. How high up the wall is the top of the ladder? Round your answer off to the nearest hundredth.

D. Distance Between Points on a Coordinate Plane

We can use a right triangle and the Pythagorean Theorem to find the distance between two points on a coordinate plane. However, such a process takes a long time because it involves plotting the points, constructing a right triangle, and determining the lengths of the legs — and that's before we can use the Pythagorean Theorem to help us find the desired distance.

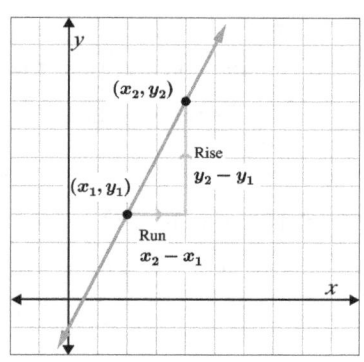

Figure 5.

There's a faster way to find the distance between two points on a coordinate plane that involves memorizing a formula. We'll take a look at where the formula comes from. Hopefully this will help you to memorize the formula.

Think back to when we learned about slope. In order to calculate the slope of a line, we must know the *rise* and *run* associated with moving from one point to another. If the points in question are labeled generically, (x_1, y_1) and (x_2, y_2) then the rise is $y_2 - y_1$, and the run is $x_2 - x_1$.

Figure 5 illustrates how this works. What if we want to find the *distance* between (x_1, y_1) and (x_2, y_2)? As Figure 5 illustrates, the run and the rise used to calculate the slope form the legs of a right triangle. If we can find the length of the hypotenuse, we'll know the distance between (x_1, y_1) and (x_2, y_2). Let's see what the Pythagorean Theorem looks like in this case.

$$A^2 + B^2 = C^2 \qquad \text{From Figure 5, replace } A \text{ with Run and } B \text{ with Rise.}$$

$$(\text{Run})^2 + (\text{Rise})^2 = C^2 \qquad \text{Now solve for } C.$$

$$\sqrt{(\text{Run})^2 + (\text{Rise})^2} = C$$

We see that the distance between two points on a coordinate plane is found with this calculation: $\sqrt{(\text{Run})^2 + (\text{Rise})^2}$.

When we use the coordinates from our generic points, (x_1, y_1) and (x_2, y_2), we have the following distance formula:

$$D = \sqrt{(x_2 - x_1)^2 + (y_2 - y_1)^2}$$

Using the distance formula to calculate distance on a coordinate plane is faster than plotting points, drawing a right triangle, and using the Pythagorean Theorem.

Let's see how this works with a couple of problems.

Example 11

Find the distance between the following pairs of points.

1. (5, 11) and (14, 23).   2. (−872, 361) and (−882, 381)

Solutions

1. (5, 11) and (14, 23)

$$D = \sqrt{(x_2 - x_1)^2 + (y_2 - y_1)^2}$$ Begin with the distance formula.

$$D = \sqrt{(14 - 5)^2 + (23 - 11)^2}$$ Plug in the coordinates from the given points.

$$D = \sqrt{9^2 + 12^2}$$ Simplify the results.

$$D = \sqrt{81 + 144}$$

$$D = \sqrt{225}$$

$$D = 15$$

Typically, there aren't any units like inches or centimeters associated with a coordinate plane, so we can simply state that the distance is 15. It's also okay to say that the distance is 15 units.

2. (−872, 361) and (−882, 381)

This problem is a good illustration of why it's a good idea to memorize the distance formula. Plotting these points on a coordinate plane certainly wouldn't be very practical.

$$D = \sqrt{(x_2 - x_1)^2 + (y_2 - y_1)^2}$$

$$D = \sqrt{(-882 - (-872))^2 + (381 - 361)^2}$$

$$D = \sqrt{(-10)^2 + 20^2}$$

$$D = \sqrt{100 + 400}$$

$$D = \sqrt{500}$$

$$D = 10\sqrt{5}$$

The distance between these two points is $10\sqrt{5}$.

Practice C — Answers

11. 8 cm   12. $6\sqrt{13}$ yds   13. About 13.67 ft

Practice D

It's your turn to use the distance formula. Make sure you give exact answers, and simplify your answers as much as possible. When you're done, turn the page to check your solutions.

14. Find the distance between $(-12, 91)$ and $(-15, 87)$.

15. Find the distance between $(901, 582)$ and $(905, 598)$

Exercises 6.3

For the following problems, use square roots to help you solve each quadratic equation.

1. $x^2 = 36$
2. $x^2 = 49$
3. $a^2 = 9$
4. $a^2 = 4$
5. $b^2 = 1$
6. $a^2 = 1$
7. $x^2 = -25$
8. $x^2 = -81$
9. $a^2 = 5$
10. $a^2 = 10$
11. $b^2 = 12$
12. $b^2 = 6$

13. $y^2 = 3$
14. $y^2 = 7$
15. $a^2 - 8 = 0$
16. $a^2 - 3 = 0$
17. $a^2 - 5 = 0$
18. $y^2 + 1 = 0$
19. $x^2 + 10 = 0$
20. $x^2 - 11 = 0$
21. $3x^2 - 27 = 0$
22. $5b^2 - 5 = 0$
23. $2x^2 = 50$
24. $4a^2 = 40$

25. $2x^2 = 24$
26. $(x - 1)^2 = 4$
27. $(x - 2)^2 = 9$
28. $(x - 3)^2 = 25$
29. $(a - 5)^2 = 36$
30. $(a + 3)^2 = 49$
31. $(a + 9)^2 = 1$
32. $(a - 6)^2 = 3$
33. $(x + 4)^2 = 5$
34. $(b + 6)^2 = 7$

For the following problems, solve the quadratic equation. Then use your calculator to find decimal approximations. Round each result to the nearest hundredth.

35. $8a^2 - 168 = 0$
36. $6m^2 - 5 = 0$
37. $0.03y^2 = 1.6$
38. $0.048x^2 = 2.01$
39. $1.001x^2 - 0.999 = 0$

For the following problems, find the missing side length. Give exact answers, not decimal approximations.

40.

18 mi

x

24 mi

41.

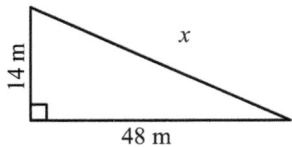

14 m

x

48 m

42.

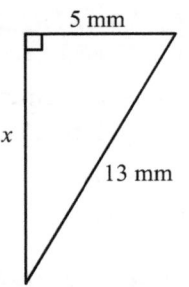

43.

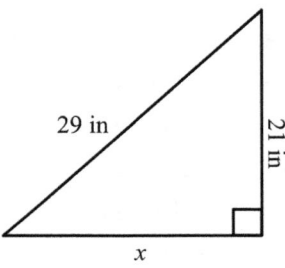

44.

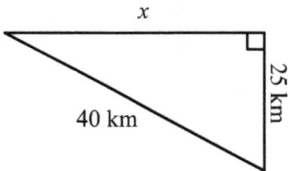

45.

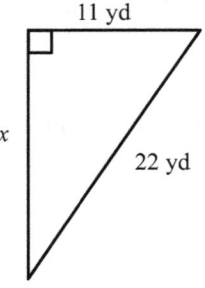

46.

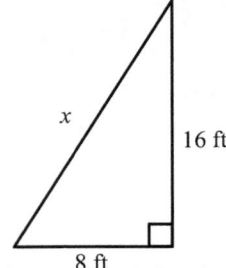

47.

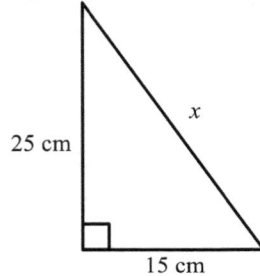

48.

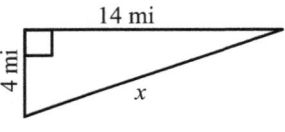

49.

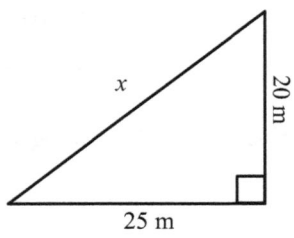

Practice D — Answers

14. 5

15. $4\sqrt{17}$

For the following problems, use the Pythagorean Theorem to help you set up and solve an equation that will allow you to answer the question. Round answers off to the nearest hundredth.

50. Krystal has built a small ramp that she likes to roll marbles down. The ramp is in the shape of a right triangle, with one leg being flat against the floor. The top of the ramp is 6 inches off the floor and the side touching the floor is 14 inches. If a marble is released from the very top of the ramp, how far will it travel before it reaches the floor?

51. Lester needs to roll a heavy rock up into the back of his pickup truck. Unfortunately, there is a ditch between the back of his truck and the heavy rock. He decides to use a sturdy board as a ramp. If the rock is sitting at the far edge of the ditch, which is 4 feet across, and if the back of the truck is 2.5 feet off the ground, then how long does the board need to be?

52. On a camping trip, Timothy and Mateo decide to live dangerously and play catch with a baseball by throwing it back and forth across a small pond. From where Timothy is standing, Mateo walks 73 feet due North and then 59 feet due West to take his place across the lake. How far must Timothy throw the ball in order for it to get to Mateo?

53. Wilhelmina is 5.5 feet tall. She decides to measure her shadow on the ground in a rather unconventional way: She has two friends stretch out a string from the top of her head to the tip of her shadow. She then uses this information to determine how long the shadow is. If the length of string joining the top of Wilhelmina's head to the tip of her shadow is 6.7 feet, then how long is her shadow?

54. When a 135-inch ladder is leaned against a building, the top of the ladder is 131 inches off the ground. How far away from the base of the wall is the bottom of the ladder?

55. When a 160-inch ladder is leaned against a building, the bottom of the ladder is 33 inches away from the base of the wall. How far off the ground is the top of the ladder?

For the following problems, find the distance between each pair of points. Give exact answers, not decimal approximations.

56. $(1, 13)$ and $(35, 17)$

57. $(71, 28)$ and $(80, 22)$

58. $(298, 523)$ and $(300, 511)$

59. $(107, -49)$ and $(115, -69)$

60. $(-139, 1)$ and $(-109, -5)$

61. $(-381, -71)$ and $(-390, -59)$

62. $(37, -1)$ and $(25, 11)$

63. $(807, -325)$ and $(828, -253)$

64. $(18, 90)$ and $(-2, 30)$

65. $(91, -13)$ and $(55, 3)$

6.4 Solving Quadratic Equations by Using the Quadratic Formula

Overview

We've used factoring to solve quadratic equations, but some equations contain expressions that are not factorable. We've also used square roots to solve quadratic equations, but this method only works if there is no first-degree term in the equation, or if the variable appears inside of a squared parentheses and nowhere else.

Here's a problem:

> Rodney throws a baseball with an initial upward velocity of 40 feet per second. If the ball is 6 feet above the ground when Rodney releases it, how long does it take until the ball hits the ground?

In order to solve this problem, we need to know a mathematical formula that can model the height of a thrown object. However, the formula by itself is not enough. The equation that the formula gives us needs to be solved, and the methods we've covered so far don't work in this situation. So we need to develop another method for solving quadratic equations. Once we've done that, we'll come back to this problem and solve it.

When you are finished working through this section, you will be able to:

♦ Understand the standard form of a quadratic equation

♦ Use the quadratic formula to solve quadratic equations

♦ Use the quadratic formula to solve application problems algebraically

A. The Standard Form of a Quadratic Equation

We've seen that a quadratic equation can be written in the form $ax^2 + bx + c = 0$, where $a \neq 0$. We also know that this is called the *standard form* of a quadratic equation. When a quadratic equation is written in standard form:

> a is the coefficient of the *quadratic* (second-degree) term
>
> b is the coefficient of the *linear* (first-degree) term
>
> c is the *constant* term

In the following equations, we will identify the values of a, b, and c.

Example 1

Identify the values of a, b, and c in the following equations.

1. $3x^2 + 5x + 2 = 0$ 3. $2y^2 + 3 = 0$ 5. $z^2 = z + 8$
2. $12x^2 - 2x - 1 = 0$ 4. $-8y^2 + 11y = 0$

Solutions

1. $3x^2 + 5x + 2 = 0$
In this equation, $a = 3$, $b = 5$, and $c = 2$.

2. $12x^2 - 2x - 1 = 0$
Remember, if a term has a minus sign in front of it then the term has a negative coefficient. In this equation, $a = 12$, $b = -2$, and $c = -1$.

3. $2y^2 + 3 = 0$
We need to be careful with this equation. There's no linear term. However, we can write the equation as $2y^2 + 0y + 3 = 0$. With this in mind, we see that $a = 2$, $b = 0$, and $c = 3$.

4. $-8y^2 + 11y = 0$
This is similar to the equation in part 3, but in this case, there's no constant term. We can write this equation as $-8y^2 + 11y + 0 = 0$, which gives us $a = -8$, $b = 11$, and $c = 0$.

5. $z^2 = z + 8$
This equation is not written in standard form because neither side of the equation is 0. In order to write this equation in standard form, we can take one of two approaches:

> *Option 1:* We can subtract z from both sides, and subtract 8 from both sides. This gives us $z^2 - z - 8 = 0$. When the equation is written this way, $a = 1$, $b = -1$, and $c = -8$.

> *Option 2:* We can subtract z^2 from both sides. This gives us $0 = -z^2 + z + 8$, which means the same thing as $-z^2 + z + 8 = 0$. In this case, $a = -1$, $b = 1$, and $c = 8$.

If neither side of the original equation is 0, then it's preferable to end up with a positive value for a, as in the first case: $z^2 - z - 8 = 0$.

Practice A

Determine the values of a, b, and c in the following quadratic equations. If an equation doesn't start out with 0 on one side, begin by rewriting it in standard form so that the first term is positive. When you are done, turn the page and check your solutions.

1. $4x^2 - 3x + 5 = 0$ 4. $x^2 - 2x = 0$ 7. $5x - 3 = -3x^2$
2. $x^2 - 5x - 1 = 0$ 5. $y^2 = 5y - 6$ 8. $2x - 11 - 3x^2 = 0$
3. $z^2 - 4 = 0$ 6. $2x^2 - 4x = -1$ 9. $y^2 = 0$

B. Using the Quadratic Formula to Solve Quadratic Equations

As we have previously noted, some quadratic equations cannot be solved by factoring or by using square roots. Now it's time to learn a formula that can be used to solve any quadratic equation, provided the equation is written in standard form.

The Quadratic Formula

If $ax^2 + bx + c = 0$, then $x = \dfrac{-b \pm \sqrt{b^2 - 4ac}}{2a}$.

Because it's a fairly complicated formula, we generally check to see if factoring or square roots can be used to solve a quadratic equation more quickly. If these methods won't work, however, then the quadratic formula is our next best option.

Example 2

Solve: $3x^2 + 5x + 2 = 0$

Solution
We will use the quadratic formula to solve this equation. The equation is in standard form, so we can see that $a = 3$, $b = 5$ and $c = 2$.

$x = \dfrac{-b \pm \sqrt{b^2 - 4ac}}{2a}$	Begin with the quadratic formula. The first step is to replace a with 3, b with 5 and c with 2.
$x = \dfrac{-5 \pm \sqrt{(5)^2 - 4(3)(2)}}{2(3)}$	Now that we have made the proper substitutions, it's time to simplify our solutions.
$x = \dfrac{-5 \pm \sqrt{25 - 24}}{6}$	Simplify the radicand: $5^2 = 25$ and $4(3)(2) = 24$ Then simplify the denominator: $2(3) = 6$
$x = \dfrac{-5 \pm \sqrt{1}}{6}$	Finish simplifying the radicand: $25 - 24 = 1$
$x = \dfrac{-5 \pm 1}{6}$	Simplify the square root: $\sqrt{1} = 1$. The radical is gone, so we can combine this result with -5. Because of the "$\pm$" symbol, we will need to use both addition and subtraction when doing so.
$x = \dfrac{-5 + 1}{6}$ and $\dfrac{-5 - 1}{6}$	Write two fractions — one for each solution.
$x = \dfrac{-4}{6}, \dfrac{-6}{6}$	
$x = -\dfrac{2}{3}, -1$	Simplify each fraction.

In the previous example, the fact that the solutions are rational numbers is an indication that the equation actually could have been solved by factoring. However, if a second-degree polynomial has a leading coefficient that is not 1 and if there isn't a GCF other than 1, then the factoring process is somewhat complicated. We don't cover this type of factoring in this book.

Now let's look at an equation that has more complicated solutions.

Example 3

Solve: $12x^2 - 2x - 1 = 0$

Solution

First we identify a, b, and c: $a = 12$, $b = -2$, and $c = -1$. Then we use these values to help us solve the equation.

$$x = \frac{-b \pm \sqrt{b^2 - 4ac}}{2a}$$

Begin with the quadratic formula and replace a, b, and c with the appropriate numbers. It's important to use parentheses when replacing variables with negative numbers.

$$x = \frac{-(-2) \pm \sqrt{(-2)^2 - 4(12)(-1)}}{2(12)}$$

Simplify the double negative in the numerator, the exponential expression, multiplications within the radical, and the expression in the denominator.

$$x = \frac{2 \pm \sqrt{4 + 48}}{24}$$

Simplify the expression within the radical.

$$x = \frac{2 \pm \sqrt{52}}{24}$$

Simplify the square root.

$$x = \frac{2 \pm \sqrt{4 \cdot 13}}{24}$$

$$x = \frac{2 \pm 2\sqrt{13}}{24}$$

Because the square root is irrational, the simplified square root still contains a radical sign. This means that we can't combine it with 2 by addition and subtraction.

$$x = \frac{2(1 \pm \sqrt{13})}{24}$$

We can, however, factor out a GCF of 2 in the numerator. This allows us to reduce the fraction.

$$x = \frac{1 \pm \sqrt{13}}{12}$$

This notation is an acceptable way to write our solutions.

$$x = \frac{1 + \sqrt{13}}{12}, \frac{1 - \sqrt{13}}{12}$$

If it is preferable to see both solutions in their entirety, we can write them out like this.

Practice A — Answers

1. $a = 4$, $b = -3$, and $c = 5$
2. $a = 1$, $b = -5$, and $c = -1$
3. $a = 1$, $b = 0$, and $c = -4$
4. $a = 1$, $b = -2$, and $c = 0$
5. $a = 1$, $b = -5$, and $c = 6$
6. $a = 2$, $b = -4$, and $c = 1$
7. $a = 3$, $b = 5$, and $c = -3$
8. $a = -3$, $b = 2$, and $c = -11$
9. $a = 1$, $b = 0$, and $c = 0$

In the previous example, if we were asked for decimal approximations, we could find them using a calculator: $\frac{1 + \sqrt{13}}{12} \approx 0.38$ and $\frac{1 - \sqrt{13}}{12} \approx -0.22$. In application problems, we often use decimal approximations to give answers.

Example 4

Use the quadratic formula to solve the following equations.

1. $2y^2 + 3 = 0$ 3. $(3x + 1)(x - 4) = x^2 + x - 2$

2. $-8x^2 + 11x = 0$

Solutions

1. $2y^2 + 3 = 0$

Because there is no linear term, we could use square roots to attempt to solve this equation. However, we'll look at how the process works with the quadratic formula. First, we identify a, b, and c: $a = 2$, $b = 0$ and $c = 3$. Next, we use the quadratic formula to solve the equation.

$x = \dfrac{-b \pm \sqrt{b^2 - 4ac}}{2a}$ Start with the quadratic formula and make the substitutions.

$x = \dfrac{-0 \pm \sqrt{0^2 - 4(2)(3)}}{2(2)}$ Simplify the resulting expressions.

$x = \dfrac{0 \pm \sqrt{-24}}{4}$

We're done! This equation has no real number solution because of the negative radicand.

2. $-8x^2 + 11x = 0$

Since there's no constant term on the left side of the equation, we could use factoring to solve. Instead, let's take a look at how it works with the quadratic formula. We begin by identifying a, b, and c: $a = -8$, $b = 11$ and $c = 0$. Then we solve the equation.

$x = \dfrac{-b \pm \sqrt{b^2 - 4ac}}{2a}$ Start with the quadratic formula and make the substitutions.

$x = \dfrac{-11 \pm \sqrt{11^2 - 4(-8)(0)}}{2(-8)}$ Simplify the resulting expressions.

$x = \dfrac{-11 \pm \sqrt{121 - 0}}{-16}$

$x = \dfrac{-11 \pm \sqrt{121}}{-16}$

$x = \dfrac{-11 \pm 11}{-16}$ There's no radical sign left, so write two separate fractions.

$x = \dfrac{-11 + 11}{-16}, \dfrac{-11 - 11}{-16}$ Simplify the solutions.

$x = 0, \dfrac{11}{8}$

3. $(3x + 1)(x - 4) = x^2 + x - 2$

For this equation, we start by writing the equation in standard form. This takes some work.

$$(3x + 1)(x - 4) = \quad x^2 + x - 2 \quad \text{The left side of the equation must be written without}$$
$$3x^2 - 12x + x - 4 = \quad x^2 + x - 2 \quad \text{parentheses. Use the FOIL method to multiply the binomials.}$$
$$3x^2 - 11x - 4 = \quad x^2 + x - 2 \quad \text{To get a positive quadratic term, we subtract } x^2 \text{ from both}$$
$$\underline{\quad - x^2 - x + 2 \quad \quad - x^2 - x + 2} \quad \text{sides. We also subtract } x \text{ and add 2.}$$
$$2x^2 - 12x - 2 = \quad \quad \quad 0$$

$$\frac{2x^2 - 12x - 2}{2} = \frac{0}{2} \quad \text{Every term is even, so divide both sides by 2 to simplify.}$$
$$x^2 - 6x - 1 = 0$$

Now we can identify a, b, and c and proceed to solve the equation. Because you know how this works now, we will just show the mathematical steps involved in solving.

$$x = \frac{-b \pm \sqrt{b^2 - 4ac}}{2a}$$

$$x = \frac{-(-6) \pm \sqrt{(-6)^2 - 4(1)(-1)}}{2(1)}$$

$$x = \frac{6 \pm \sqrt{36 + 4}}{2}$$

$$x = \frac{6 \pm \sqrt{40}}{2}$$

$$x = \frac{6 \pm \sqrt{4 \cdot 10}}{2}$$

$$x = \frac{6 \pm 2\sqrt{10}}{2}$$

$$x = \frac{2(3 \pm \sqrt{10})}{2}$$

$$x = 3 \pm \sqrt{10}$$

Practice B

Now it's your turn to solve quadratic equations using the quadratic formula. Take your time. When you're done, turn the page and check your solutions.

10. $2x^2 + 3x - 7 = 0$

11. $5a^2 - 2a - 1 = 0$

12. $6y^2 + 5y = -3$

13. $3m^2 = 2m$

C. Application Problems

Now that you have had some practice using the quadratic formula, it's time to look at some application problems. We'll use the same five-step approach we've used before.

Example 5

The length of a rectangle is 4 cm more than the width. The area of the rectangle is 37 cm². What are the dimensions of the rectangle? Round values off to the nearest thousandth.

Solution

Step 1: In this problem, the length of the rectangle is defined based on the width. We can let w represent the width, and then $w + 4$ will represent the length.

Step 2: We know that for a rectangle, Area = (Length)(Width). We also know the area of this rectangle is 37 cm². Using this information and the variable expressions from Step 1, we have our equation: $37 = (w + 4)(w)$

Step 3: Let's solve the equation.

$37 = (w + 4)(w)$	Use the distributive property to write the expression on the right side without parentheses.
$37 = w^2 + 4w$	The quadratic term is positive, so leave it on the right side.
$\underline{-37 = \quad\quad -37}$	Subtract 37 from both sides.
$0 = w^2 + 4w - 37$	Because the quadratic expression is not factorable, use the quadratic formula to solve.

$$w = \frac{-b \pm \sqrt{b^2 - 4ac}}{2a}$$

$$= \frac{-4 \pm \sqrt{(4)^2 - 4(1)(-37)}}{2(1)}$$

$$= \frac{-4 \pm \sqrt{164}}{2}$$

$$= \frac{-4 \pm 2\sqrt{41}}{2}$$

$$= \frac{-4}{2} \pm \frac{2\sqrt{41}}{2}$$

$$= -2 \pm \sqrt{41} \qquad \text{The instructions indicate that a rounded-off decimal is expected. Using a calculator, find an approximation for } \sqrt{41}.$$

$$\approx -2 \pm 6.403$$

A distance cannot be negative, so we only keep the positive solution: $w \approx -2 + 6.403 = 4.403$. Because the length is given by $w + 4$, we find that as well: $4.403 + 4 = 8.403$.

Step 4: Check the solution:

$Area$ = (Length)(Width)

37 = (8.403)(4.403)

37 = 36.998409   Because we're using solutions that are decimal approximations, the numbers don't match exactly. However, they are very close, so we can be confident in our solution.

Step 5: The rectangle is about 8.403 cm long and 4.403 cm wide.

Quadratic equations are useful in many scientific applications. In physics, for example, it's often necessary to model the height of an object that is thrown or launched into the air. To do this, we can use the following information:

The height, h, in feet of an object that is thrown or launched is given by the formula $h = -16\,t^2 + v_0\,t + h_0$

In this formula, t represents the time in seconds since the object is launched, v_0 represents the initial upward velocity in feet per second, and h_0 represents the initial height of the object in feet.

Now it's time to tackle the problem from the beginning of this section.

Example 6

Rodney throws a baseball with an initial upward velocity of 40 feet per second. If the ball is 6 feet above the ground when Rodney releases it, how long does it take until the ball hits the ground? Round your answer off to the nearest thousandth.

Solution

Step 1: We are figuring out the time until the ball hits the ground. Let t represent time in seconds.

Step 2: We know the formula for the height of an object that is thrown or launched:

$h = -16\,t^2 + v_0\,t + h_0$

When the ball hits the ground, the height is 0. We have the initial upward velocity of 40 ft/sec and the initial height of 6 feet. By substituting these values into our equation, we have:

$0 = -16\,t^2 + 40t + 6$

Practice B — Answers

10. $x = \dfrac{-3 \pm \sqrt{65}}{4}$

11. $a = \dfrac{1 \pm \sqrt{6}}{5}$

12. no real number solutions

13. $m = 0, \dfrac{2}{3}$

Step 3: This equation has 0 on one side, so we start with the quadratic formula and go from there.

$$t = \frac{-b \pm \sqrt{b^2 - 4ac}}{2a}$$

$$= \frac{-40 \pm \sqrt{(40)^2 - 4(-16)(6)}}{2(-16)}$$

$$= \frac{-40 \pm \sqrt{1984}}{-32} \qquad \text{Using a calculator, find an approximation for } \sqrt{1984}.$$

$$\approx \frac{-40 \pm 44.5421149}{-32}$$

Now use a calculator to find the solutions:

$$\frac{-40 + 44.5421149}{-32} \approx -0.142 \text{ and } \frac{-40 - 44.5421149}{-32} \approx 2.642$$

A negative answer doesn't make sense in this case, so we only report the positive solution:

$$t = 2.642$$

Step 4: Check the solution.

$$0 = -16(2.642)^2 + 40(2.642) + 6$$

$$0 = -111.682624 + 105.68 + 6$$

$$0 = -0.002642 \qquad \text{These numbers are slightly off because we're using a decimal approximation for our solution, but they're very close. We can be confident in our solution.}$$

Step 5: The baseball is in the air for about 2.642 seconds.

Practice C

Now, you get to tackle a couple of application problems. Use the quadratic formula to help you solve each problem. Round your answers off to the nearest thousandth. When you are done check your solutions.

14. The length of a rectangle is 3 feet less than four times its width. The area of the rectangle is 100 square feet. Find the dimensions of the rectangle.

15. A football is punted into the air with an initial velocity of 75 feet per second. If the ball is 4 feet off the ground when it is kicked, how much time passes before the football hits the ground? (Hint: use the formula $h = -16t^2 + v_0 t + h_0$)

Exercises 6.4

For the following problems, solve the equations using the quadratic formula.

1. $x^2 - 2x - 3 = 0$

2. $x^2 + 5x + 6 = 0$

3. $y^2 - 5y + 4 = 0$

4. $a^2 + 4a - 21 = 0$

5. $a^2 + 12a + 20 = 0$

6. $b^2 - 4b + 4 = 0$

7. $b^2 + 4b + 4 = 0$

8. $x^2 + 10x + 25 = 0$

9. $2x^2 - 5x - 3 = 0$

10. $6y^2 + y - 2 = 0$

11. $4x^2 - 2x - 1 = 0$

12. $3y^2 + 2y - 1 = 0$

13. $5a^2 - 2a - 3 = 0$

14. $x^2 - 3x + 1 = 0$

15. $x^2 - 5x - 4 = 0$

16. $(x + 2)(x - 1) = 1$

17. $(a + 4)(a - 5) = 2$

18. $(x - 3)(x + 3) = 7$

19. $(b - 4)(b + 4) = 9$

20. $x^2 + 8x = 2$

21. $y^2 = -5y + 4$

22. $x^2 = -3x + 7$

23. $x^2 = -2x - 1$

24. $x^2 + x + 1 = 0$

25. $a^2 + 3a - 4 = 0$

26. $y^2 + y = -4$

27. $b^2 + 3b = -2$

28. $x^2 + 6x + 8 = -x - 2$

29. $x^2 + 4x = 2x - 5$

30. $6b^2 + 5b - 4 = b^2 + b + 1$

31. $4a^2 + 7a - 2 = -2a + a$

32. $(2x + 5)(x - 4) = x^2 - x + 2$

33. $(x - 4)^2 = 3$

34. $(x + 2)^2 = 4$

35. $(b - 6)^2 = 8$

36. $(3 - x)^2 = 6$

37. $3(x^2 + 1) = 2(x + 7)$

38. $2(y^2 - 3) = -3(y - 1)$

39. $-4(a^2 + 2) + 3 = 5$

40. $-(x^2 + 3x - 1) = 2$

For the following exercises, use the quadratic formula to solve each problem. Round answers off to the nearest thousandth. For problems involving launched or thrown objects, use the formula $h = -16t^2 + v_0 t + h_0$.

41. The length of a rectangle is 7 meters more than the width. If the area of the rectangle is 84 square meters, what are the dimensions of the rectangle?

42. The length of a rectangle is 3 miles less than twice the width. If the area of the rectangle is 62 square miles, what are the dimensions of the rectangle?

43. The length of a rectangle is 12 millimeters more than thrice the width. If the area of the rectangle is 314 square millimeters, what are the dimensions of the rectangle?

44. Kendrick launches a model rocket into the air at an initial velocity of 180 feet per second. If the rocket is 2 feet off the ground when it is launched, how much time passes before the rocket hits the ground?

45. Rosie tees off from a tee box 10.75 feet above the green. She hits the ball with an initial upward velocity of 84 feet per second, and the ball lands on the green. How much time passes before the ball hits the ground? Hint: let 10.75 represent the initial height of the golf ball, so that 0 represents the height of the green.

46. Dennis shoots a rock from his slingshot with an initial upward velocity of 60 feet per second. If the rock is 4.5 feet above the ground when it leaves the slingshot, how much time passes before it hits the ground?

Practice C — Answers

14. The rectangle has a length of about 18.556 feet and a width of about 5.389 feet.

15. The football hits the ground after about 4.740 seconds.

6.5 Graphing Quadratic Equations

Overview

At this point, you should feel comfortable using factoring, square roots, or the quadratic formula to help you solve quadratic equations. The last method for solving that we'll cover involves graphing. Before we get there, however, we need to learn how to graph quadratic equations.

In Chapter 2, we graphed linear equations in two variables. As the word *linear* suggests, the solutions of a linear equation make a straight line when graphed. In this section, we find and graph solutions of quadratic equations. These solution points will *not* make a straight line.

Here's a problem like that:

> Graph: $y = x^2 - 6x + 4$

After we cover the basics of graphing quadratic equations, we'll come back to this problem. By working through this section, you will be able to:

- ◆ Locate the vertex of a parabola given a graph, or given an equation
- ◆ Graph a quadratic equation in two variables by hand
- ◆ Graph a quadratic equation in two variables with a calculator

A. Locating the Vertex of a Parabola

You will recall that a polynomial in one variable is *quadratic* if the highest power of the variable is 2. In this book, we won't examine equations containing two variables in which both variables are raised to the power of 2, so we'll define a quadratic equation in two variables as follows:

> A **quadratic equation in two variables** is an equation that can be written in the form $y = ax^2 + bx + c$, where the coefficients a, b and c are all real numbers and $a \neq 0$
>
> .

Informally, we can also define a quadratic equation in two variables as an equation that can be written as: $y = \{$quadratic polynomial containing $x\}$.

Whenever a quadratic equation in two variables is graphed on a coordinate plane, we see a nonlinear shape called a **parabola**, as in Figure 1

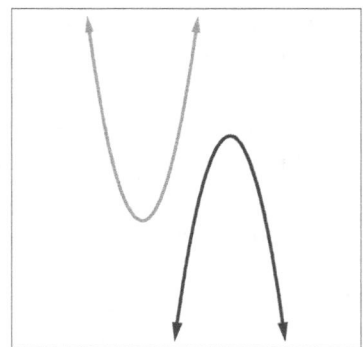

Figure 1.

If a parabola opens upward, then the lowest point is called the vertex, and there is no highest point. If a parabola opens downward, the highest point is called the vertex, and there is no lowest point. See Figure 2.

When a parabola is graphed on a coordinate plane, we can name the vertex as an ordered pair.

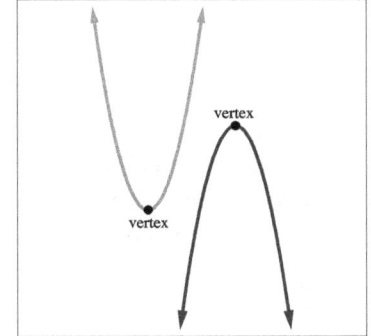

Figure 2.

Example 1

In Figure 3, the graph for the equation $y = x^2 - 4x - 1$ is given. Find the vertex.

Solution

We can identify several points that lie on this parabola. However, because this parabola opens upward, the only point we're interested in is the *lowest* point. The vertex is $(2, -5)$.

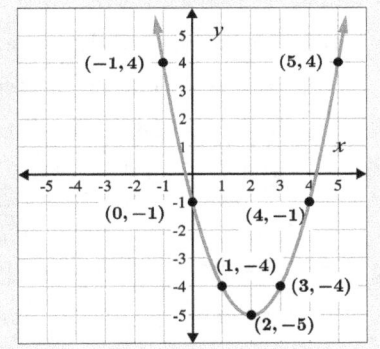

Figure 3.

We know that the standard form of a quadratic polynomial containing the variable x is $ax^2 + bx + c$. Similarly, the **standard form** of a quadratic equation in two variables is $y = ax^2 + bx + c$.

When we're given an equation in this form, the following formula is useful in locating the vertex algebraically:

If $y = ax^2 + bx + c$, then the x-coordinate of the vertex is $\frac{-b}{2a}$.

We can remember this formula by thinking about what the quadratic formula would look like if the numerator did not contain a "±" symbol or square root expression.

Once we know the x-coordinate of the vertex, we can plug it into the equation and find the corresponding y-coordinate. The following example shows how this is done.

Example 2

Find the vertex of the graph whose equation is: $y = x^2 - 8x + 17$.

Solution

We haven't talked about graphing quadratic equations yet, but that's okay. We have a formula that will give us the x-coordinate of the vertex.

Looking at the equation $y = x^2 - 8x + 17$, we can see that $a = 1$, $b = -8$ and $c = 17$.

$$\frac{-b}{2a} = \frac{-(-8)}{2(1)}$$ In the formula for the x-coordinate of the vertex, replace a with 1 and b with -8.

$$= \frac{8}{2}$$ Simplify the fraction.

$$= 4$$

The x-coordinate of the vertex is 4. Now we can find the y-coordinate.

$$y = x^2 - 8x + 17$$ Replace x with 4.

$$= (4)^2 - 8(4) + 17$$ Now simplify the result.

$$= 16 - 32 + 17$$

$$= 1$$

Now that we know both coordinates of the vertex, we can write the answer. The vertex is $(4, 1)$.

Practice A

Now you get a chance to locate some vertices, which is the plural word for "vertex." For each problem, locate the vertex of the parabola whose information is given. When you are done, turn the page and check your solutions.

1. Graph:

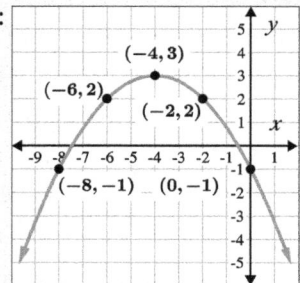

2. Equation: $y = x^2 + 6x + 2$

3. Equation: $y = -2x^2 + 16x - 20$

B. Graphing by Hand: Equation → Table → Graph

In Chapter 2, we graphed a lot of linear equations in two variables. Before we learned graphing short-cuts, we used this four-step process to construct a graph for a linear equation that was solved for y:

1. Pick three numbers to use as x-coordinates.

2. Plug each of these numbers into the equation and find the corresponding y-coordinates.

3. Plot the three points.

4. If the three points form a straight line, draw the graph. If the three points don't form a straight line, go back and check the work in the previous steps.

Now our graphs won't be linear. They'll be parabolas. We need to locate at least *five* points to draw a parabola. However, not just any five points will do. We need to locate the vertex and then find at least two points on either side of the vertex.

With this in mind, we modify Step 1:

Pick *five* numbers to use as x-coordinates. Find the x-coordinate of the vertex, and then pick two numbers that are less than this value and two numbers greater than this value.

We'll also modify Step 4 slightly to make sure the points form a parabola instead of a straight line. Let's work through a couple of examples to see how this works. We'll start with the problem from the beginning of this section.

Example 3

Graph: $y = x^2 - 6x + 4$

Solution

Step 1: We begin by locating the x-coordinate of vertex.

$$\frac{-b}{2a} = \frac{-(-6)}{2(1)} \qquad \text{Replace } a \text{ with } 1 \text{ and } b \text{ with } -6.$$

$$= \frac{6}{2} \qquad \text{Simplify.}$$

$$= 3$$

Now we know that the x-coordinate of the vertex is 3. We also need to pick two numbers that are less than 3 and two numbers that are greater than 3. We will pick 1 and 2, as well as 4 and 5.

Step 2: Using the five numbers from Step 1 as x-coordinates, we use the equation $y = x^2 - 6x + 4$ to help us calculate the corresponding y-coordinates in the table below.

x	*Equation with x-coordinate Plugged In*	y	*(x, y)*
1	$y = (1)^2 - 6(1) + 4$	-1	$(1, -1)$
2	$y = (2)^2 - 6(2) + 4$	-4	$(2, -4)$
3	$y = (3)^2 - 6(3) + 4$	-5	$(3, -5)$
4	$y = (4)^2 - 6(4) + 4$	-4	$(4, -4)$
5	$y = (5)^2 - 6(5) + 4$	-1	$(5, -1)$

Step 3: In Figure 4, we plot the five points on a coordinate plane.

Step 4: Looking at the points we've plotted, we verify that they form a parabola. The last thing to do is to draw the graph in Figure 5.

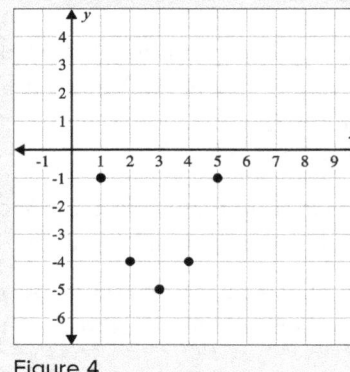

Figure 4.

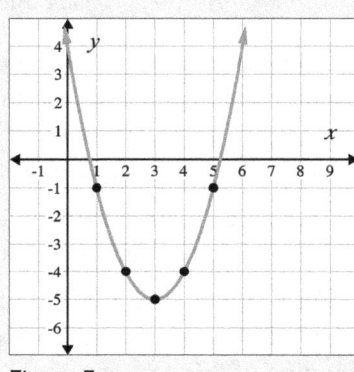

Figure 5.

Example 4

Graph: $y = 2x^2 - 6$

Solution

Step 1: With this equation, it's tempting to think that $a = 2$ and $b = -6$, but that's not the case. The polynomial containing x doesn't have a first-degree term, which means that we can only see values for a and c. If we rewrite the equation with a first-degree term containing x, we can more easily see the values of a, b and c: $y = 2x^2 + 0x - 6$. Now we calculate the x-coordinate of the vertex:

$$\frac{-b}{2a} = \frac{-(0)}{2(2)}$$
$$= \frac{0}{4}$$
$$= 0$$

Incidentally, whenever $b = 0$, the x-coordinate of the vertex will also be 0. In addition to using 0 as an x-coordinate, we need to pick two numbers that are less than 0 and two numbers that are greater than 0. We'll pick -2 and -1, as well as 1 and 2.

Step 2: Using the five numbers from Step 1 as x-coordinates, we use the equation $y = 2x^2 - 6$ to help us calculate the corresponding y-coordinates.

x	*Equation with x-coordinate Plugged In*	y	*(x, y)*
–2	$y = 2(-2)^2 - 6$	2	$(-2, 2)$
–1	$y = 2(-1)^2 - 6$	–4	$(-1, -4)$
0	$y = 2(0)^2 - 6$	–6	$(0, -6)$
1	$y = 2(1)^2 - 6$	–4	$(1, -4)$
2	$y = 2(2)^2 - 6$	2	$(2, 2)$

Step 3: In Figure 6, we plot the points on a coordinate plane.

Step 4: We verify that the points form a parabola, and then in Figure 7, we draw the graph.

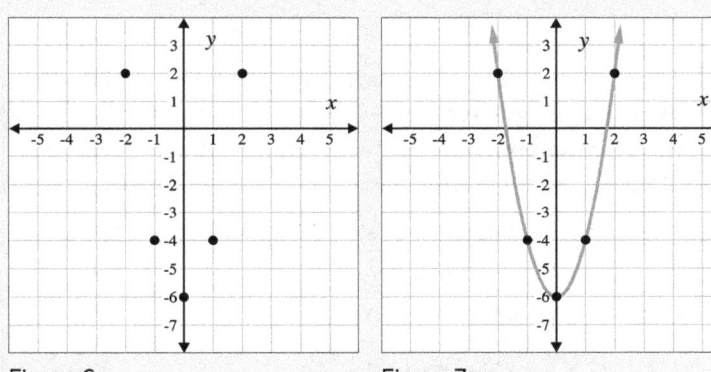

Figure 6.   Figure 7.

Look at the values of a in each of the last two examples. In Example 3, $a = 1$, while in Example 4, $a = 2$. Now look at the shape of the parabolas. The parabola in Example 4 goes up more steeply than the parabola in Example 3. In other words, the parabola in Example 4 is *narrower* than the parabola in Example 3. In general, the farther away the value of a is from zero, the narrower the parabola will be.

Practice A — Answers

1. $(-4, 3)$ 2. $(-3, -7)$ 3. $(4, 12)$

Let's see what happens when a is negative.

Example 5

Graph: $y = -\frac{1}{2}x^2 - x + 2$

Solution

Start by calculating the x-coordinate of the vertex:

$$\frac{-b}{2a} = \frac{-(-1)}{2\left(-\frac{1}{2}\right)}$$

$$= \frac{1}{-1}$$

$$= -1$$

In addition to using -1 as an x-coordinate, we need two numbers that are less than -1 and two numbers that are greater than -1. Notice that the coefficient of the term containing x^2 is a fraction, $-\frac{1}{2}$. Because denominator of this fraction is 2, we should choose even numbers as x-coordinates. By doing so, the y-coordinates will come out as integers. We pick -4 and -2, as well as 0 and 2.

Step 2: Use the equation $y = -\frac{1}{2}x^2 - x + 2$ to help calculate the y-coordinates.

x	*equation with x-coordinate plugged in*	y	(x, y)
-4	$y = -\frac{1}{2}(-4)^2 - (-4) + 2$	-2	$(-4, -2)$
-2	$y = -\frac{1}{2}(-2)^2 - (-2) + 2$	2	$(-2, 2)$
-1	$y = -\frac{1}{2}(-1)^2 - (-1) + 2$	$2\frac{1}{2}$	$\left(-1, 2\frac{1}{2}\right)$
0	$y = -\frac{1}{2}(0)^2 - (0) + 2$	2	$(0, 2)$
2	$y = -\frac{1}{2}(2)^2 - (2) + 2$	-2	$(2, -2)$

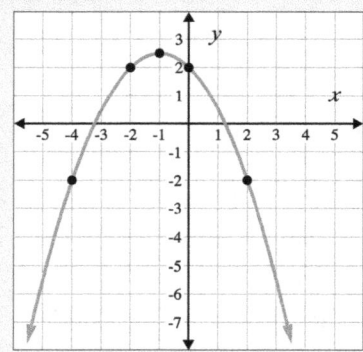

Figure 8.

We can't avoid the fraction coordinate in the vertex, but at least the other points have integer coordinates.

Steps 3 and 4: Plot the points, verify that they form a parabola, and then draw the graph in Figure 8.

In Example 5, the parabola opens downward. In general, any time the value of a is negative, the parabola will open downward.

Practice B

It's your turn to construct a few graphs. Use the process shown in the examples above to graph each equation. When you are done, turn the page and check your solutions.

4.   Graph: $y = x^2 + 8x + 13$ 5.   Graph: $y = -2x^2 + 4x + 1$ 6.   Graph: $y = \frac{1}{3}x^2 - 2$

C. Graphing with a Calculator

We can graph any equation containing the variables x and y on our graphing calculator as long as that equation is solved for y. Now let's graph a couple of parabolas on our graphing calculator.

Example 6

Use a graphing calculator to graph $y = 0.7x^2 - 3.9x + 2$.

Solution

Begin by entering $0.7x^2 - 3.9x + 2$ in the Y_1 equation. On an older calculator, the expression might look like this: $0.7x^2 - 3.9x + 2$.

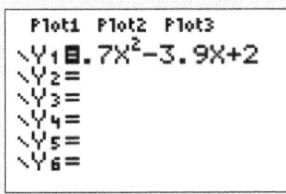

Press the GRAPH button. Your viewing window may need to be adjusted. In this image, the viewing window is $\{-10 < x < 10$ and $-10 < y < 10\}$.

If your graph looks different than the image above, check the settings on your viewing window.

Example 7

Use a graphing calculator to graph $1.9x^2 + 8.4x + y = 0$.

Solution

This equation must be solved for y before we can graph it on a calculator. After we subtract $1.9x^2$ and $8.4x$ from both sides, the equation is in the proper form: $y = -1.9x^2 - 8.4x$.

Begin by entering $-1.9x^2 - 8.4x$ in the Y_1 equation.

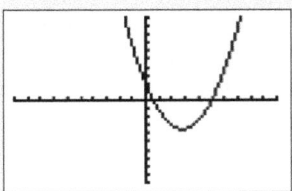

Press the GRAPH button. The viewing window for this image is $\{-10 < x < 10$ and $-10 < y < 10\}$.

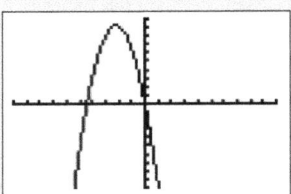

In addition to viewing graphs of equations that are solved for y, we've used the table feature of our graphing calculators to view ordered pairs given by these equations. This feature gives us a convenient way to find or check ordered pairs for equations that we need to graph by hand.

Example 8

Use the table feature of your graphing calculator to verify that the ordered pairs in Example 3 are correct.

Solution

Begin by entering $x^2 - 6x + 4$ in the Y_1 equation.

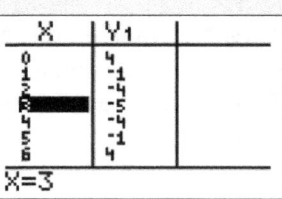

Press the [TABLE] button, which you access with [2nd] + [GRAPH].
Use the up or down arrow keys as necessary to locate the correct ordered pairs.

The table on our calculator shows us the same ordered pairs we found by hand in Example 3.

Practice C

Use your graphing calculator to graph the following equations. Turn the page to check your work.

7. Graph: $y = 2.1\,x^2 + 3.9x + 0.6$

8. Graph: $0.13\,x^2 - 5.2x + y = 0$

Practice B — Answers

4.

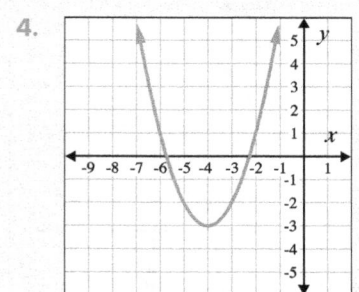

5.

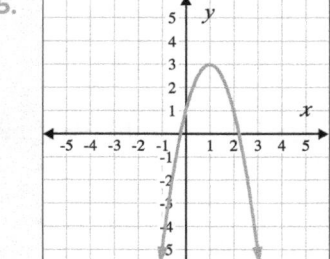

6.

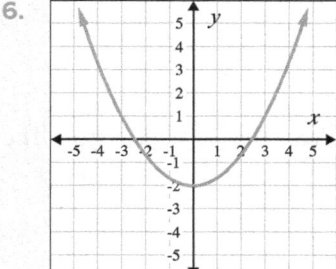

Exercises 6.5

Find the vertex of the following graphs.

1.

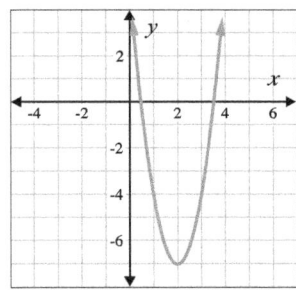

3.

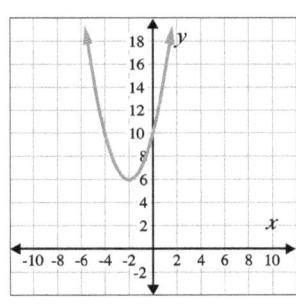

5.

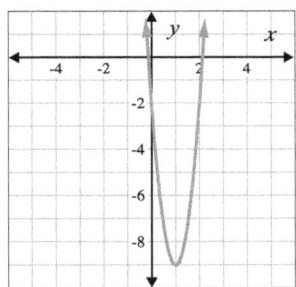

2.

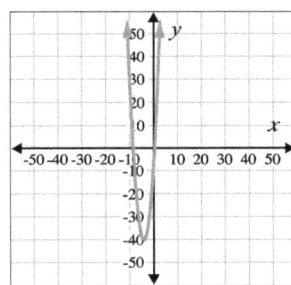

4.

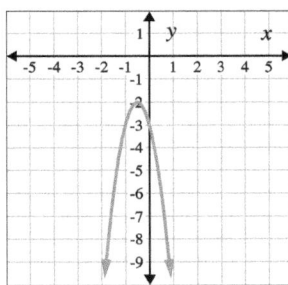

For the following problems, find the vertex of the graph whose equation is given below.

6. $y = 4x^2 - 24x + 26$

7. $y = 2x^2 - 12x + 26$

8. $y = x^2 - 10x + 23$

9. $y = 4x^2 + 8x + 7$

10. $y = 2x^2 + 6x + 3.5$

Use the process shown in Part B of this section to graph the following equations.

11. $y = -x^2 - 1$

12. $y = x^2 + 5$

13. $y = x^2 - 4$

14. $y = x^2 + 3$

15. $y = x^2 + 8x + 10$

16. $y = x^2 - 3x + 4$

17. $y = -x^2 + 4x + 1$

18. $y = -2x^2 - 4x + 3$

19. $y = 10x^2 - 10x + 10$

20. $y = 9x^2$

21. $y = \frac{1}{3}x^2 + \frac{2}{3}x$

22. $y = \frac{1}{2}x^2$

Use your graphing calculator to graph the following equations.

23. Graph $y = -1.72x^2 + 4.5x - 1.8$

24. Graph $y = -4.07x^2 - 2.26x - 0.33$

25. Graph $y = -.08x^2 - 0.31x$

26. Graph $y = 5.08x^2 - 0.2x + 10$

27. Graph $y = 2.66x^2 - 2.4x + 1.3$

Practice C — Answers

7. $\{-10 < x < 10 \text{ and } -10 < y < 10\}$

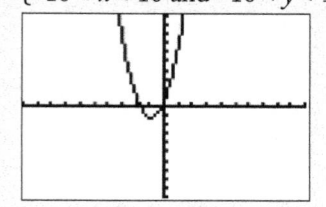

8. $\{-10 < x < 50 \text{ and } -10 < y < 60\}$

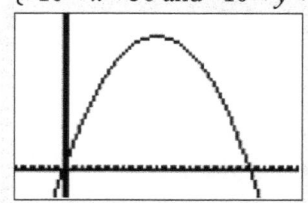

6.6 Solving Quadratic Equations by Graphing

Overview

You now know how to use factoring, square roots, and the quadratic formula to solve quadratic equations by hand. We'll conclude our study of elementary algebra by solving quadratic equations with a graphing calculator.

Here is a problem like that:

> The future net profit for Jasper's Jelly Beans can be modeled by the equation $y = 1.8x^2 + 1.6x + 15$, while the future net profit for Devonte's Donuts is given by the equation $y = 22.7x + 43.9$. In both of these equations y represents net profit, in thousands of dollars, and x represents the number of years from now. Use a graphing calculator to help you determine when the companies will have equal net profits. Then, discuss the circumstances under which it will be preferable to own each company.

This problem gives us a system of two equations. One of these equations is a quadratic equation in two variables. We can solve this system graphically by using graphing calculators. Before we do that, however, we need to understand how to solve quadratic equations by graphing.

One method of solving graphically involves calculating points of intersection. We did something similar to this in Chapter 3. Another method involves calculating zeros. Once we have had a chance to look at both of those methods, we'll return to this problem.

As a result of studying this section, you should be able to:

◆ Use a graphing calculator to solve quadratic equations by calculating points of intersection

◆ Use a graphing calculator to solve quadratic equations by calculating zeros

◆ Use a graphing calculator when solving application problems involving quadratic equations

A. Calculating Points of Intersection

In section 3.1, we solved a system of two linear equations by graphing both equations and finding the point of intersection. We also saw how to use a modification of this method in order solve a single equation by using the intersect feature on our graphing calculator.

Now we'll use this approach to solve quadratic equations.

Example 1

Use the given graphs to help you solve the equations.

1. $x^2 - 6x + 8 = 3$

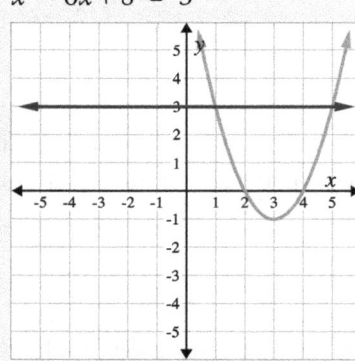

2. $x^2 + 2x - 4 = 2x - 3$

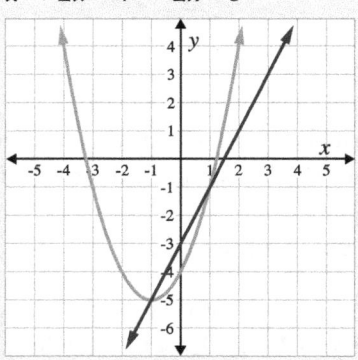

Solutions

1. $x^2 - 6x + 8 = 3$

The solutions of this equation are the x-values that cause the left and right sides of this equation to take on the same value. We can determine these solutions by analyzing the graphs of $y = x^2 - 6x + 8$ and $y = 3$. Notice that the points of intersection are $(1, 3)$ and $(5, 3)$. The x-coordinates of these points are the solutions we are seeking: $x = 1, 5$.

We can check both solutions algebraically.

$$\begin{aligned} x^2 - 6x + 8 &= 3 \\ (1)^2 - 6(1) + 8 &= 3 \\ 1 - 6 + 8 &= 3 \\ 3 &= 3 \checkmark \end{aligned}$$

$$\begin{aligned} x^2 - 6x + 8 &= 3 \\ (5)^2 - 6(5) + 8 &= 3 \\ 25 - 30 + 8 &= 3 \\ 3 &= 3 \checkmark \end{aligned}$$

2. $x^2 + 2x - 4 = 2x - 3$

Again, we need to determine the x-values that cause both sides of the equation to take on the same value. Figure 2 shows the graphs of $y = x^2 + 2x - 4$ and $y = 2x - 3$. The points of intersection are $(-1, -5)$ and $(1, -1)$. The x-coordinates of these points are our solutions: $x = -1, 1$.

Let's verify these solutions:

$$\begin{aligned} x^2 + 2x - 4 &= 2x - 3 \\ (-1)^2 + 2(-1) - 4 &= 2(-1) - 3 \\ 1 - 2 - 4 &= -2 - 3 \\ -5 &= -5 \checkmark \end{aligned}$$

$$\begin{aligned} x^2 + 2x - 4 &= 2x - 3 \\ (1)^2 + 2(1) - 4 &= 2(1) - 3 \\ 1 + 2 - 4 &= 2 - 3 \\ -1 &= -1 \checkmark \end{aligned}$$

When graphs aren't provided, we can use our graphing calculator to do most of the work for us. Here are the steps for working through this process:

1. Store the left side of the equation on the calculator as Y_1 and store the right side of the equation as Y_2.

2. View the graph, and make sure the viewing window allows you to see the point(s) of intersection.

3. Have the calculator find the point(s) of intersection — the steps for using the "intersect" feature are detailed in section 3.1

4. Write the x-coordinate of each point of intersection as the solution(s) of the original equation.

Example 2

Solve: $x^2 - 2.1x - 7.4 = 1.7x - 3.2$. Round the solution(s) off to the nearest thousandth.

Solution

Step 1: Begin by storing the left side of the equation as Y_1 and the right side of the equation as Y_2.

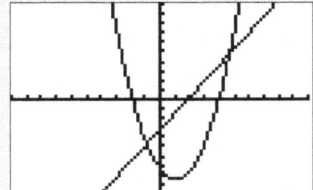

Step 2: View the graph. The viewing window needs to be set so that the points of intersection are visible. (In this image, viewing window: $-10 < x < 10, -10 < y < 10$)

Step 3: There are two points of intersection. Use your calculator find both of them. Use the procedure outlined in Section 3.1, and remember that the "guess" step in this procedure is used to tell the calculator which point of intersection you're looking for.

The intersection point on the left is $(-0.895, -4.721)$.

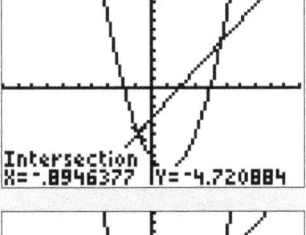

The intersection point on the right is $(4.695, 4.781)$.

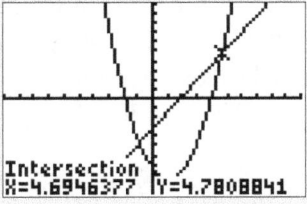

Step 4: Because the original equation only contained the variable *x*, we are only interested in the *x*-coordinates of the points of intersection. These are the solutions of the equation:

$$x \approx -0.895, 4.695$$

Step 5: We can check each of the solutions we found by substituting for *x* in the original equation. First we check −0.895.

$$x^2 - 2.1x - 7.4 = 1.7x - 3.2$$

$$(-0.895)^2 - 2.1(-0.895) - 7.4 = 1.7(-0.895) - 3.2$$

$$0.801025 + 1.8795 - 7.4 = -1.5215 - 3.2$$

$$-4.719475 = -4.7215$$

The numbers on either side of this equation are slightly different, but they are close enough to let us know that this solution is reasonable. Next, we check 4.695:

$$x^2 - 2.1x - 7.4 = 1.7x - 3.2$$

$$(4.695)^2 - 2.1(4.695) - 7.4 = 1.7(4.695) - 3.2$$

$$22.043025 - 9.8595 - 7.4 = 7.9815 - 3.2$$

$$4.783525 = 4.7815$$

This solution is also reasonable.

In future examples, we won't show how the solutions are checked, but when you are solving equations on your own, it's always a good idea to check your solutions.

Example 3

Solve: $0.05 x^2 - 8 = -0.029 x^2 + x + 5$. Round the solutions to the nearest thousandth.

Solution

Step 1: Store the left and right sides of the equation as Y_1 and Y_2, respectively.

```
Plot1  Plot2  Plot3
\Y1◼0.05X²−8
\Y2◼-0.029X²+X+5
\Y3=
\Y4=
\Y5=
\Y6=
```

Step 2: The viewing window from the previous example doesn't allow us to see both points of intersection. Here is what the graph looks like when the viewing window has been set at
    $-10 < x < 30, -10 < y < 20$

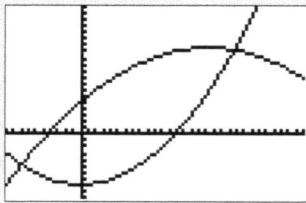

Step 3: The intersection point on the left is $(-7.975, -4.820)$. The intersection point on the right is $(20.633, 13.287)$.

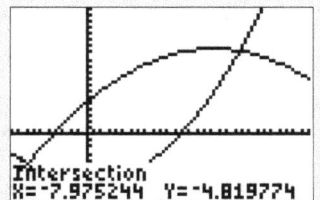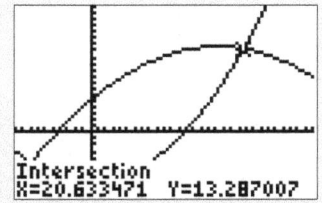

Step 4: Finish by writing the solutions of the original equation: $x \approx -7.975, 20.633$.

In the previous example, both sides of the equations were quadratic expressions, so it's not surprising that the graphs in Step 2 were parabolas. In some cases, it isn't possible to see both of the graphs in Step 2.

Example 4

Solve: $0.7x^2 + 2.8x - 5 = 0$. Round the solutions to the nearest thousandth.

Solution

Step 1: Store the left and right sides of the equation.

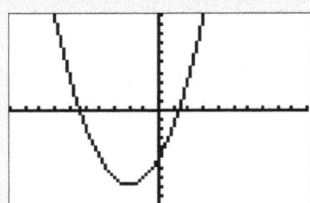

Step 2: A viewing window of $-10 < x < 10, -10 < y < 10$ works well for this step.

Only one graph is visible — the parabola. This is because the other graph, $Y_2 = 0$, is a horizontal line passing through 0 on the y-axis. The other graph is lying on top of the x-axis.

Step 3: The intersection point on the left is $(-5.338, 0)$. The intersection point on the right is $(1.338, 0)$.

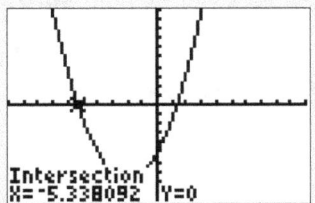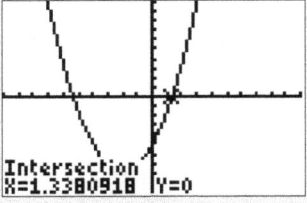

Step 4: Finish by writing the solutions of the original equation: $x \approx -5.338, 1.338$.

Sometimes it is not possible to find solutions of a quadratic equation graphically.

Example 5

Solve: $x^2 + 3.1x - 1.6 = 0.5x - 7.2$. Round the solutions to the nearest thousandth.

Solution

Step 1: Store the left and right sides of the equation.

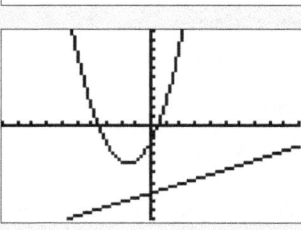

Step 2: Use a viewing window of $-10 < x < 10, -10 < y < 10$.

We have to stop now because these two graphs will *never* intersect. This means that the equation doesn't have any real-number solutions.

If we use the quadratic formula to solve this equation, our answer will include a negative number inside of the square root symbol — another indication that the equation has no real-number solutions.

Practice A

It's time for you to solve a few equations on your own. Use your graphing calculator to solve these equations by finding points of intersection. Round your solutions to the nearest thousandth. When you're done, turn the page to check your answers.

1. $-1.42x^2 - 9.7x - 8.2 = 0.78x + 2.4$ 3. $0.3x^2 + 2.8 = -0.13x^2 - 0.1x - 2.6$

2. $0.46x^2 - 1.3x - 5 = -0.08x^2 + 0.27x + 7$

B. Finding Zeros

We have seen that if an equation containing the variables x and y is written in the form

$$y = \{\text{expression containing } x\}$$

then we can say that y is the dependent variable and x is the independent variable because the value of y depends on the value assigned to x. We can also say that y is a *function* of x.

A **zero** of a function is any value assigned to x that causes the function to take on a value of 0. For example, suppose $y = x + 10$. In this case, y is a function of x. If we assign a value of -10 to x, then y takes on a value of 0:

$$y = -10 + 10$$
$$y = x + 10$$
$$y = 0$$

Therefore, -10 is a zero of this function. In fact, -10 happens to be the *only* zero of this function.

Now think back to what we learned about intercepts in Chapter 2. If a point on the graph of an equation has a y-coordinate of 0, that point is an x-intercept of the graph. In this example, the point $(-10, 0)$ is an x-intercept of the graph of $y = x + 10$. In fact, the point $(-10, 0)$ happens to be the *only* x-intercept of the graph of $y = x + 10$.

The following rule summarizes the relationship between zeros and x-intercepts:

> If n is a zero of $y = $ {expression containing x}, then the point $(n, 0)$
> is an x-intercept of the graph of $y = $ {expression containing x}

In essence, the word *zero* in this case is synonymous with the word *x-intercept*. When applied to quadratic equations, the rule looks like this:

> If $an^2 + bn + c = 0$, then $(n, 0)$ is an x-intercept of the graph of $y = ax^2 + bx + c$.

Using this fact, we can follow this procedure to solve quadratic equations in one variable.

1. Make sure the equation is in standard form: $ax^2 + bx + c = 0$.

2. Graph $y = ax^2 + bx + c$.

3. Find the x-intercept or intercepts of the graph — the zeros of the function.

4. Write the solution or solutions of the equation.

Example 6

Use the given graph to help you solve the equation: $-3x^2 + 30x - 72 = 0$

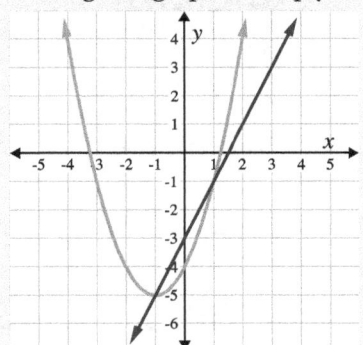

Solution

This equation is in standard form. That means that the solutions of the equation zeros of the equation $y = -3x^2 + 30x - 72$. In other words, the solutions are the x-intercepts of the graph of $y = -3x^2 + 30x - 72$. By looking at the graph, we can see that the x-intercepts are $(4, 0)$ and $(6, 0)$. This means that the zeros are 4 and 6. We can now write our solutions: $x = 4, 6$.

Our graphing calculators can be used to find zeros of a function. To see how that's done, let's take another look at the equation we saw in Example 4 above.

Example 7

Solve: $0.7x^2 + 2.8x - 5 = 0$. Round the solutions to the nearest thousandth.

Solution

Step 1: This quadratic equation is already in standard form, with 0 on one side.

Step 2: We need to find the zeros of $y = 0.7x^2 + 2.8x - 5$.
Put this in as the Y_1 equation.

Now graph the equation.
We use a viewing window of $-10 < x < 10, -10 < y < 10$ here.

Step 3: Here are the steps for finding the x-intercept or zero on the left. First, press [CALC]. Do this by pressing [2nd] and then [TRACE][ENTER]. Then select option 2, Zero.

The calculator prompts you to set a left boundary. Move the blinking icon to the left of the x-intercept and press [ENTER]. The calculator will not analyze any points to the left of this boundary.

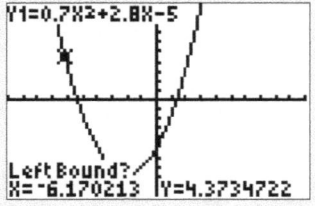

The calculator prompts you to set a right boundary. Move the blinking icon to the right of the x-intercept and press [ENTER]. The calculator will not analyze any points to the right of this boundary.

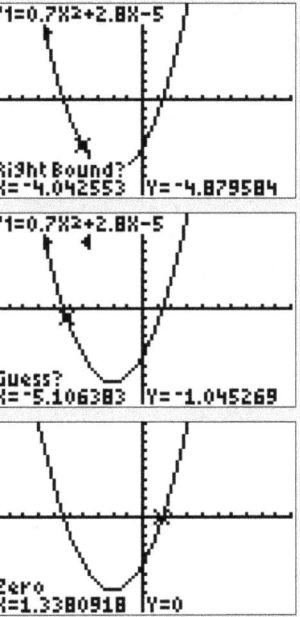

Finally, the calculator prompts you to "guess." Move the blinking icon so that it is close to the x-intercept, and press [ENTER].

Using the same procedure, find the x-intercept or zero on the right.

This time, the calculator displays the y-coordinate as 0. The x-coordinate, 1.338, matches the other solution we found in Example 3.

Step 4: Finish by writing the solutions of the original equation: $x \approx -5.338, 1.338$.

With this method, the equation must have 0 on one side before we can use our graphing calculators.

Example 8

Solve: $0.05 x^2 - 8 = -0.029 x^2 + x + 5$. Round the solutions to the nearest thousandth.

Solution

This as the same equation we solved in Example 2, but this time, we'll solve it by calculating zeros.

Step 1: This quadratic equation isn't in standard form, so we need to get 0 on one side of the equation.

$$
\begin{array}{rcl}
0.05 x^2 - 8 &=& -0.029 x^2 + x + 5 \\
\underline{+\ 0.029 x^2 - x - 5} &=& \underline{+\ 0.029 x^2 - x - 5} \\
0.079 x^2 - x - 13 &=& 0
\end{array}
$$

Step 2: We need to find the zeros of $y = 0.079 x^2 - x - 13$. Put this in as the Y_1 equation.

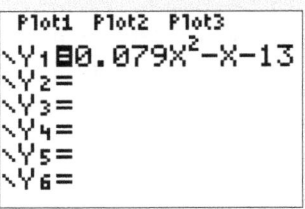

Practice A — Answers

1. $x \approx -6.171, -1.210$

2. $x \approx -3.479, 6.387$

3. no real-number solutions

Now graph the equation.
We use a viewing window of $-15 < x < 25, -20 < y < 5$ here.

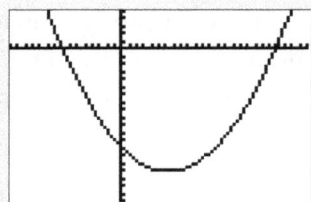

Step 3: The zero on the left is -7.975.
The zero on the right is 20.633.

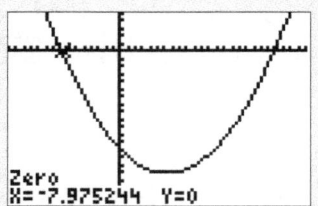 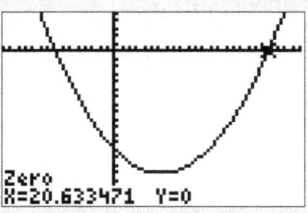

Step 4: Finish by writing the solutions of the original equation: $x \approx -7.975, 20.63$.

Not surprisingly, these solutions match what we found in Example 2.

Example 9

Solve: $2.03\,x^2 + 9.14x + 13.61 = 0$. Round the solutions to the nearest thousandth.

Step 1: This quadratic equation is already in standard form, so we move on to Step 2.

Step 2: We need to find the zeros of $y = 2.03\,x^2 + 9.14x + 13.61$.
Put this in as the Y_1 equation.

Now graph the equation.
We use a viewing window of $-10 < x < 10, -10 < y < 10$.

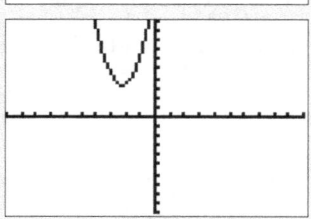

Step 3: At this point, we must stop. This graph doesn't have any x-intercepts. If we attempt to calculate a zero using a graphing calculator, we get an error message. Therefore, the equation has no real-number solutions.

Practice B

Now it's your turn. Use your graphing calculator to solve these equations by finding zeros. Round your solutions to the nearest thousandth. Turn the page when you're ready to check your solutions.

4. $-1.3x^2 - 4.9x + 5.1 = 0$

6. $3.7x^2 - 0.5x + 2 = 2.6x^2 - 9.8x - 11.4$

5. $0.5x^2 - 3.7x + 8.6 = 0$

C. Application Problems

Now that we know how to use a graphing calculator to solve quadratic equations by calculating points of intersection and calculating zeros, we're ready to look at some application problems. We'll use the same five-step approach for application problems, starting with the one from the beginning of this section.

Example 10

The future net profit for Jasper's Jelly Beans can be modeled by the equation $y = 1.8x^2 + 1.6x + 15$, while the future net profit for Devonte's Donuts is given by the equation $y = 22.7x + 43.9$. In both equations, y represents net profit, in thousands of dollars, and x represents the number of years from now. Use a graphing calculator to determine when the companies will have equal net profits. Then, discuss the circumstances under which it will be preferable to own each company.

Solution

Step 1: This problem defines the variables for us: x represents the number of years from now, and y represents net profit, in thousands of dollars.

Step 2: We are given the following system:

$$\begin{cases} y = 1.8x^2 + 1.6x + 15 \\ y = 22.7x + 43.9 \end{cases}$$

We use substitution. In the second equation, we replace y with $1.8x^2 + 1.6x + 15$. This generates the following equation:

$$1.8x^2 + 1.6x + 15 = 22.7x + 43.9$$

By solving the equation, we will find the number of years from now when the net profits of the two companies will be equal.

Step 3: Now we solve the equation by calculating points of intersection.

Store the left side of the equation as Y_1 and the right side of the equation as Y_2.

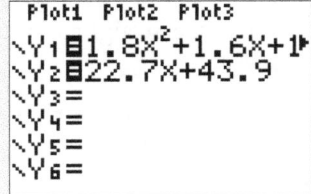

View the graph.

Use a viewing window of $-10 < x < 15, -30 < y < 400$.

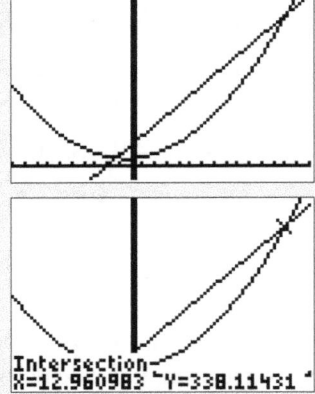

The intersection point on the left is $(-1.239, 15.780)$. The intersection point on the right is $(12.961, 338.114)$.

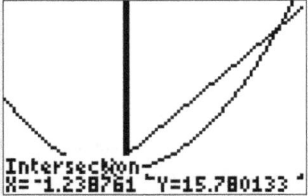

Write the solutions of the equation: $x \approx -1.239, 12.961$.

Step 4: It's time to check the solutions.

The first solution, -1.239, doesn't make sense within the context of this problem because all of the given information pertains to *future* net profit. These equations may or may not be valid when calculating the net profit 1.239 years in the past. The directions also ask us to determine when these companies *will* have equal not profits, not when they *had* equal net profits in the past.

Let's check the second solution algebraically.

$$1.8x^2 + 1.6x + 15 = 22.7x + 43.9$$
$$1.8(12.961)^2 + 1.6(12.961) + 15 = 22.7(12.961) + 43.9$$
$$302.3775378 + 20.7376 + 15 = 294.2147 + 43.9$$
$$338.1151378 = 338.1147$$

The second solution, 12.961, is reasonable.

Step 5: Now we translate the solution into a written statement:

The companies will have equal net profits 12.961 years from now — in about 13 years, in other words.

We're also directed to discuss the circumstances under which it will be preferable to own each company. Let's look at the graph from Step 3 again.

The net profit for Jasper's Jelly Beans, $y = 1.8x^2 + 1.6x + 15$, corresponds with the parabola. The net profit for Devonte's Donuts, $y = 22.7x + 43.9$, corresponds with the straight line. The straight line is *above* the parabola for x-values that are between 0 and 12.961. For x-values greater than 12.961, the straight line is *below* the parabola. Based on this information, we can state the following:

It will be preferable to own Devonte's Donuts for the next 13 years. After that, it will be preferable to own Jasper's Jelly Beans.

Let's try one more application problem.

Example 11

During an economic crisis, the amount of cash on the balance sheet of an investment firm can be modeled by the equation $y = -17.9x^2 + 42.3x + 600$ where y represents cash in thousands of dollars, and x represents the number of months elapsed. After how many months will the firm be out of cash?

Solution

Step 1: The problem defines the variables for us: x represents the number of months elapsed during the economic crisis, and y represents cash on the balance sheet in thousands of dollars.

Step 2: When the investment firm runs out of cash, there will be $0 of cash on the balance sheet. Therefore, in the given equation, we replace y with 0.

$$0 = -17.9x^2 + 42.3x + 600$$

Alternatively, this equation can be written so that 0 appears on the right side.

$$-17.9x^2 + 42.3x + 600 = 0$$

Step 3: Now we solve the equation by calculating zeros.
Store the left side of the equation as Y_1.

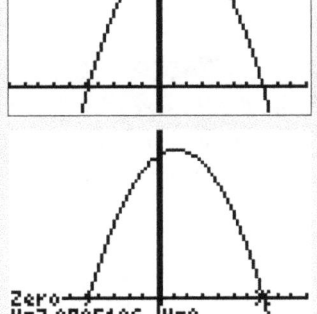

View the graph.
Use a viewing window of $-10 < x < 10, -100 < y < 700$.

We can disregard the zero or x-intercept on the left because negative solutions don't make sense within the context of the problem. The zero on the right is 7.091

Write the solution: $x \approx 7.091$.

Practice B — Answers

4. $x \approx -4.619, 0.849$ 5.   no real number solutions.   6. $x \approx -6.612, -1.842$

Step 4: Check the solution algebraically.

$$-17.9\,x^2 + 42.3x + 600 = 0$$

$$-17.9\,(7.091)^2 + 42.3(7.091) + 600 = 0$$

$$-900.0528299 + 299.9493 + 600 = 0$$

$$-0.1035299 = 0$$

Our solution is reasonable.

Step 5: Translate the solution into a written statement:

After about 7 months, the investment firm will be out of cash.

Practice C

Now you can give it a try. Use your graphing calculator to help you with the following application problems. When you're ready to check your answers, turn the page

7. Dikembe's Doggie Delights and Carlita's Canine Cuisine both sell packages of Tristan's Treats. The number of packages of Tristan's Treats sold by Dikembe's Doggie Delights can be modeled by the equation $y = 11x + 173$, and the number sold by Carlita's Canine Cuisine can be modeled by the equation $y = 0.74\,x^2 + 132$. In both equations, y represents the number of packages of Tristan's Treats sold, and x represents the number of months from now. Determine when both stores will sell the same number of packages of Tristan's Treats.

8. When Pete the Punter kicked a football, the height of the football was given by the equation $h = -16\,t^2 + 73.49t + 3.17$. h represents the height of the football in feet, and t represents elapsed time in seconds. How much time elapsed when the football hit the ground?

Exercises 6.6

For the following problems, use the given graph to help you solve the equation. When you are finished, you will find the odd-numbered solutions at the end of the book.

1. $2\,x^2 + 6x - 3 = 5$

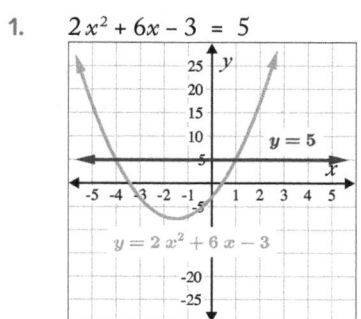

2. $-x^2 + 12x - 37 = -2$

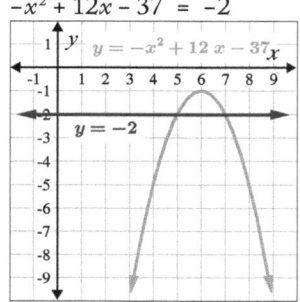

3. $0.5\,x^2 + 3x - 4 = 4$

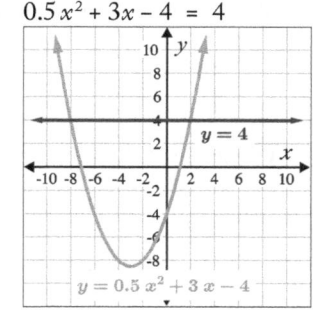

4. $x^2 - 10x + 18 = -7$

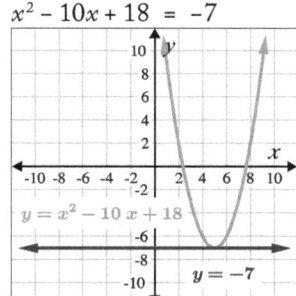

7. $0.5x^2 + 4x + 6 = 2x - 6$

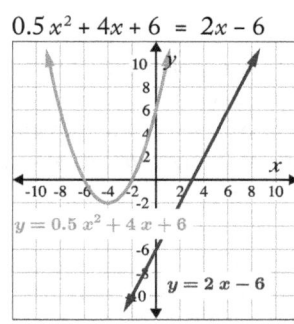

10. $\frac{1}{8}x^2 + \frac{1}{4}x - 6 = 0$

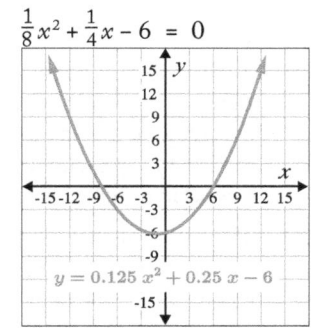

5. $x^2 - 9x + 9 = -2x + 3$

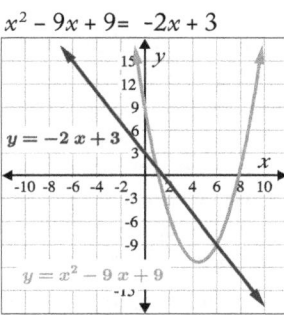

8. $x^2 + x - 6 = 0$

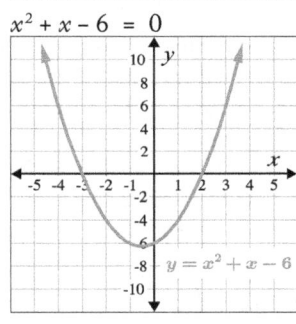

6. $-2x^2 - 3x + 1 = x + 7$

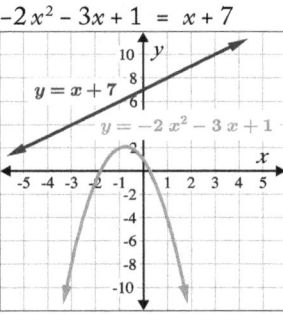

9. $-2x^2 + 30x - 108 = 0$

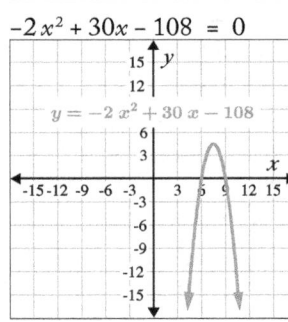

For the following problems, use a graphing calculator to help you solve the equations. Round answers off to the nearest thousandth.

11. $x^2 - 4x + 1 = 8$

12. $-x^2 - 3x + 7 = 4$

13. $-2.79n^2 + 4.92n + 4.65 = -0.38n + 2.81$

14. $1.67a^2 - 3.92 = -1.59a + 0.83$

15. $5.21t^2 + 15.37t + 2.79 = -0.81t - 5.76$

16. $0.6v^2 - 1.5v - 3.9 = -1.7v^2 - 0.9v + 6.6$

17. $4.8x^2 - 16.1x + 7.9 = -0.2x^2 - 0.9x - 5.7$

18. $-0.17g^2 + 2.9g + 19.6 = 0.23g^2 - 3.6g - 9.8$

19. $0.07m^2 + 1.35m + 11.8 = -0.05m^2 - 1.03m + 7.2$

20. $x^2 - 7x + 4 = 0$

21. $x^2 + 3x - 1 = 0$

22. $h^2 + 9h + 24 = 0$

23. $-3.1p^2 + 24.9p - 7.8 = 0$

24. $4.8y^2 + 32.7y + 17.2 = 0$

25. $-2.8w^2 + 6.7w - 5.4 = 0$

26. $-0.06x^2 - 0.11x + 7.46 = 0$

27. $0.09x^2 + 1.28x + 2.35 = 0$

28. $-6.72c^2 + 31.99c - 41.32 = 0$

Use a graphing calculator to help you solve the following application problems.

29. The number of people who read Hector's college football blog each week can be modeled by the equation $y = 11x + 723$. The number of people who read Wendy's college football blog each week can be modeled by the equation $y = 1.7x^2 - 5.8x + 517$. In both equations, y represents the number people who read the blog each week, and x represents the number of weeks from now. When will both blogs have the same number of readers?

30. The gross income for Jorge's Yoga Center can be modeled by the equation $y = 1.2x^2 + 3.1x + 91$ and the gross income for Briana's Gym can be modeled by the equation $y = 0.57x^2 - 1.13x + 152$. In both equations, y represents gross income in thousands of dollars, and x represents the number of years from now. When will both businesses have the same gross profit?

31. Scientists at the Pipe Dream Institute estimate that once a functional time machine is invented, the number of people who can be transported into the future can be modeled by the equation $y = 2.7x^2 + 5.1x + 600$. They also estimate that the number of people who will *want* to be transported into the future can be modeled by the equation $y = 4.2x^2 + 7.3x + 275$. In these equations, y represents the number of people who can or want to be transported into the future, and x represents the number of years after the invention of the functional time machine. If these scientists are correct, how many years after a functional time machine is invented will the number of people who can be transported into the future be equal to the number of people who want to be transported into the future?

32. A model airplane made of balsa wood is thrown from the top of a building. This situation can be modeled by the equation $y = -2x^2 + 3x + 34$, where y represents how high in feet the airplane is above the ground, and x represents the elapsed time in seconds after the airplane is thrown. How much time elapses before the airplane hits the ground?

33. A grocery store owner puts out a stack of "buy one, get one free" ice cream coupons for people to take. This situation can be modeled by the equation $y = -3.24x^2 - 1.79x + 200$, where y represents the number of coupons left in the stack, and x represents the number of hours since the owner put out the stack. How much time will it take for all of the coupons to be taken?

34. A math professor estimates that the percentage of people who are scared of graphing calculators can be modeled by the equation $y = -0.02x^2 - 0.04x + 23.1$. In this equation, y represents the percentage of people who are scared of graphing calculators, and x represents the number of years since 2015. If this equation is accurate, in what year will 0% of people be scared of graphing calculators?

Practice C — Answers

7. In about 18 months, both stores will sell the same number of packages of Tristan's Treats.

8. About 4.636 seconds had elapsed when the football hit the ground.

Solutions to Odd-Numbered Exercises

Chapter 1: Solving Linear Equations and Inequalities

1.1 Solving Linear Equations

1. conditional
3. identity
5. contradiction
7. $m = 7$
9. $y = -17$
11. $x = -14$
13. $g = -287$
15. $x = -443$
17. $y = -18.059$
19. $p = \frac{3}{5}$
21. $x = \frac{11}{12}$
23. $t = -\frac{7}{6}$
25. $x = 14$
27. $x = 8$
29. $x = 14$

31. $a = -16$
33. $p = -18$
35. $a = -4$
37. $x = 7$
39. $k = -42$
41. $x = 6$
43. $k = 42$
45. $x = 768$
47. $m = -56$
49. $f = -6386$
51. $k = 0.06$
53. $y = 9.453$
55. $m = -\frac{10}{3}$
57. $h = \frac{21}{8}$
59. $x + 13 = -20$;
    Solution: $x = -33$

61. $-37 = x - 18$;
    Solution: $x = -19$
63. $x - (-4) = 11$;
    Solution: $x = 7$
65. $45 = 2x$;
    Solution: $x = 22.5$
67. $8x = 62$;
    Solution: $x = \frac{31}{4}$
69. The teacher graded 556 final exams last year.
71. He averaged 25.3 points per game.
73. It takes light 240 minutes, or 4 hours, to travel from the sun to Neptune.
75. The radius of the earth is about 6,378 km.
77. The Ducks won 33 games.

1.2 Solving Multi-Step Equations

1. $x = 5$
3. $a = 7$
5. $y = -5$
7. $x = -3$
9. $y = -2$
11. $x = 36$
13. $y = 24$
15. $m = -44$
17. $k = -5$
19. $x = -8$
21. $k = -14$
23. $y = -5$
25. $y = 20$
27. $k = -1$
29. contradiction
31. $x = -3$

33. $x = \frac{19}{14}$
35. $k = 3$
37. $m = 2$
39. $n = \frac{10}{9}$
41. $w = \frac{39}{8}$
43. $x = -1$
45. $x = 9$
47. $x = 6$
49. $x = 7$
51. $5x = x - 2$;
    Solution: $x = -\frac{1}{2}$
53. $10x - 4 = 66$;
    Solution: $x = 7$
55. $\frac{11}{15}(x + 2) = 8$;
    Solution: $x = \frac{98}{11}$
57. $2x + 2 = \frac{1}{2}x - 3$;

Solution: $x = -\frac{10}{3}$
59. $\frac{x + 7}{2} = 22$;
    Solution: $x = 37$
61. $\frac{x - 11}{15} - 1 = 5$;
    Solution: $x = 101$
63. Last year's output was 50 items.
65. There are 15 ml of alcohol and 69 ml of water in the solution.
67. Each beaker will hold $263\frac{1}{3}$ ml of chloride solution.
69. The odd integers are -15 and -13.
71. This problem has no solution.
73. The building is 73 feet long and 57 feet wide.
75. Each leg is 42 inches long, and the base is 21 inches long.

1.3 Solving Literal Equations

1. solved for y

3. not solved

5. not solved

7. solved for d

9. $n = 4 - m$

11. $b = -a + 3c + d - 2f$

13. $c = 2a - 3b - 11$

15. $p = \dfrac{7r}{q}$

17. $b = \dfrac{2d}{a}$

19. $b = \dfrac{15}{8}a^2 c$

21. $\Delta = \dfrac{\nabla\ \Phi}{\square}$

23. $m = \dfrac{21 - 8t}{-12k}$

Equivalently: $m = -\dfrac{7}{4k} + \dfrac{2t}{3k}$

25. $\Phi = \dfrac{2\Delta - 3\square}{5\nabla}$

Equivalently: $\Phi = \dfrac{2\Delta}{5\nabla} - \dfrac{3\square}{5\nabla}$

27. $y = 3x + 7$

29. $y = 3x + 8$

31. $y = -\dfrac{9}{4}x - 7$

33. $y = 3.7x - 12.5$

35. $y = -\dfrac{5}{6}x - \dfrac{3}{7}$

37. $a = 4b - 9$

39. $k = -\dfrac{11}{3}c$

41. $q = 3r$

43. $g = \dfrac{5f + 4}{8}$

Equivalently: $g = \dfrac{5}{8}f + \dfrac{1}{2}$

45. $P = \dfrac{8t - Q}{6}$

Equivalently: $P = \dfrac{4}{3}t - \dfrac{1}{6}Q$

47. $j = \dfrac{\Delta\Phi - \square}{9}$

Equivalently: $j = \dfrac{\Delta\Phi}{9} - \dfrac{\square}{9}$

1.4 Solving Linear Inequalities

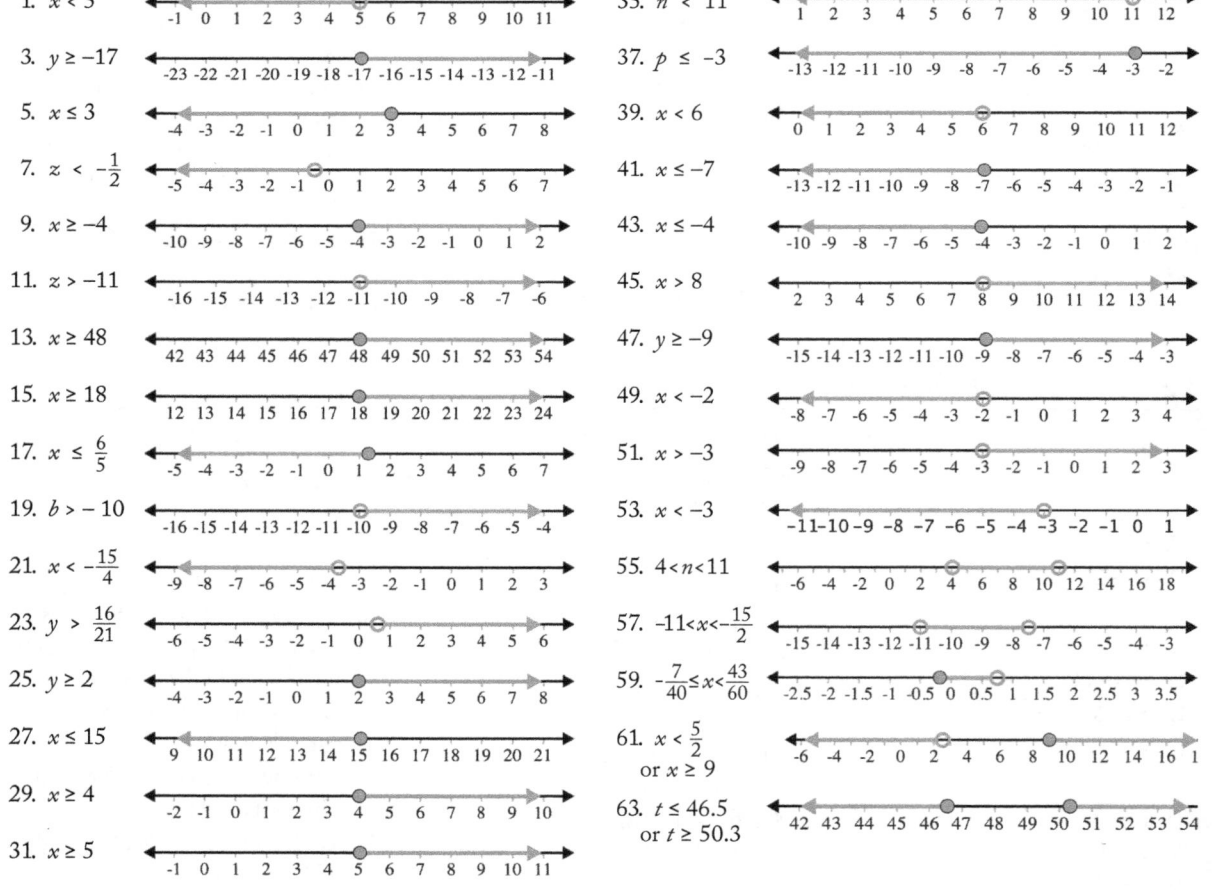

1. $x < 5$

3. $y \geq -17$

5. $x \leq 3$

7. $z < -\dfrac{1}{2}$

9. $x \geq -4$

11. $z > -11$

13. $x \geq 48$

15. $x \geq 18$

17. $x \leq \dfrac{6}{5}$

19. $b > -10$

21. $x < -\dfrac{15}{4}$

23. $y > \dfrac{16}{21}$

25. $y \geq 2$

27. $x \leq 15$

29. $x \geq 4$

31. $x \geq 5$

33. $y < -3$

35. $n < 11$

37. $p \leq -3$

39. $x < 6$

41. $x \leq -7$

43. $x \leq -4$

45. $x > 8$

47. $y \geq -9$

49. $x < -2$

51. $x > -3$

53. $x < -3$

55. $4 < n < 11$

57. $-11 < x < -\dfrac{15}{2}$

59. $-\dfrac{7}{40} \leq x < \dfrac{43}{60}$

61. $x < \dfrac{5}{2}$
or $x \geq 9$

63. $t \leq 46.5$
or $t \geq 50.3$

65. $2x + 1 > -3$; Solution: $x > -2$

67. $x + 25 \leq 57 - x$; Solution: $x \leq 16$

69. $x + 51 - 4x > 18$; Solution: x < 11

71. $2(x - 8) \geq 4x + 7$; Solution: $x \leq -\frac{23}{2}$

73. Maggie's golf score must be less than or equal to 78.

75. If more than 9.5 months pass, then the amount of Tommy's debt will be less than the amount of Juan's debt.

77. The width of the closet must be greater than 0 feet and less than or equal to 6 feet.

Chapter 2: Linear Models

2.1 Introduction to Linear Models

1. a) The total cost is $7.30.
   b) Your friends can afford to purchase 4 burritos.
3. a) $240 was available at the beginning of the month.
   b) $84 was available after three weeks.
5. a) It takes 7 weeks.
   b) The boxer weighs 164.5 pounds after three weeks.
   c) The boxer will be in the "super middleweight" weight class after 6 weeks.
7. a) The cost is $15.40.
   b) About 5.45 gallons of gasoline can be purchased.
9. a) 952 gallons of water will be in the tank.
   b) 600 gallons of water will remain after 25 minutes.
   c) The tank will be empty after 43.75 minutes.

11. a) Andre will be 2,420 miles from his destination.
    b) After 12 days of riding, he will be 1,540 miles from his destination.
    c) He will reach his destination after 26 days of riding.
13. a) 19,500 fish will be in Loch Lake.
    b) 22,000 fish will be in the lake after 8 years.
15. Table 3
17. Table 1
19. Table 5
21. Graph 4
23. Graph 3
25. Graph 6
27. Graph 8

2.2 Tables and Graphs

1.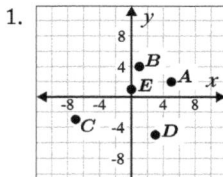

3.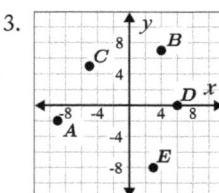

5.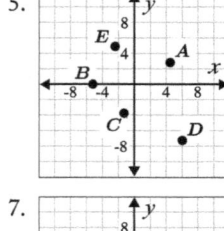

7.

9. quadrant IV
11. y-axis
13. quadrant II
15. quadrant III
17. quadrant II
19. x-axis
21. origin (both axes)
23. quadrant I
25.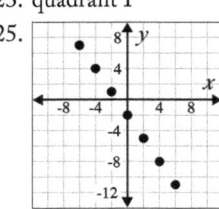

exact linear relationship

27.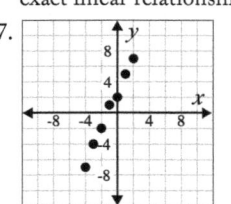

approximately linear relationship

29.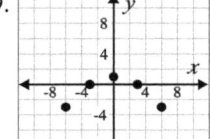

relationship is not linear

31.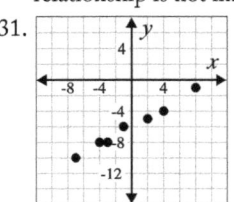

approximately linear relationship

33.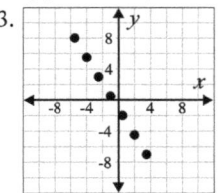

exact linear relationship

35.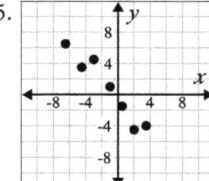

approximately linear relationship

37.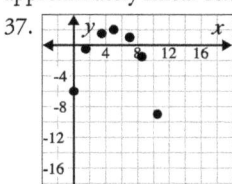

relationship is not linear

39. a) no, b) no, c) yes

41. a) yes, b) yes, c) no

43. a) no, b) no, c) no

45.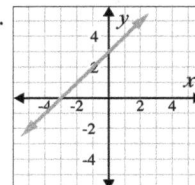

47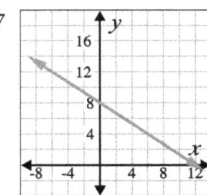

49.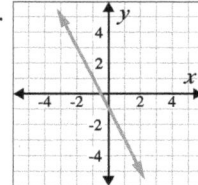

51.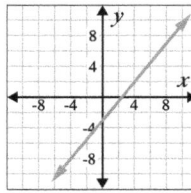

53.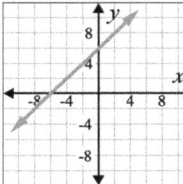

55.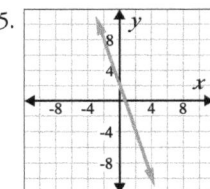

57.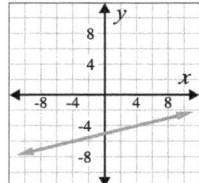

59.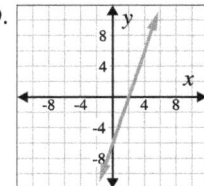

61.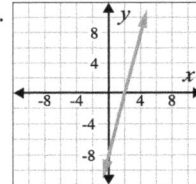

63.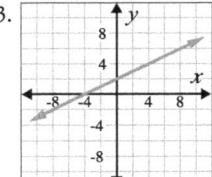

2.3 Slope

1. 2

3. 1

5. −3

7. −3

9. $\frac{1}{3}$

11. $-\frac{3}{2}$

13. $\frac{2}{5}$

15. 0

17. undefined (no slope)

19. $m = 1$

21. $m = -2$

23. $m = \frac{2}{3}$

25. $m = -5$

27. $m = -\frac{1}{7}$

29. $m = 0$

31. $m = 2$

33. undefined (no slope)

35. $m = \frac{2}{3}$

37. 2

39. $-\frac{5}{4}$

41. −1

43. −3

45. 1

47. 4

49. 0

51. undefined (no slope)

53. $-\frac{6}{7}$

55. Slope = $\frac{5}{4}$ The temperature in the car increases by 5 degrees when the elapsed time increases by 4 minutes.

57. Slope = $-\frac{7}{2}$ The temperature goes down by 7 degrees when elevation increases by 2 thousand feet.

2.4 Slope-Intercept Form

1.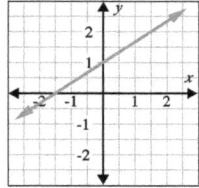

3.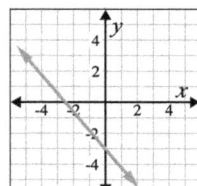

5.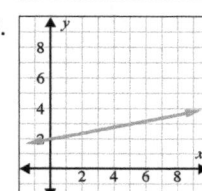

7.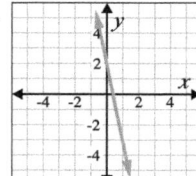

25.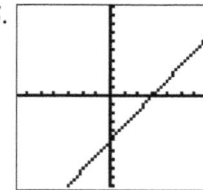

9.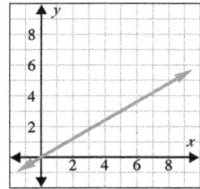

11.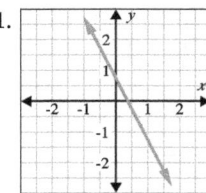

13.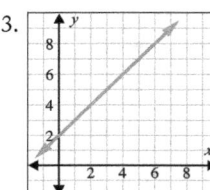

15.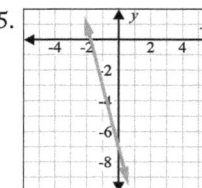

27.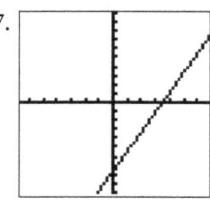

17.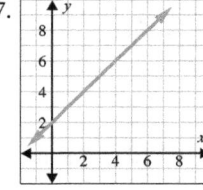

19.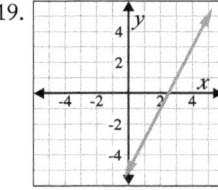

21.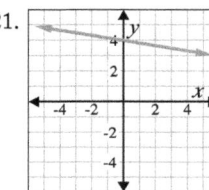

23.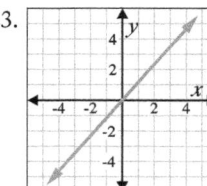

29.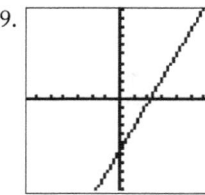

31. a) d, the number of days since Brittany started working out; n, the number of sit-ups she can do (choice of variables may vary)

b) $n = 3d + 15$

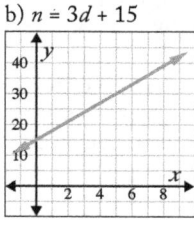

c) 78 sit-ups

33. a) t, the time (in minutes) that has passed; g, the number of gallons of water remaining (choice of variables may vary)

b) $g = -12t + 42$

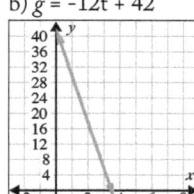

c) 6 gallons remain

35. a) c, the cost of the party; g, the number of guests (choice of variables may vary)

b) $c = 7.80g + 44.95$

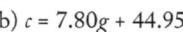
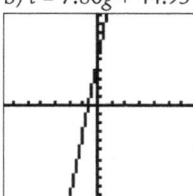

c) \$185.35

37. $y = \frac{4}{5}x - 2$

39. $y = 4x$

41. $y = \frac{5}{3}x + 2$

43. $y = \frac{5}{4}x$

45. $y = 6x + 5$

47. $y = -4x - 7$

2.5 Standard Form

1.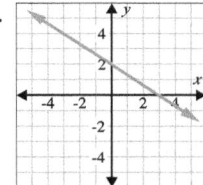

3.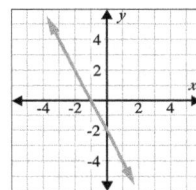

5.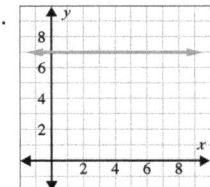

7.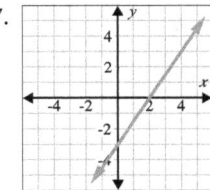

9.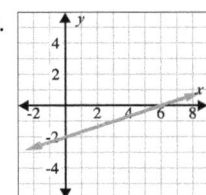

11.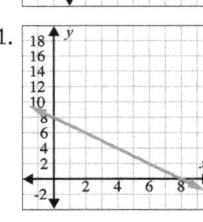

13.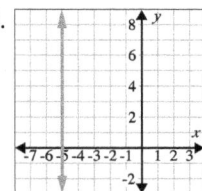

15.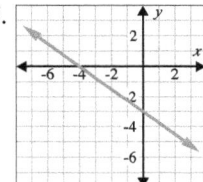

17.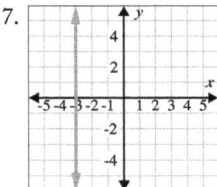

19.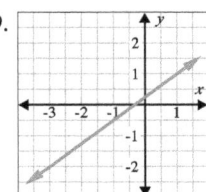

21.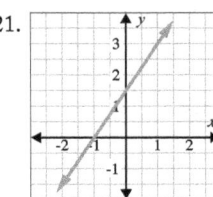

23.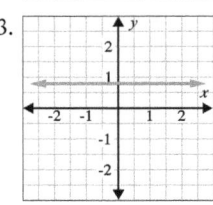

25.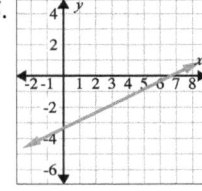

27.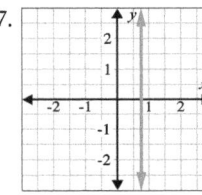

29. x = the number of general admission tickets, y = the number of preferred-seat tickets; $15x + 25y = 1200$

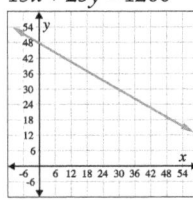

31. x = the number of large pizzas, y = the number of family-size pizzas; $3x + 5y = 45$

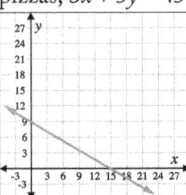

33. x = the number of paperbacks, y = the number of hardbacks; $1.5x + 2.5y = 30$

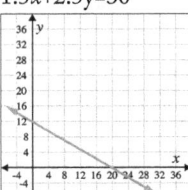

35. $4x+y=4$

37. $y = -\frac{1}{6}x$

39. $y=5$

41. $7x - 4y=28$

43. $x = -6$

45. $-2x+3y=1$

47. $3x -4y=24$

49. $-4x + y = -7$

51. $-3x+4y=4$

53. $2x+5y=-6$

55. $-15x+12y=-10$

2.6 Ordered Pairs and Equations

1. $y = 3x + 4$

3. $y = 8x + 1$

5. $y = -6x -1$

7. $y = -\frac{3}{2}x$

9. $y - 1 = -2(x - 5)$

11. $y - 3 = 5(x + 10)$

13. $y + 6 = \frac{3}{4}(x + 12)$

15. $y + 10 = -\frac{8}{3}(x - 3)$

17. $y = x + 5$

19. $y = 8x -32$

21. $y = -x + 6$

23. $y = -2x + 1$

25. $y = -\frac{6}{7}x + 3$

27. $y = \frac{5}{2}x - 13$

29. $y = -\frac{1}{9}x + 4$

31. $y = 5$

33. $y = 321$

35. $y = -57$

37. $y = \frac{8}{5}x$

39. $y = x + 3$

41. $y = -3x - 23$

43. $y = 3$

45. $x = 4$

47. $y = -4x + 25$

49. $y = -\frac{9}{5}x + \frac{2}{5}$

51. $y = \frac{9}{7}x - 6$

53. Variables: x = number of party guests (people), y = number of cookies baked. Equation: $y = 3x+4$

55. Variables: x = number of police officers parked on the side of the highway (people), y = number of passing drivers who are speeding (people per minute). Equation: $y = -8x + 73$

57. Variables: x: temperature (degrees), y = number of athletes who complain (people). Equation: $y = \frac{1}{3}x - 18$

2.7 Line of Best Fit

1.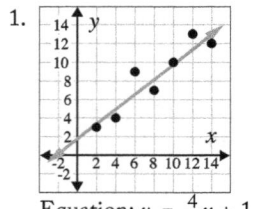
Equation: $y = \frac{4}{5}x + 1\frac{4}{5}$

3.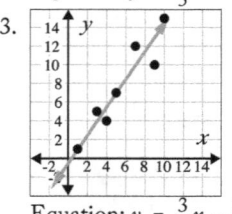
Equation: $y = \frac{3}{2}x - \frac{1}{2}$

5.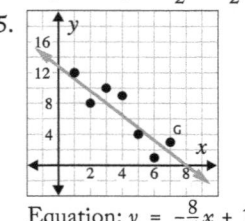
Equation: $y = -\frac{8}{5}x + 12\frac{4}{5}$

7.
Equation: $y = -\frac{3}{4}x + 13\frac{1}{2}$

9.
Equation: $y = 2x + 6$

11.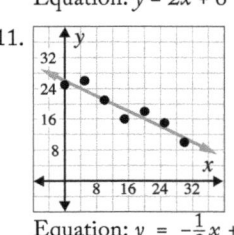
Equation: $y = -\frac{1}{2}x + 26$

13.
Equation: $y = \frac{13}{10}x + 9$

15. $y = -9.429x + 99.429$

17. $y = 0.6x + 24.986$

19. $y = 0.858x + 101.334$

21. $y = -0.182x + 55.357$

23. $y = 0.329x + 2.5$

25. $y = -7.057x + 120.571$

27. a) If he studies for 2 hours, Phillip's test score will be 77.

   b) If he studies for 5 hours, Phillip's test score will be 99.5.

29. a) In 2026, there will be 952 gym memberships

    b) In 2040, there will be 1,820 gym memberships

31. a) In 2024, the portfolio will be worth \$335,800.

    b) In 2033, the portfolio will be worth \$584,200.

33. Equation: $y = 0.109x + 1.707$

    a) In 2023, impressionable youth in the U.S. will watch 4.2 hours of television per day.

    b) In 2036, impressionable youth in the U.S. will watch 5.6 hours of television per day.

35. Equation: $y = 23.971x + 161.238$

a) In 2026, the University of Oregon will spend \$568,745 on football.

b) In 2045, the University of Oregon will spend \$1,024,194 on football.

2.8 Parallel and Perpendicular Lines

1. b, e

3. a, c, d

5. a, e

7. $y = 3x - 16$

9. $y = -4x - 9$

11. $y = \frac{1}{2}x + 1$

13. $y = \frac{4}{3}x - 7$

15. $y = -\frac{2}{9}x - 3$

17. $y = \frac{2}{3}x + \frac{13}{3}$

19. $y = -2x + 15$

21. $y = -\frac{3}{4}x - 1$

23. $y = 2x - 19$

25. $y = -x + 12$

27. $y = -\frac{2}{7}x + 8$

29. c, e

31. a, c, d

33. b, e

35. $y = -\frac{1}{2}x + 5$

37. $y = -\frac{1}{4}x - 12$

39. $y = \frac{1}{7}x + 35$

41. $y = -5x - 51$

43. $y = \frac{4}{5}x - 17$

45. $y = -\frac{3}{4}x + \frac{23}{2}$

47. $y = -2x + 3$

49. $y = 3x + 10$

51. $y = \frac{5}{6}x - 5$

53. $y = -2x - 2$

55. $y = \frac{3}{4}x + 9$

Chapter 3: Systems of Linear Equations

3.1 Using Graphing to Solve Systems

1. a) no, b) no, c) yes

3. a) no, b) no, c) no

5. Solution: (3, 4)

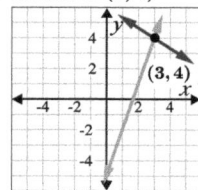

7. Solution: (4, –3)

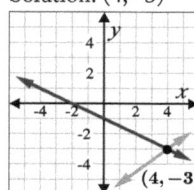

9. (–3, – 2)

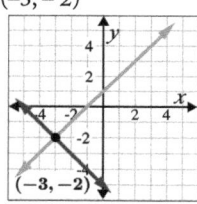

11. (–1, 2)

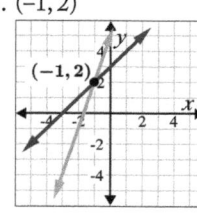

13. $\left(\frac{12}{13}, -\frac{36}{13}\right)$

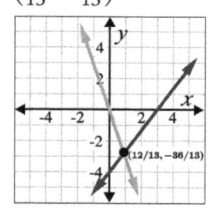

15. (– 6, 6)

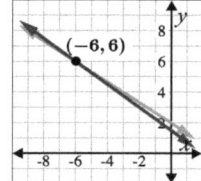

17. no solution (inconsistent)

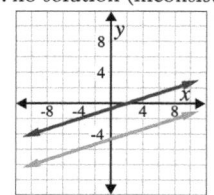

19. infinitely many solutions (dependent)

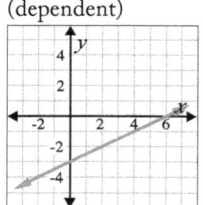

21. Solution: (3.834, 3.711)

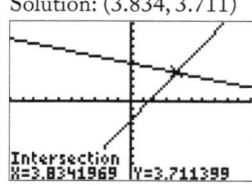

23. Solution: (3.933, –1.301)

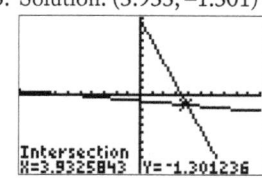

25. Solution: (–0.231, –1.577)

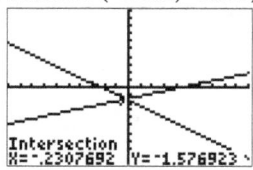

27. Solution: (–0.266, –3.752)

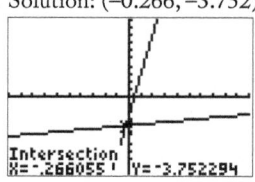

29. Solution: $x \approx 2.280$

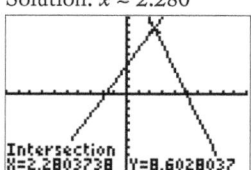

31. Solution: $x = 6.25$

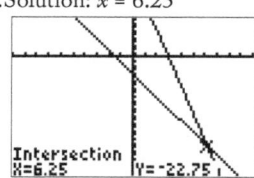

33. Solution: $x \approx 1.847$

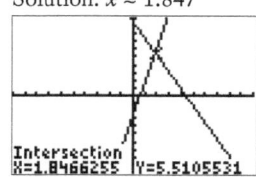

35. Solution: (7, 12)

X	Y₁	Y₂
4	21	6
5	18	8
6	15	10
7	12	12
8	9	14
9	6	16
10	3	18

X=7

37. Solution: $(-4, -2)$

X	Y1	Y2
-7	-1	-3.5
-6	-1.333	-3
-5	-1.667	-2.5
-4	**-2**	**-2**
-3	-2.333	-1.5
-2	-2.667	-1
-1	-3	-.5

X= -4

39. Solution: $(3, -6)$

X	Y1	Y2
0	6	-8
1	2	-7.333
2	**-2**	**-6.667**
3	-6	-6
4	-10	-5.333
5	-14	-4.667
6	-18	-4

X=3

3.2 Using Substitution to Solve Systems

1. $(1, 3)$
3. $(-1, 1)$
5. $(2, 2)$
7. $\left(4, -\frac{1}{3}\right)$
9. infinitely many solutions (dependent)
11. $(1, -2)$
13 no solution (inconsistent)
15. $(-1, -5)$

17. $\left(\frac{15}{4}, 4\right)$
19. $(-1, -3)$
21. $(4, -1)$
23. no solution (inconsistent)
25. $(1, -1)$
27. The two numbers are 14 and 8.
29. The two numbers are 18 and 6.
31. System: $\begin{cases} 38 = 2L + 2W \\ L = 2W - 2 \end{cases}$
    Answer: The box is 12 inches long

and 7 inches wide.
33. System: $\begin{cases} x = 4y \\ x + y = 14,250 \end{cases}$
    Answer: 11,400 people attended on Sunday and 2,850 people attended on Thursday.
35. System: $\begin{cases} x = 3y \\ 4x + 7y = 551 \end{cases}$
    Answer: 87 6.1-inch sandwiches were sold and 29 12.1-inch sandwiches were sold.

3.3 Using Elimination to Solve Systems

1. $(5, 6)$
3. $(2, 2)$
5. $(0, -6)$
7. $(2, -2)$
9. infinitely many solutions (dependent)
11. $(1, 1)$
13. $(-2, 2)$
15. $(-4, 3)$
17. $(-1, 6)$

19. infinitely many solutions (dependent)
21. $(1, 1)$
23. no solution (inconsistent)
25. infinitely many solutions (dependent)
27. no solution (inconsistent)
29. $(0, 4)$
31. $\left(-\frac{258}{7}, \frac{519}{7}\right)$
33. infinitely many solutions (dependent)

35. System: $\begin{cases} 5c + 7g = 41 \\ 10c + 2g = 46 \end{cases}$
    Answer: Cherries cost $4 per pound and grapes cost $3 per pound.
37. System: $\begin{cases} 3x + 5y = 7,350 \\ 2x + 9y = 9,150 \end{cases}$
    Answer: It costs $1,200 to have dinner with the governor and $750 to have dinner with the local congressperson.

3.4 Using Systems of Equations to Solve Mixture Problems

1. The mixture contains 6 pounds of green grapes and 8 pounds of red grapes.
3. The compound contains 12 grams of Chemical A and 30 grams of Chemical B.
5. 163 students and 37 non-students purchased tickets for the performance.
7. There were 85 pennies and 68 nickels in the well.
9. There are 28 nickels and 16 quarters in the hat.
11. She should use 17.6 ml of Solution A and 4.4 ml of Solution B

13. a) No — the mixture cannot be created with Solution A and Solution B. (When solving algebraically, one of the answers is negative.)
    b) The mixture can be made with 0.4 ml of Solution A and 1.6 ml of Solution C. Or, the mixture can be made with 0.8 ml of Solution B and 1.2 ml of Solution C.
15. Victor should use 20 ml of pure water and 80 ml of the 20% acid solution.
17. The mixture should contain 9 pounds of brazil nuts and 3 pounds of walnuts.
19. The mixture should contain 26 quarts (6.5 gallons) of apple juice and 14 quarts (3.5 gallons) of grape juice.

Chapter 4: Polynomials

4.1 Addition and Subtraction of Polynomials

1. $4x$

3. $2m$

5. $-y$

7. $-7hp$

9. $37\,a^2\,b$

11. $63x - 33$

13. $x - 11$

15. $39xy - 27x + 10y + 1$

17. $7\,x^2 - 4x - 7$

19. $-16\,x^2\,y$

21. $x^4 - 3\,x^3 - 8\,x^2 + 6x + 1$

23. $a^3\,b^2\,c + 7\,a^2\,b^2$

25. $7\,x^2\,y^3 + 3\,x^3\,y^2 + 1$

27. $7a + 18$

29. $-28b + 18$

31. $31x + 8$

33. $10a$

35. $32y^2 + 72x$

37. $5r - 53s$

39. $-24a - 16$

41. $-20n + 33$

43. $7\,x^2 - 5x - 7$

45. $6\,y^2 - 19xy + 2x$

47. $6\,x^2 - 33x - 1$

49. $9\,a^2\,b + 13a\,b^2 - ab$

51. $a^3 - 11\,a^2 - 7a - 46$

53. $12x - 9$

55. $7x - 10$

57. $26x + 5$

59. $-20x - 18$

61. $-6\,x^2 - 34x - 4$

4.2 Multiplication of Polynomials

1. $x^2 + 6x$

3. $m^2 - 4m$

5. $3\,x^2 + 6x$

7. $6\,a^2 - 30a$

9. $15\,x^2 + 12x$

11. $2\,b^2 - 2b$

13. $15\,x^4 + 12\,x^2$

15. $20\,a^7 + 12\,a^6 + 8\,a^5$

17. $-5\,x^3 - 10\,x^2$

19. $6\,x^4\,y^3 - 12\,x^3\,y$

21. $2\,b^6\,x^3 - 11\,b^5\,x^2$

23. $18\,y^7 - 27\,y^6 + 72\,y^5 + 9\,y^4 - 54\,y^3$

25. $a^2 + 6a + 8$

27. $y^2 - 9$

29. $i^2 + 2i - 15$

31. $6\,a^2 - 20a + 6$

33. $18\,y^2 + 93y + 110$

35. $-x^2 - x + 12$

37. $x^3 + x^2 + 2x + 2$

39. $6\,x^4 - 7\,x^2 - 5$

41. $12\,x^5\,y^7 + 30\,x^3\,y^5 + 12\,x^3\,y^3 + 30xy$

43. $4\,a^2 - 28a - 32$

45. $x^3 + 5\,x^2 + 4x$

47. $y^5 - 5\,y^4 + 6\,y^3$

49. $5\,y^8 + 40\,y^7 + 35\,y^6$

51. $10\,x^6\,y^5 - 5\,x^5\,y^4 - 6\,x^4\,y^3 + 3\,x^3\,y^2$

53. $8\,c^3 + 40c + 88$

55. $24\,a^5\,b^9 + 42\,a^4\,b^{11} + 12\,a^3\,b^{13} + 84\,a^3\,b^3$

57. $x^3 - 6\,x^2 - 10x + 21$

59. $21\,a^7 - 22\,a^5 - 15\,a^3 - 7\,a^2 - 2a - 2$

61. $10\,a^3 + 8\,a^3\,b + 4\,a^2\,b^2 + 5\,a^2\,b - b^2 - 8a - 4b - 2ab$

63. $x^2 + 2x + 1$

65. $a^2 + 4a + 4$

67. $-9\,x^2 + 30x - 25$

69. $n^3 + 12\,n^2 + 48n + 64$

4.3 Rules of Exponents

1. d^{12}

3. $3^8\,y^3$

5. $9\,d^2\,b^3$

7. $25\,p^8\,q^6$

9. $64\,r^5\,w^{10}$

11. $7\,z^8$

13. $(v - u)^3$

15. $14\,(3 + c^2)^5$

17. c^{3w+7}

19. p

21. $9\,x^6$

23. $4\,x^4\,y^{39}$

25. $20\,r^2\,st$

27. $(7 + y)^4$

29. y^{p-q}

31. d^{7-n}

33. a. 1   b. 7

35. a. 1   b. −1

37. $20\,y^{11}$

39. $144\,x^4\,y^5\,z^3$

41. $32\,x^3\,y^3$

43. $2\,x^3\,y^7$

45. $\dfrac{15}{2}\,z^{20}$

47. $4x^5z^3\,(z-x^4)^7\,(x+z)^2$

4.4 Power Rules of Exponents

1. w^{12}

3. $n^{\blacktriangle\,\Phi}$

5. a^5c^5

7. $8\,a^3$

9. $-32\,v^5$

11. $81j^4$

13. $49\,y^6$

15. $125\,x^{18}$

17. $10,000\,a^{12}\,b^{36}$

19. $x^8\,y^{12}\,z^{20}$

21. $x^{15}\,y^{10}\,z^{20}$

23. $a^{32}\,b^{56}\,c^{48}\,d^{64}$

25. 1

27. $\dfrac{1}{64}\,c^{20}\,d^{16}\,e^8\,f^{18}$

29. $x^6\,y^8$

31. $108\,m^8\,t^{24}$

33. $x^{26}\,y^{14}\,z^8$

35. x^8

37. $m^4\,n^4\,p^{12}$

39. $1000\,x^8\,y^7\,z^{33}$

41. $25\,x^{14}\,y^{22}$

43. $\dfrac{27\,a^3\,b^3}{64\,x^3\,y^3}$

45. $\dfrac{27\,a^6\,b^9}{c^{12}}$

47. $\dfrac{27\,x^9\,y^{24}}{8\,z^{12}}$

49. $\dfrac{16\,n^{36}\,p^4}{x^{28}}$

51. $\dfrac{x^8\,(y-1)^{12}}{(x+6)^4}$

53. x^{n+6}

55. $2^\nabla\,a^\nabla\,b^\nabla$

57. $2\,m^{\Delta-\nabla}$

59. $\dfrac{2^\square\,x^{\Delta\square}}{y^{\nabla\square}}$

4.5 Negative Exponents

1. $\dfrac{1}{x^2}$

3. $\dfrac{1}{b^{12}}$

5. $\dfrac{1}{n^5}$

7. $\dfrac{1}{128}$

9. $-\dfrac{1}{49}$

11. $\dfrac{1}{25}$

13. $\dfrac{1}{k^2\,m}$

15. $\dfrac{1}{x^7y^{35}}$

17. $\dfrac{y^{22}}{81\,x^8}$

19. $\dfrac{81}{n^{20}\,m^{32}}$

21. $\dfrac{1}{(x-5)^{13}}$

23. $\dfrac{x^3}{y^2}$

25. $\dfrac{a^4}{b}$

27. $\dfrac{w\,x^3\,z^2}{y^4}$

29. $\dfrac{x^9}{r^2\,w^5\,y^6\,z}$

31. $\dfrac{5\,x^2\,y^2}{z^5}$

33. $\dfrac{7\,a^2\,(a-4)^3}{b^6\,c^7}$

35. $\dfrac{7\,(w+1)^3}{(w+2)^2}$

37. $\dfrac{(x^2+3)^3}{(x^2-1)^4}$

39. $\dfrac{10}{x^4}$

41. $\dfrac{12}{a^{10}}$

43. $\dfrac{c^6}{5\,a^2\,b^{17}}$

45. $\dfrac{35\,a^3\,c^2}{b^8}$

47. $\dfrac{1}{(a-4)^7}$

49. $\dfrac{12xz^5}{y^7}$

51. $\dfrac{200}{a^{19}\,b^6}$

53. a

55. $\dfrac{3\,b^3\,c^5}{a^3}$

57. $\dfrac{4bc^9\,y^2}{a^2\,d^3\,z^8}$

59. $\dfrac{4\,a^3}{b^2\,c}$

61. $\dfrac{3\,x^3}{y^2\,z^4}$

63. $\dfrac{3}{a^7\,b^5}$

65. $\dfrac{(y-6)^{12}}{(x+3)^2}$

67. $8\,m^6\,n^{15}$

69. $\dfrac{h^{10}\,j^{30}\,p^5}{k^{20}}$

71. 640

73. -381

75. $5\,a^{-3}\,x^4\,y^3$

77. $2bc\,d^{-11}$

4.6 Applications of Exponent Rules

1. 54 cubic feet.
3. 3.06 cubic meters.
5. 0.64 cubic yards.
7. 3,097,600 square yards.
9. 15,500,031 cubic inches.
11. 37.43 square yards.
13. Veronica needs about 556.88 ft^2 of carpet.
15. The exhibit contains about 3,030 m^3 of water.
17. 5.89×10^3
19. 2.5×10^{23}
21. 1×10^{-10}
23. 1.7×10^{-11}
25. 7.35×10^{12}
27. 3.36×10^3
29. 5.1×10^{11}
31. 1.03×10^{-18}
33. 3.16×10^2
35. 7.4×10^5
37. 2.61×10^{-8}
39. 5,980,000,000,000,000,000,000,000

41. 30,000,000
43. 100,000,000,000,000,000,000
45. 0.000316
47. 0.0000015
49. 602,200,000,000,000,000,000,000
51. 0.000054054
53. 6×10^9
55. 3.6×10^5
57. 6×10^{-6}
59. 9.61×10^{-2}
61. 1.21×10^{12}
63. 1.0404×10^{-34}
65. 8×10^{17}
67. 7.5×10^{-48}
69. 3.125×10^{12}
71. 3.08×10^{15}
73. 6×10^{-7}
75. 4×10^{30}

Chapter 5 Factoring

5.1 Factoring out the Greatest Common Factor

1. 5
3. 3
5. $2xy$
7. $-2c^2d^3$
9. $2x + 5$
11. $3x + 4$
13. $3b^3 + 4b + 5$
15. $2x + 7$
17. $a - b$
19. $2x - 4y + z$
21. $4(4x + 3)$

23. $18(f - 2)$
25. $10y(3y + 1)$
27. $-7(x^2 + 3)$
29. $5x(a^2x + 2)$
31. $8(2x^2 + x - 3)$
33. $-3y^3(5y^2 + 8y - 3)$
35. $13x^2y^5(c^5 - 2c^3 - 3)$
37. $(x - 9)(a + b)$
39. $(9a - b)(2w - 7x)$
41. $(4x + 3)(-5x + 3)$ or
$-(4x + 3)(5x - 3)$

43. $(7q^3 - 15)(21q + 19)$
45. $12n^3(5x + 1)(2n^2 - 1)$
47. $(2n - 5)(3n^4 + 7)$
49. $(y + 1)(x + 3)$
51. $(r + 4s)(a - 5b)$
53. $(9k - 2h)(4a - 3b)$
55. not factorable by grouping
57. not factorable by grouping
59. $(n + 6)(2n - 5m)$
61. $(x + 2y)(8x - 1)$

5.2 Factoring Trinomials of the Form x2 + bx + c

1. $(x + 3)(x + 1)$
3. $(x + 3)(x + 4)$
5. $(y + 6)(y + 2)$
7. $(y - 4)(y - 1)$
9. $(a + 4)(a - 1)$
11. $(x - 7)(x + 3)$

13. $(y + 8)(y + 2)$
15. not factorable (prime)
17. $(y - 7)(y - 1)$
19. $(a + 6)(a - 5)$
21. $(a - 10)(a - 2)$
23. $(x + 6)(x + 7)$

25. $(x + 5)(x + 8)$
27. not factorable (prime)
29. $(b + 8)(b + 7)$
31. $4(x + 2)(x + 1)$
33. $5(y^2 - 14y + 88)$
35. $x(x + 4)(x + 2)$

37. not factorable (prime)
39. $x^2(x + 7)(x + 2)$
41. $-4a(a - 7)(a - 3)$
43. $-2y^2(n - 8)(n + 3)$
45. $y^3(y + 6)(y + 7)$

5.3 Special Cases for Factoring

1. $(a + 3)(a - 3)$
3. $(x + 4)(x - 4)$
5. not factorable (prime)
7. not factorable (prime)
9. $3(x + 3)(x - 3)$
11. $(2a + 5)(2a - 5)$
13. $(6y + 5)(6y - 5)$
15. $3(2a + 5)(2a - 5)$
17. $2(2y + 5)(2y - 5)$
19. $(xy + 5)(xy - 5)$
21. $(x^2y^2 + 3a)(x^2y^2 - 3a)$
23. $b^2(2a + 3)(2a - 3)$

25. not factorable (prime)
27. $(x^2 + y^2)(x + y)(x - y)$
29. $(a^4 + y)(a^4 - y)$
31. $(b^3 + x^2)(b^3 - x^2)$
33. $(5 + a)(5 - a)$ or $-(a + 5)(a - 5)$
35. $32(2 + x)(2 - x)$ or $-32(x + 2)(x - 2)$
37. $(a^2 + b^2)(a + b)(a - b)$
39. $(x^6 + y^6)(x^3 + y^3)(x^3 - y^3)$
41. $ac^2(a + 5)(a - 5)$
43. $(2x - 1)(x + 5)(x - 5)$
45. $(3x + 7)(2x + 3)(2x - 3)$
47. $(x + 4)^2$

49. $(a + 2)^2$

51. not factorable (prime)

53. $(m - 8)^2$

55. $(b + 11)^2$

57. $(x - 25)^2$

59. $7a(n - 3)^2$

61. $-5j(p + 4)^2$

63. $(3x + 1)^2$

65. $(4a - 3)^2$

5.4 Simplifying Rational Expressions

1. $\dfrac{1}{n - 13}$

3. $\dfrac{x^4}{5x + 7}$

5. $\dfrac{-9}{p^3(p + 6)(p - 2)}$ or $\dfrac{-9}{p^5 + 4p^4 - 12p^3}$

7. $\dfrac{r + 9}{r - 4}$

9. $\dfrac{g + 3}{g - 6}$

11. $\dfrac{x - 20}{x + 11}$

13. $\dfrac{a + 1}{a - 1}$

15. $\dfrac{2(n - 5)}{5(n - 1)}$ or $\dfrac{2n - 10}{5n - 5}$

17. $\dfrac{(x^2 + 9)(x - 3)}{x + 8}$ or $\dfrac{x^3 - 3x^2 + 9x - 27}{x + 8}$

19. $\dfrac{(3x + 1)(x + 2)}{x + 10}$ or $\dfrac{3x^2 + 7x + 2}{x + 10}$

21. all real numbers except 0

23. all real numbers except 11

25. all real numbers except −9 and 12

27. all real numbers except −11 and 11

29. all real numbers except 0 and −7

31. all real numbers except 0, −11 and 2

33. $\dfrac{x - 7}{x - 5}$;

Original expression: all real numbers except −7 and 5;

Simplified expression: all real numbers except 5

35. $\dfrac{-9x}{x - 6}$;

Original expression: all real numbers except −11 and 6;

Simplified expression: all real numbers except 6

37. $\dfrac{x + 10}{x - 5}$;

Original expression: all real numbers except −6 and 5;

Simplified expression: all real numbers except 5

39. $\dfrac{2(x + 3)}{7}$ or $\dfrac{2x + 6}{7}$;

Original expression: all real numbers except 0 and −5

Simplified expression: all real numbers

Chapter 6: Quadratic Equations

6.1 Solving Quadratic Equations by Factoring

1. 4

3. $0, -7$

5. $4, 8$

7. $-5, -4$

9. $-27, 1$

11. $3, \frac{6}{5}$

13. $-\frac{5}{6}, \frac{6}{11}$

15. $-\frac{5}{4}, \frac{7}{2}$

17. 8

19. $0, 4$

21. $-2, 2$

23. $-10, 10$

25. $-5, 5$

27. $0, -1, 1$

29. $-8, 8$

31. $-6, 6$

33. $-3, 3$

35. 4

37. $-11, -3$

39. $-5, 2$

41. $3, 4$

43. $1, 2$

45. $-4, 10$

47. $-12, 4$

49. $-4, -1$

51. $0, 3, 6$

53. $0, 4, 5$

55. The manufacturer should set the price for fuel injectors at \$8 or \$14.

57. a) The population of Falls City is now 12,036.

b) The population will be 12,000 3 and 12 years from now.

59. a) The ball's initial height is 48 feet.

b) After one second in the air, it is at 64 feet.

c) It hits the ground after 3 seconds.

61. The length of the rectangle is 8 feet. The width is 1 feet.

63. The length of the rectangle is 8 feet and the width is 12 feet.

65. The base of the triangle is 7 inches. The height is 4 inches.

67. There are two sets of consecutive integers whose product is 72: $-9, -8$ or $8, 9$.

69. There are two sets of odd, consecutive integers whose product is 143: $-13, -11$ or $11, 13$.

71. The integer is either 2 or -5.

73. The walkway should be 2 feet wide, which is pretty narrow for a walkway.

6.2 Understanding and Simplifying Square Roots

1. $8, -8$

3. $5, -5$

5. $12, -12$

7. $100, -100$

9. $8.43, -8.43$

11. $6.86, -6.86$

13. $\frac{1}{7}, -\frac{1}{7}$

15. $\frac{11}{15}, -\frac{11}{15}$

17. $0.4, -0.4$

19. 7

21. -6

23. -13

25. $-\frac{11}{13}$

27. not real

29. not real

31. 0.9

33. $x \geq -4$

35. $h \geq 11$

37. $x \geq -\frac{8}{7}$

39. $y \leq 3$

41. $5\sqrt{3}$

43. $2\sqrt{6}$

45. $3\sqrt{15}$

47. $6\sqrt{3}$

49. $6\sqrt{7}$

51. $10\sqrt{10}$

53. $15\sqrt{5}$

6.3 Solving Quadratic Equations Using Square Roots

1. $x = \pm 6$

3. $a = \pm 3$

5. $b = \pm 1$

7. no real-number solutions

9. $a = \pm\sqrt{5}$

11. $b = \pm 2\sqrt{3}$

13. $y = \pm\sqrt{3}$

15. $a = \pm 2\sqrt{2}$

17. $a = \pm\sqrt{5}$

19. no real-number solutions

21. $x = \pm 3$

23. $x = \pm 5$

25. $x = \pm 2\sqrt{3}$

27. $x = 5, -1$

29. $a = 11, -1$

31. $a = -8, -10$

33. $x = -4 \pm \sqrt{5}$

35. $a = \pm 4.58$

37. $y = \pm 7.30$

39. $x = \pm 1.00$

41. 50 m

43. 20 in

45. $11\sqrt{3}$ yd

47. $5\sqrt{34}$ cm

49. $5\sqrt{41}$ m

51. Equation: $4^2 + 2.5^2 = C^2$;

Answer: The board should be at least 4.72 feet long.

53. Equation: $5.5^2 + B^2 = 6.7^2$;
Answer: The shadow is about 3.83 feet long.

55. Equation: $33^2 + B^2 = 160^2$;
Answer: The top of the ladder is about 156.56 inches off the ground.

57. $3\sqrt{13}$

59. $4\sqrt{29}$

61. 15

63. 75

65. $4\sqrt{97}$

6.4 Solving Quadratic Equations Using the Quadratic Formula

1. $x = -1, 3$

3. $y = 1, 4$

5. $a = -10, -2$

7. $b = -2$

9. $x = -\frac{1}{2}, 3$

11. $x = \frac{1}{4} - \frac{\sqrt{5}}{4}, \frac{1}{4} + \frac{\sqrt{5}}{4}$

13. $a = -\frac{3}{5}, 1$

15. $x = \frac{5}{2} - \frac{\sqrt{41}}{2}, \frac{5}{2} + \frac{\sqrt{41}}{2}$

17. $a = \frac{1}{2} - \frac{\sqrt{89}}{2}, \frac{1}{2} + \frac{\sqrt{89}}{2}$

19. $b = -5, 5$

21. $y = -\frac{5}{2} - \frac{\sqrt{41}}{2}, -\frac{5}{2} + \frac{\sqrt{41}}{2}$

23. $x = -1$

25. $a = -4, 1$

27. $b = -2, -1$

29. no real-number solutions

31. $a = -1 + \frac{\sqrt{6}}{2}, -1 - \frac{\sqrt{6}}{2}$

33. $x = 4 - \sqrt{3}, 4 + \sqrt{3}$

35. $b = 6 - 2\sqrt{2}, 6 + 2\sqrt{2}$

37. $x = \frac{1}{3} - \frac{\sqrt{34}}{3}, \frac{1}{3} + \frac{\sqrt{34}}{3}$

39. no real-number solutions

41. The rectangle is about 13.311 meters long and about 6.311 meters wide.

43. The rectangle is about 37.272 millimeters long and about 8.424 millimeters wide.

45. The ball hits the ground after 5.375 seconds.

6.5 Graphing Quadratic Equations

1. Vertex: (2, -7)

3. Vertex: (−2, 6)

5. Vertex: (1, −9)

7. Vertex: (3, 8)

9. Vertex: (−1, 3)

11.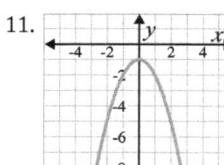

13.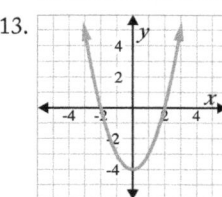

15.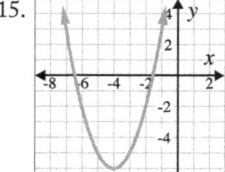

17.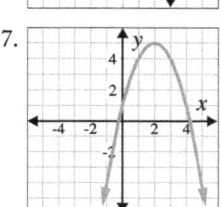

19.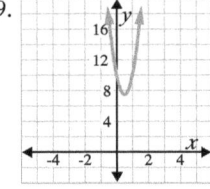

23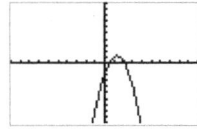

27.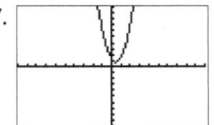

25.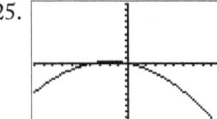

21.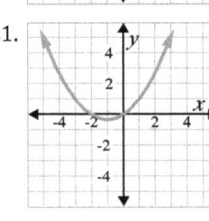

6.6 Solving Quadratic Equations Using a Graphing Calculator

1. $x = -4, 1$

3. $x = -8, 2$

5. $x = 1, 6$

7. no real number solutions

9. $x = 6, 9$

11. $x \approx -1.317, 5.317$

13. $n \approx -0.300, 2.199$

15. $t \approx -2.430, -0.675$

17. no real number solutions

19. $m \approx -17.663, -2.170$

21. $x \approx -3.303, 0.303$

23. $p \approx 0.327, 7.706$

25. no real number solutions

27. $x \approx -12.056, -2.166$

29. Both blogs will have the same number of readers in about 17 weeks.

31. These numbers will be equal about 14 years after the functional time machine is invented.

33. All of the coupons will be taken in about 7.585 hours.

CPSIA information can be obtained
at www.ICGtesting.com
Printed in the USA
LVHW01s0401181117
556666LV00001B/3/P